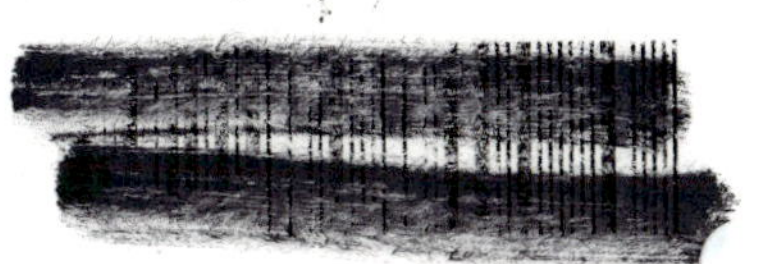

AF616473

Two week loan

Please return on or before the last date stamped below.
Charges are made for late return.

WITHDRAWN FROM STOCK

INNOVATIONS IN GIS 7

GIS and Geocomputation

INNOVATIONS IN GIS 7

GIS and Geocomputation

Edited by
Peter Atkinson and David Martin
Department of Geography
University of Southampton

G
70.2
.G4

First published 2000 by Taylor & Francis
11 New Fetter Lane, London EC4P 4EE

Simultaneously published in the USA and Canada
by Taylor & Francis
29 West 35th Street, New York, NY 10001

Taylor & Francis is an imprint of the Taylor & Francis Group

© 2000 Taylor & Francis

except where stated in the text, and Fig 16.1 © Commonwealth of Australia, taken from *National Wilderness Inventory Handbook of Procedures, Content and Usage* (2nd ed.) R. Lesslie & M. Maslen. Australian Heritage Commission: Canberra
http://www.environment.gov.au/heritage/anlr/index.htm

Printed and bound in Great Britain by TJ International Ltd, Padstow, Cornwall

Publisher's Note

This book has been prepared from camera-ready copy provided by the editors

All rights reserved. No part of this book may be reprinted or reproduced or utilised in any form or by any electronic, mechanical, or other means, now known or hereafter invented, including photocopying and recording, or in any information storage or retrieval system, without permission in writing from the publishers.

British Library Cataloguing in Publication Data
A catalogue record for this book is available from the British Library

Library of Congress Cataloging in Publication Data
A catalogue record for this book has been requested

ISBN 0-7484-0928-9

Contents

Preface vii

Contributors viii

GISRUK Committees xii

1 Introduction 1
Peter Atkinson and David Martin

PART I: MODELLING SPACE-TIME AND CYBERSPACE 9

2 An introduction to TeleGeoMonitoring: problems and potentialities 11
Robert Laurini

3 Models and queries in a spatio-temporal GIS 27
Baher El-Geresy and Christopher Jones

4 A representation of relationships in temporal spaces 41
Christophe Claramunt and Bin Jiang

5 Multi-agent simulation: computational dynamics within GIS 55
Michael Batty and Bin Jiang

6 A parameterised urban cellular model combining spontaneous and self-organising growth 73
Fulong Wu

7 Testing space-time and more complex hyperspace geographical analysis tools 87
Ian Turton, Stan Openshaw, Chris Brunsdon, Andy Turner and James Macgill

PART II: ZONATION AND GENERALIZATION 101

8 Automated zone design in GIS 103
David Martin

9 Designing zoning systems for flow data 115
Seraphim Alvanides, Stan Openshaw and Oliver Duke-Williams

10 Propagating updates between geographic databases with different scales 135
Thierry Badard and Cécile Lemarié

11 A simple and efficient algorithm for high-quality line labelling 147
Alexander Wolff, Lars Knipping, Marc van Kreveld, Tycho Strijk and Pankaj K. Agarwal

12 Modelling knowledge for automated generalisation of categorical maps – a constraint based approach 161
Alistair Edwardes and William Mackaness

13 Preserving density contrasts during cartographic generalisation 175
Nicolas Regnauld

PART III: SPATIAL INFORMATION AND ACCURACY 187

14 Applying signal detection theory to spatial data 189
Brian Lees and Susan Hafner

15 Localized areal disaggregation for linking agricultural census data to remotely sensed land cover data 205
Alessandro Gimona, Alistair Geddes and David Elston

16 A fuzzy modelling approach to wild land mapping in Scotland 219
Steffen Fritz, Linda See and Steve Carver

17 Evaluating the derivation of sub-urban land-cover data from Ordnance Survey Land-Line.Plus® 231
Michael Hughes and Peter Fisher

18 Interpolating elevation with locally-adaptive kriging 241
Christopher D. Lloyd and Peter M. Atkinson

19 Assessing spatial similarity in geographic databases 255
Alia Abdelmoty and Baher El-Geresy

Index 271

Preface

This volume is the seventh in a series based on the GIS Research UK (GISRUK) Conference Series. The original titles in the series reflected well the diversity of contributions to the GISRUK conferences, and of GIS research more generally. Over the seven years of its existence, GISRUK has seen an enormous amount of change in GIS research and publication, with areas which were initially represented by one or two pioneering papers now reflected in entire sessions and even in separate conferences. Last year's volume represented something of a shift in focus for the *Innovations* series, being more of a thematic volume dealing specifically with the integration of information infrastructures with geographic information technology. GISRUK '99 was characterised by lively debate about the relationship between GIS research and the emerging field of GeoComputation, and in this volume the thematic focus is continued as we have selected a range of papers which are all concerned in some way with computational issues in GIS.

GISRUK is not a formal organization, but rather a series of independently organized conferences sharing a common identity and objectives, under the guidance of a national steering committee. The objectives of the GISRUK meetings are:

- To act as a focus for GIS research in the UK
- To provide a mechanism for the announcement and publication of GIS research
- To act as an interdisciplinary forum for the discussion of research ideas
- To promote active collaboration amongst researchers from diverse parent disciplines
- To provide a framework in which postgraduate students can see their work in a national context

GISRUK '99, held at the University of Southampton on 14–16 April 1999, drew around 160 delegates from many different countries and disciplines. In common with previous GISRUK conferences, the format of the academic content included a mixture of full and short paper presentations, a poster display and panel discussions. Again continuing a model established in 1998, a conference prize was given for the best postgraduate paper, and this forms the basis for Chapter 11 of this volume. The breadth of interest in GISRUK, and the goodwill that the series clearly has from so many quarters should serve to ensure that it remains one of the key meeting points for GIS research, and an environment in which future research directions will emerge and develop.

Contributors

Alia Abdelmoty
School of Computing, University of Glamorgan, Pontypridd CF37 1DL, United Kingdom
E-mail: aiabdel@glam.ac.uk

Pankaj K. Agarwal
Center for Geometric Computing, Department of Computer Science, Duke University, Durham, NC, USA
E-mail: pankaj@cs.duke.edu

Seraphim Alvanides
School of Geography, University of Leeds, Leeds, LS2 9JT, United Kingdom
E-mail: s.alvanides@geog.leeds.ac.uk

Peter Atkinson
Department of Geography, University of Southampton, Southampton, SO17 1BJ, United Kingdom
E-mail: pma@soton.ac.uk

Thierry Badard
Institut Géographique National, Service de la Recherche/ Laboratoire COGIT, 2 à 4 Avenue Pasteur, 94165, Saint Mandé Cedex, France
E-mail: Thierry.Badard@ign.fr

Michael Batty
CASA, University College London, 1-19 Torrington Place, London, WC1E 6BT, United Kingdom
E-mail: m.batty@ucl.ac.uk

Chris Brunsdon
Department of Geography, University of Newcastle, Newcastle-upon-Tyne, NE1 7RU, United Kingdom
E-mail: Chris.Brunsdon@ncl.ac.uk

Steve Carver
School of Geography, University of Leeds, Leeds, LS2 9JT, United Kingdom
E-mail: S.Carver@geog.leeds.ac.uk

Christophe Claramunt
Department of Computing,The Nottingham Trent University, Nottingham, NG1 4BU, United Kingdom
E-mail: clac@doc.ntu.ac.uk

Oliver Duke-Williams
School of Geography, University of Leeds, Leeds, LS2 9JT, United Kingdom
E-mail: oliver@geog.leeds.ac.uk

Alistair Edwardes
Department of Geography, University of Edinburgh, Drummond Street, Edinburgh, EH8 9XP, United Kingdom
E-mail: aje@geo.ed.ac.uk

Baher El-Geresy
School of Computing, University of Glamorgan, Pontypridd, CF37 1DL, United Kingdom
E-mail: bageresy@glam.ac.uk

David Elston
Environmental Modelling Unit, Biomathematics and Statistics Scotland, Craigiebuckler, Aberdeen, AB15 8HQ, United Kingdom
E-mail: d.elston@mluri.sari.ac.uk

Peter Fisher
Department of Geography, University of Leicester, Leicester, LE1 7RH, United Kingdom
E-mail: pff1@le.ac.uk

Steffen Fritz
School of Geography, University of Leeds, Leeds, LS2 9JT, United Kingdom
E-mail: pgsf@geog.leeds.ac.uk

Alessandro Gimona
Land Use Science Group, MLURI, Craigiebuckler, Aberdeen, AB15 8HQ, United Kingdom
E-mail: a.gimona@mluri.sari.ac.uk

Alistair Geddes
Land Use Science Group, MLURI, Craigiebuckler, Aberdeen, AB15 8HQ, United Kingdom
E-mail: a.geddes@mluri.sari.ac.uk

Susan Hafner
Division of Water Resources, CSIRO, Canberra, ACT2601, Australia

Michael Hughes
Environmental Change Research Centre, University College London, 26 Bedford Way, London, WC1H 0AP, United Kingdom
E-mail: m.hughes@geography.ucl.ac.uk

Bin Jiang
CASA, University College London, 1-19 Torrington Place, London, WC1E 6BT, United Kingdom
E-mail: b.jiang@ucl.ac.uk

Christopher Jones
School of Computing, University of Glamorgan, Pontypridd, CF37 1DL, United Kingdom
E-mail: cbjones@glam.ac.uk

Lars Knipping
Sender Freies Berlin, Forckenbeckstraße 52, D-14199 Berlin, Germany
knipping@altus.de

Marc van Kreveld
Department of Computer Science, Utrecht University, The Netherlands
E-mail: marc@cs.uu.nl

Robert Laurini
Laboratoire d'Ingénierie des Systèmes d'Information, Université Claude Bernard Lyon 1 INSA de Lyon, F-69621 Villeurbanne Cedex, France
E-mail: Laurini@if.insa-lyon.fr

Brian Lees
Department of Geography, Australian National University, Canberra, ACT 0200, Australia
E-mail: Brian.Lees@anu.edu.au

Cécile Lemarié
Institut Géographique National, Service de la Recherche/ Laboratoire COGIT, 2 à 4 Avenue Pasteur , 94165 Saint Mandé Cedex, France

Christopher D. Lloyd
School of Geosciences, Queens University of Belfast, Belfast, BT7 1NN, United Kingdom
E-mail: c.lloyd@qub.ac.uk

James Macgill
School of Geography, University of Leeds, Leeds, LS2 9JT, United Kingdom
E-mail: J.Macgill@geog.leeds.ac.uk

William Mackaness
Department of Geography, University of Edinburgh, Drummond Street, Edinburgh, EH8 9XP, United Kingdom
E-mail: wam@geo.ed.ac.uk

David Martin
Department of Geography,University of Southampton, Southampton, SO17 1BJ, United Kingdom
E-mail: djm1@soton.ac.uk

Stan Openshaw
School of Geography, University of Leeds, Leeds, LS2 9JT, United Kingdom
E-mail: stan@geog.leeds.ac.uk

Nicolas Regnauld
Department of Geography, University of Edinburgh, Drummond Street, Edinburgh, EH8 9XP, United Kingdom
E-mail: nrr@geo.ed.ac.uk

Linda See
School of Geography, University of Leeds, Leeds, LS2 9JT, United Kingdom
E-mail: L.See@geog.leeds.ac.uk

Tycho Strijk
Department of Computer Science, Utrecht University, The Netherlands
E-mail: tycho@cs.uu.nl

Andy Turner
School of Geography, University of Leeds, Leeds, LS2 9JT
E-mail: andyt@geog.leeds.ac.uk

Ian Turton
School of Geography, University of Leeds, Leeds, LS2 9JT, United Kingdom
E-mail: Ian@geog.leeds.ac.uk

Alexander Wolff
Institut für Mathematik und Informatik, Ernst-Moritz-Arndt-Universität, Jahnstraße 15a, D-17487 Greifswald, Germany
E-mail: awolff@mail.uni-greifswald.de

Fulong Wu
Department of Geography, University of Southampton, Southampton, SO17 1BJ, United Kingdom
E-mail: F.Wu@soton.ac.uk

GISRUK Committees

GISRUK National Steering Committee

Steve Carver	University of Leeds
Jane Drummond	University of Glasgow
Bruce Gittings(Chair)	University of Edinburgh
Peter Halls	University of York
Gary Higgs	University of Leeds
Zarine Kemp	University of Kent
David Kidner	University of Glamorgan
David Martin	University of Southampton
George Taylor	University of Newcastle

GISRUK'99 Local Organising Committee

Nigel Arnell
Peter Atkinson
Chris Lloyd
David Martin (Chair)
Jim Milne
Shane Murnion
David Sear
Andrew Trigg
David Wheatley
Fulong Wu

GISRUK'99 Sponsors

The GISRUK Steering Committee are enormously grateful to the following organisations who generously supported GISRUK'99:

The Association for Geographic Information (AGI)
Taylor and Francis Ltd
John Wiley and Sons
The Quantitative Methods Research Group of the RGS-IBG
Blackwell Publishers
Whitbread plc
Ordnance Survey
The GeoData Institute, University of Southampton
Smallworld Systems Ltd

1

Introduction

Peter Atkinson and David Martin

1.1 GEOCOMPUTATION

GeoComputation was unleashed upon the world of geographical analysis in 1996 when Bob Abrahart organised and ran the first International Conference on GeoComputation at the University of Leeds. The conference was a resounding success and has been repeated every year since: at the University of Otago, New Zealand in 1997, the University of Bristol in 1998 and Mary Washington College, Frederiksberg, Virginia, USA in 1999. Appropriately, this year's conference will be hosted once again by Bob Abrahart at the University of Greenwich.

The success of the conference series, the proliferation of literature resulting from and related to the series (e.g. Longley *et al.*, 1998), and the promotion of GeoComputation by its various supporters, has meant that GeoComputation is now much more than a conference series. It is regarded as a discipline in its own right, just like GIS. However, unlike GIS, GeoComputation is still in its infancy and in these early days, several authors have taken the time to proffer their own view of what GeoComputation really is. For example, Mark Gahegan (1999), in a guest editorial of *Transactions in GIS*, emphasizes the enabling technology and defines four significant advances in computer science that have enabled GeoComputation:

1. Computer architecture and design (i.e. parallel processing)
2. Search, classification, prediction and modelling (e.g. artificial neural networks)
3. Knowledge discovery (i.e. data mining tools)
4. Visualisation (e.g. replacement of statistical summaries with graphics)

This view of GeoComputation as enabled by technology is similar to the view of astronomy enabled by the telescope (MacMillan, 1998). It is clear that GIS itself has played a large part in this enabling process.

In the opening chapter of Longley *et al.* (1998), Longley (1998) describes GeoComputation, largely from the perspective of the 21 invited contributors to the book. Longley's emphasis is on data mining and data visualization, but particularly on dynamic spatial modelling as described by Burrough (1998) in the same book. Dynamic spatial modelling is also known as distributed spatial process modelling, and encompasses cellular automata and computational fluid dynamics.

It is not surprising given Longley's emphasis on dynamics and process that he levels some criticism at Openshaw (1998) for promoting purely empiricist, inductive approaches to geographical analysis. However, such criticism is interesting since it is surely Prof. Stan Openshaw who has done most to initiate, develop and promote GeoComputation as a discipline in its own right and also to define what GeoComputation is. Prof. Openshaw's argument is essentially a reaction against classical statistical methods that are clearly not applicable to spatial data where the assumption of data independence (lack of statistical correlation) is invalid. The message could be interpreted as follows:

1. *do not* apply classical statistics to geographical data as though they were statistically independent
2. *do not* 'generalise out' the geography with global statistics (e.g. global mean)
3. *do not* use stationary models (e.g. spatially constant mean or variogram) for geographical data
4. *do not* rely on model-based statistics for inference when the power of the computer can let spatial data speak for themselves

These are compelling arguments for geographers, whose very discipline is concerned with variation across space. Why, as geographers, would we want to throw away the geography? However, it is important to realize that this 'call to geographers' to be more geographical does not imply that that there is no place for (e.g. statistical) models in geography. The geographically weighted regression (GWR) developed by Brunsdon *et al.* (1996) is a good example of a statistical approach (linear regression) being adapted to emphasize the geography. Similarly, in the field of dynamic spatial modelling, analytical and stochastic approaches very much fit in with Openshaw's view.

The relationship between GeoComputation and GIS is unclear. For example, is GeoComputation a reaction against (often off-the-shelf) GIS systems that place constraints on what researchers can and cannot do? Also, is GeoComputation to be regarded as an equal parallel discipline to GIS, or does one envelope the other (and if so, which one?). To confuse things further, much of what we now regard as GeoComputation was covered in Burrough's (1986) original book on the principles of *GIS* (e.g. dynamic spatial modelling).

Our own view of GeoComputation is that is what geographers are doing with the wealth of data and computer processing power that is now available to them. It will be defined, as time passes, by what GeoComputation researchers do. In this volume, we have collected those papers presented at the seventh annual GISRUK conference held at the University of Southampton in April 1999 with the largest GeoComputational elements. However, we acknowledge that there are many aspects of GeoComputation that are not covered, and others that are included that would be better described by the term GIS.

1.2 MODELLING SPACE-TIME AND CYBERSPACE

In Chapter 2, **Robert Laurini** introduces the concept of TeleGeoMonitoring, which he characterises as a new field, concerned with the development of GIS which function in real time by use of data supplied by sensors at fixed locations and in vehicles using telecommunications. The objective of telegeomonitoring systems is to allow short and longer-term decision making based on data collected in real time about the workings of complex geographical systems. The chapter introduces the reader to a range of alternative telegeomonitoring applications, and outlines the principal issues concerning the architecture and use of these systems, concluding with a challenging research agenda.

As increasing amounts of data become available for manipulation within GIS, and as GIS are now often implemented for long-term monitoring, the capabilities of GIS for handling temporal data are increasingly being exploited. In Chapter 3, **Baher El-Geresy** and **Christopher Jones** provide an exhaustive review and classification of conceptual models for temporal GIS (TGIS). This review focuses on the type of query that the TGIS is intended to handle (e.g. What, Where and When corresponding to space, feature and event), but also introduces the How and Why views.

In Chapter 4, **Christophe Claramunt** and **Bin Jiang** address the representation of relationships in temporal spaces. A longstanding weakness of existing GIS tools has been their inability to cope adequately with temporal data, and one of the contributing reasons for this slow development has undoubtedly been the additional complexity involved in modelling the relationships between objects which have both temporal and geographical existence. The chapter briefly reviews work in this field, and proposes a new reasoning and computational approach for the integration of time and space information within an integrated framework. The discussion clearly illustrates the additional sophistication that is required for the incorporation of the temporal dimension within the conceptual approaches to GIS with which we may be familiar.

Michael Batty and **Bin Jiang's** Chapter 5 concerns a rather different aspect of the incorporation of a temporal component into a GeoComputational framework, in that it deals with a series of multi-agent models which operate in a cellular space in simulated time. They begin by setting out the ways in which their multi-agent modelling approach differs from cellular automata (CA) approaches, although sharing a similar cellular model of geographical space. The approach is demonstrated by reference to a route finding problem, in which shortest paths are 'discovered' by the model agents rather than being directly computed. These ideas are then extended to deal with directed networks such as river systems, and with the modelling of visual fields.

In Chapter 6, **Fulong Wu** discusses the GeoComputational tool of cellular automata (CA) as applied to urban evolution. He argues that while CA has become a popular tool for simulating urban growth, many questions remain unanswered in relation to its reliability in the real world. Fulong describes a parametrised approach to urban simulation that combines two different processes of growth. First, spontaneous growth is simulated that is independent of the evolving state of the CA and, second, self-organised growth is simulated that is directly controlled by the state of the CA. Flexibility in balancing the weight of each modelled process in the overall evolution is clearly an advantage of CA. Fulong argues that the dual approach is most appropriate for simulating urban land use change.

The time-space complexity so clearly revealed by Chapter 4 has also provided an obstacle to the identification of patterns in data. In Chapter 7, **Ian Turton, Stan Openshaw, Chris Brunsdon, Andy Turner** and **James Macgill** examine several time-space pattern searching procedures based on the concept of data mining. They consider the unique aspects of searching for pattern in geographical spaces, and evaluate the performance of a range of GeoComputational approaches including pattern searching algorithms using simulated time-space data. Although the results are preliminary, and based on synthetic data, they demonstrate a good degree of success in the identification of spatial and space-time clustering.

1.3 ZONATION AND GENERALIZATION

Two chapters address the computation of zoning systems, a problem for which a variety of algorithms are available, but for which practical implementation with real-world sized problems has only recently become tractable. The emergence of large volumes of digital boundary data for small geographical areas, together with GIS structures for the management of these data provide the essential prerequisites for the practical application of such approaches.

In Chapter 8, **David Martin** introduces the topic of zonation and zone design in a chapter that reviews previous work and focuses on his own involvement with the application of zone design tools to the 2001 UK Census output geography. The author explains that the zone design problem can be conceptualized as one part of the popular modifiable areal unit problem (MAUP), the other part being the aggregation problem. Zone design is a complex problem requiring optimization of some goal under constraints. Several approaches to automated zone design are reviewed ranging from the first, and simplest, automated zoning procedure (AZP) to the zone design implemented for the forthcoming 2001 UK census.

In Chapter 9, **Seraphim Alvanides, Stan Openshaw** and **Oliver Duke-Williams** present a further application of automated zone design principles, this time applied to the analysis of flow data. Flow data, for example relating to migration flows or journeys to work, present even more complex zone design requirements than static data such as census counts, but the chapter demonstrates that an automated solution can be implemented, and a working application is presented, applied to journey to work flows.

The second section addresses several technical issues which may be grouped together by their cartographic orientation. These are characterised by the fact that they deal with the creation of cartographic products from existing digital spatial data, and by their use of inherently geographical computation techniques in order to overcome a range of traditional cartographic problems. There is a frequent tendency to separate research in automated cartography from more general purpose GIS research, although many cartographic operations require the kinds of explicitly spatial processing which is central to GIS. Increasingly, the data used in real-world GIS implementations are derived from national topographic reference datasets, which have been devised and developed for cartographic purposes. Thus, techniques for appropriate handling of generalization, intelligent update and scale changes should also be of concern to GIS researchers and users.

In Chapter 10, **Thierry Badard** and **Cécile Lemarié** deal with the issue of propagating updates between geographic databases with different scales. The problem is made particularly complex by the need to update a users' own database from more recent reference data at a different scale, while preserving those characteristics which are part of the users' own modifications to the data. They examine the nature of the relationships that may exist between the objects in the two databases and devise a logical and geographical schema for the implementation of an update process, discussing its applications to major products of IGN, the French national mapping agency.

In Chapter 11, **Alexander Wolff, Lars Knipping, Marc van Kreveld, Tycho Strijk** and **Pankaj Agarwal** propose a new algorithm for the placement of labels on cartographic lines. The technique again represents a high degree of geographical computation, while being a simple and efficient approach to a longstanding problem.

The original version of this chapter won the prize for the best postgraduate presentation at the GISRUK'99 Conference.

Alistair Edwardes and **William Mackaness** deal with a rather different cartographic problem in Chapter 12, that of modelling knowledge for automated generalization of categorical maps. They take the example of soil mapping, in which polygons that are distinct at one scale are merged with others, with a resultant impact on legend classes, as the scale changes. They adopt an object-oriented programming approach in which empirical case-based evaluation of categorical map generalisation provides the knowledge to define the ontology of entities that can exist in a generalised categorical coverage.

Nicolas Regnauld is again concerned with the visual appearance and information content of cartographic products in Chapter 13, in particular, with the preservation of density contrasts during map generalization. The chapter deals with the preservation of building density in maps of urban areas, and examines several approaches to the measurement of density at the level of individual buildings and land parcels. The resultant approach operates at three levels. Parcel density is used to segment the urban area into homogeneous districts. Districts are then classified according to their density levels, to preserve contrasts across the urban area. The lowest level, within-parcel analysis considers factors such as the alignment of buildings to roads to inform the local operation of the generalization process.

1.4 SPATIAL INFORMATION AND ACCURACY

The application of artificial neural networks (ANNs) was identified by Gahegan (1999) as a component of the enabling technology that has catapulted GeoComputation to the forefront of geographical analysis. The key advantages of such approaches over more classical statistical approaches (e.g. regression and maximum likelihood classification) are that the relations can be non-linear, no model is assumed for the data distribution and prediction is usually fast and accurate once training is complete. In Chapter 14, **Brian Lees** and **Susan Hafner** exploit these advantages to map areas of the Liverpool Plains in Australia that are subject to salinization. The problem was essentially to resolve different sources of spatial variation given a remotely sensed image of the region of interest using signal processing (ANN) approaches. The apparent success of the ANN, applied to this complex problem, further supports its increasingly widespread use for prediction.

Chapter 15, by **Alessandro Gimona**, **Alistair Geddes** and **David Elston**, deals with the important problem of combining data from different sources in a GIS. Their particular problem relates to the June Agricultural and Horticultural Census (JAHC) data provided by Scottish Office Agricultural Environment and Fisheries Department (SOAFED). These data are not 'GISable' because no location data are provided with agricultural information. The idea, therefore, was to link these data with land cover data provided by the Macaulay Land Use Research Institute (MLURI). This involved transformation of the data via a disaggregation procedure. The synergy achieved in the research through the merging of complementary datasets was very much enabled by GIS.

Often, the real world does not sit comfortably with the traditional hard classification model (e.g. eco-gradients) and, today, fuzzy approaches are increasingly the norm. In Chapter 16, **Steffen Fritz**, **Linda See** and **Steve Carver** describe an approach to wilderness mapping based on fuzzy set theory. The very concept of wilderness is difficult to define because it varies with subjective interpretation.

Therefore, the tools of fuzzy set theory are very much suited to the mapping of wilderness. The authors describe a case study of mapping wilderness in Scotland. Users were asked to respond to an internet questionnaire which was used to classify respondents into different behavioural groups and to define wilderness per individual. The various responses were then used to build a fuzzy model of wilderness and this, in turn, was used to produce a fuzzy wilderness map.

Remote sensing of land cover and land use is a theme that is continued by **Michael Hughes** and **Peter Fisher** in Chapter 17. In a neat case study, the authors compare the classification of land cover in a 1 km^2 area of suburban Leicester using two different data sources. The first source is traditional aerial photograph interpretation and the second is Ordnance Survey Land-Line Plus data. The comparison showed that for the two classes of paved surface and built surface the agreement was very good. However, for the woody vegetation class severe differences existed. The reasons for the differences were clear: the view from the air is very different to the representation of vegetation provided by the OS. The authors suggest that a combination of both data sources might be the way ahead.

Christopher Lloyd and Peter Atkinson, in Chapter 18, develop a geostatistical approach to local estimation (mapping) based on a non-stationary variogram model. Traditional geostatistical approaches are based on a stationary covariance function or variogram model. In words, this means that geographical variation is modelled as being the same from place to place. This kind of model is adopted for statistical inference, that is, to allow estimation. However, where data are sufficiently numerous (as is often the case with elevation data) then non-stationary models can be beneficial, and certainly fit better our belief that spatial variation is not the same everywhere. In Chapter 18, the non-stationary geostatistical approach is applied to digital elevation data.

Alia Abdelmoty and **Baher El-Geresy**, in Chapter 19, provide a computer science view of the problem faced by Gimona *et al.* (this volume) when they attempted to merge data from different sources in a GIS. The problem is generic, and indeed is a fundamental raison d'être for GIS. If GIS are good at one thing, it is organising spatial data into a common framework such that they can be handled simultaneously. However, while this is a fundamental operation in GIS, achieving the transformations needed in such a way as to maintain the integrity and meaning of the features in the original data requires complex formal procedures. The review and classification provided by the authors includes aspects of spatial equivalence (e.g. object-based and relation-based types) and topological equivalence (which is also extended to cover orientation equivalence).

1.5 CONCLUSION

In this opening chapter we have attempted to describe GeoComputation and to introduce the various Chapters of this volume. In simple terms, GeoComputation is the application of computational processing to *geographical* problems. For problems to be truly geographical they should have a spatial component, and its is this spatial or geographical component that makes GeoComputation different to other forms of computation. Spatial analysis problems require for their solution spatial modelling and spatial computational processing. Such explicitly spatial models and processing can be entirely new, leading to better approaches to prediction and greater understanding.

We believe that GeoComputation and GIS have a long and exciting future in geography and related disciplines. We hope that the chapters of this book help to disseminate ideas and encourage further research relating to these themes.

1.6 REFERENCES

Burrough, P.A., 1998, Dynamic modelling and geocomputation. In *Geocomputation: a Primer*, edited by Longley, P.A., Brooks, S.M., McDonnell, R. and MacMillan, W., (Chichester: Wiley), pp. 165-191.

Burrough, P.A., 1986, *Principles of Geographical Information Systems for Land Resources Assessment*, 1st edition, (Oxford: Oxford University Press).

Gahegan, M., 1999, What is geocomputation?. *Transactions in GIS*, **3**, pp. 203-206.

Brunsdon, C., Fotheringham, A.S., and Charlton, M.E., 1996, Geographically weighted regression: a method for exploring spatial nonstationarity. *Geographical Analysis*, **28**, pp. 281-298.

Longley, P.A., 1998, Foundations. In *Geocomputation: a Primer*, edited by Longley, P.A., Brooks, S.M., McDonnell, R. and MacMillan, W., (Chichester: Wiley), pp. 3-15.

Longley, P.A., Brooks, S.M., McDonnell, R. and MacMillan, W., 1998, *Geocomputation: a Primer*, (Chichester: Wiley).

MacMillan, W., 1998, Epilogue. In *Geocomputation: a Primer*, edited by Longley, P.A., Brooks, S.M., McDonnell, R. and MacMillan, W., (Chichester: Wiley), pp. 257-264.

Openshaw, S., 1998, Building Automated Geographical Analysis and Explanation Machines. In *Geocomputation: a Primer*, edited by Longley, P.A., Brooks, S.M., McDonnell, R. and MacMillan, W., (Chichester: Wiley), pp. 95-115.

PART I

Modelling Space-Time and Cyberspace

2

An introduction to TeleGeoMonitoring: problems and potentialities

Robert Laurini

2.1 INTRODUCTION

For several years, new geographical and urban applications have been emerging in which communications and real time have been very important characteristics. In other words, we no longer have to deal with applications for which conventional cartography is the target, but in which spatial aspects are involved in management in real time or with very strong temporal constraints, such as mission-critical applications. For this kind of problem, the intensive usage of positioning systems (such as GPS) is the key element, allowing vehicles to know their position and to exchange any kind of information by any kind of telecommunication means.

For instance, the management of a fleet such as police or rapid delivery vehicles needs continually to know the exact position of all vehicles. By means of GPS and embarked or on-board computers, they can communicate with a control centre, which in turn can send them other information. In this family of applications, the control centre is equipped with huge electronic panels, which represent the moving objects' locations and trajectory superimposed over a base map. In order to fulfil this task, the vehicles must have embarked computers, connected to GPS and to the control centre. In some applications, a control centre does not exist, and all vehicles exchange information regarding their positions only between themselves.

So, we can see novel applications, engendered by geoprocessing and telecommunications, all possessing common characteristics and implying adapted research. We think that we are facing a new discipline called telegeomonitoring (Laurini, 1998). Telegeomonitoring, as a child of Geographical Information Systems and telecommunications, can be considered as a new discipline characterised by positioning systems, cartography, the exchange of information between different sites and real time spatial decision making.

Applications including traffic monitoring, fleet management, environmental planning, transportation of hazardous materials and surveillance of technological and natural risks, all have in common not only some functional similarities, but also several aspects of computer architecture such as centralised, co-operative or federated. A description of those applications will be given in order to compare them. In addition, some generic problems arising in the interoperability of several telegeomonitoring systems will be briefly addressed.

The goal of this chapter will be to analyse some application classes, to construct a grid to compare them, to define a new architecture, and to examine research tracks in telegeomonitoring. This chapter is the result of work and discussions in the laboratory, especially with A. Boulmakoul, T. Tanzi and S. Servigne (Boulmakoul and Laurini, 1998; Boulmakoul *et al.* 1997a, 1997b, Tanzi *et al.*, 1998; Tanzi, 1998 and Tanzi and Servigne, 1998).

2.2 DESCRIPTION OF SOME APPLICATIONS

Let us examine some applications such as traffic management on toll motorways, fleet management, hazardous materials transportation, pollution, and major risk monitoring.

2.2.1 Traffic management on toll motorways (Tanzi, 1998; Tanzi and Servigne, 1998)

The objective of this application is to maximise the number of vehicles passing along the motorway with optimal security. A control centre consists of a computer receiving information on traffic from sensors. In addition, in order to secure rescue operations, all vehicles such as those belonging to the police, maintenance, rescue vehicles, towing trucks, and so on, are equipped with GPS. More precisely, those vehicles are also connected to the control centre via some telecommunication system.

During the day, and particularly at peak hours, the number of vehicles using the motorway can vary. Moreover, weather, road works and recovering after accident can also influence traffic conditions. So, for determining the number and location of rescue vehicles, it is necessary to run simulations in order to forecast traffic for the coming days. A database populated in real time with information coming from sensors allows all the operations of traffic monitoring, and the determination of the rescue team. The control centre receives all information necessary to follow the traffic and to manage the motorway. In order to assist managers to make sound decisions, huge synoptics are used and animated in real time.

In Figure 2.1, the schema is given showing the architecture (Tanzi, 1998) in which we can see that any motorway vehicle is linked to two satellites (GPS and INMARSAT) in order to communicate not only their position, but also to exchange any information with the control centre. See also Tanzi *et al.* (1997) and Tanzi and Servigne (1998) for details.

A crucial issue in this architecture is that animals such as wild boars or deer may damage the sensors distributed along the roads. As a consequence, as missing information cannot be easily detected, the diagnosis of systematically erroneous or biased information is very difficult. So regular inspection is necessary and must be made by competent personnel.

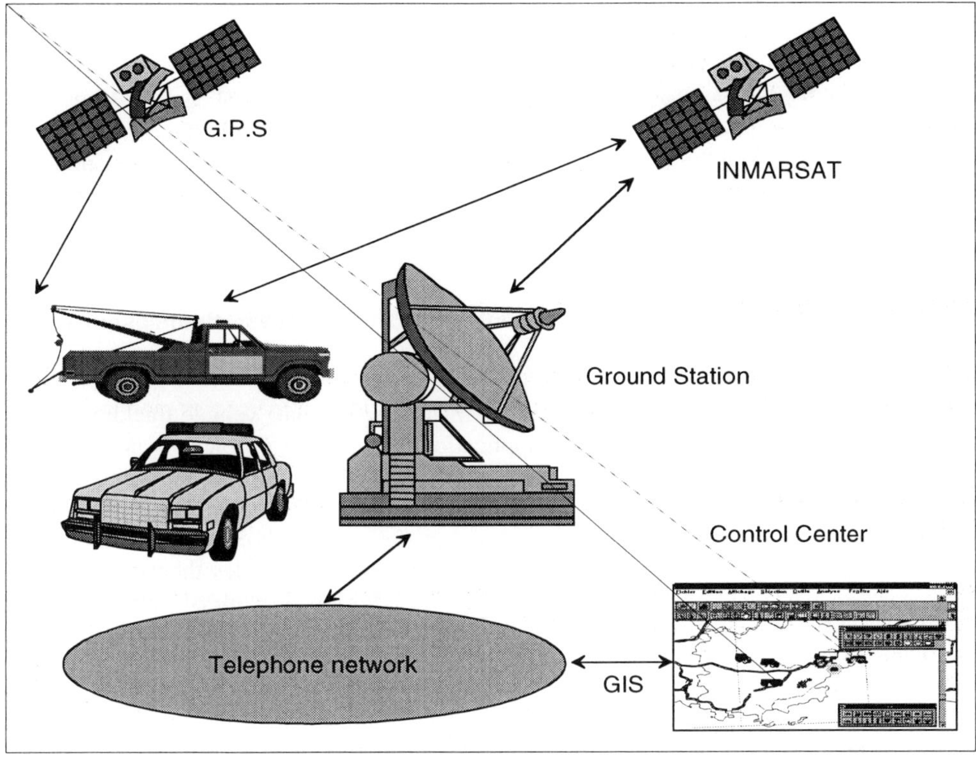

Figure 2.1 Schema of a system for the monitoring of toll motorway traffic (Tanzi, 1998)

2.2.2 Fleet management

In the previous application, the objective was to manage an infrastructure (toll motorway) characterised by the circulation of undifferentiated vehicles along a totally known linear space; now in the fleet management application, the scope is to monitor specific vehicles circulating along totally different trajectories, possibly exiting from the supervised space. By fleets of vehicle, we mean:

- Road vehicles such as police, civil protection, taxis, buses, post-office vehicles, and vehicles for any kind of delivery;
- Besides road vehicles, we may add boats, aircraft, rockets, submarines, etc.;
- We can also add trains, tramways, metros, etc. although their characteristics and constraints are a little different.

In this category, we can also add rally vehicles (such as Paris-Dakar), and yacht competitions, which have some similarities from a telegeomonitoring point of view. In this class of applications, all the vehicles have some freedom to select their trajectory, each of them being followed by the system; because here the important issue is to know, without interruption, the location and the route followed by each of them.

Whereas in the motorway application, the control centre was only monitoring a few vehicles over a demarcated space, in the fleet application, the control centre must follow all vehicles. The consequence is that each of them must have a computer connected to GPS and regularly exchange information with the control centre.

In most of those applications, the vehicles send information to the control centre, which perhaps sends back information sometimes in a vocal form. But it is possible that the embarked computers can communicate not only with the control centre, but also between each other, so that each vehicle can know the position of all of the others.

2.2.3 Hazmat transportation (Boulmakoul and Laurini, 1998; Boulmakoul *et al.*, 1997b)

From an application point of view, hazardous materials (hazmat) vehicle tracking can be seen as a mixture of the two previous applications. Here the problem is no longer to monitor some vehicles in a known system (motorway), but to follow all vehicles in a region, possibly having different trajectories selected from a graph. The optimal path must take into account not only road characteristics (height, width of tunnels and bridges, population exposed to risk, works in progress), but also the vehicles themselves due to their size and the type of materials transported. In addition, in case of very dangerous accidents, the other vehicles which were initially planned to pass through, must immediately be informed about the accident and must modify their path accordingly.

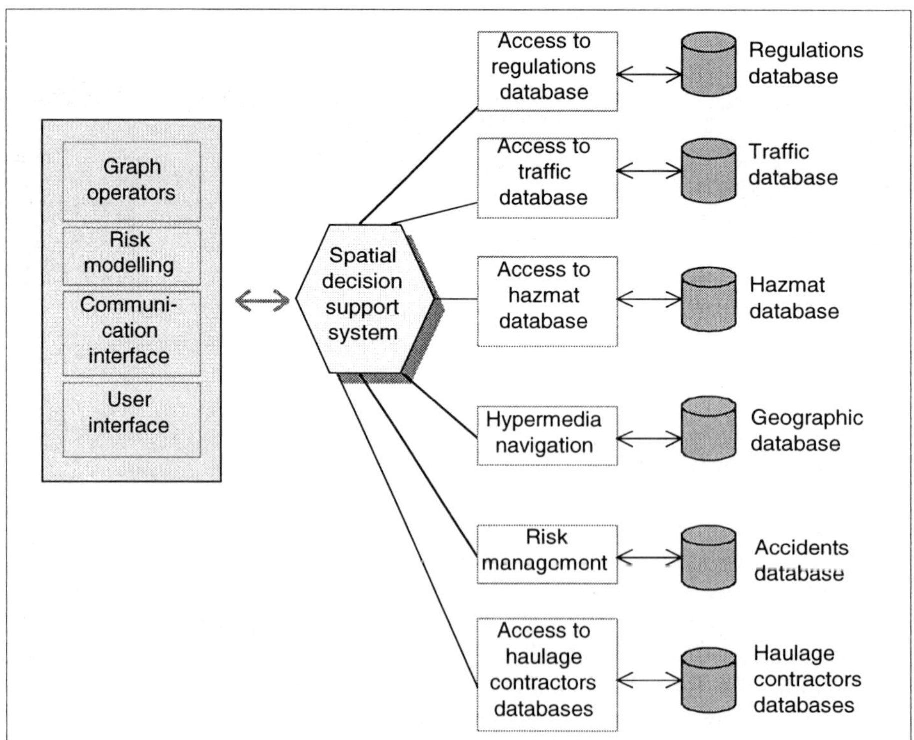

Figure 2.2 Functional description of a system for hazmat transportation (Boulmakoul and Laurini, 1998)

We therefore see that path computation must be done in real-time in connection with the control centre. In order to reach this objective, all vehicles have to be equipped with an embarked computer linked with GPS. For details, please refer to Boulmakoul *et al.* (1997a; 1997b) and Boulmakoul and Laurini (1998). Figures 2.2 and 2.3 which come from the previous papers present the functional and physical architectures. Figure 2.2 (functional architecture) depicts the system as a spatial decision-support system consisting of several components, and of real time accesses to several databases. In Figure 2.3, some physical elements are mentioned, the names of which are given as acronyms from Figure 2.2; here also the key element is the network allowing the connections to all databases. The previous architecture can be extended to other vehicles transporting hazardous materials such as boats and trains.

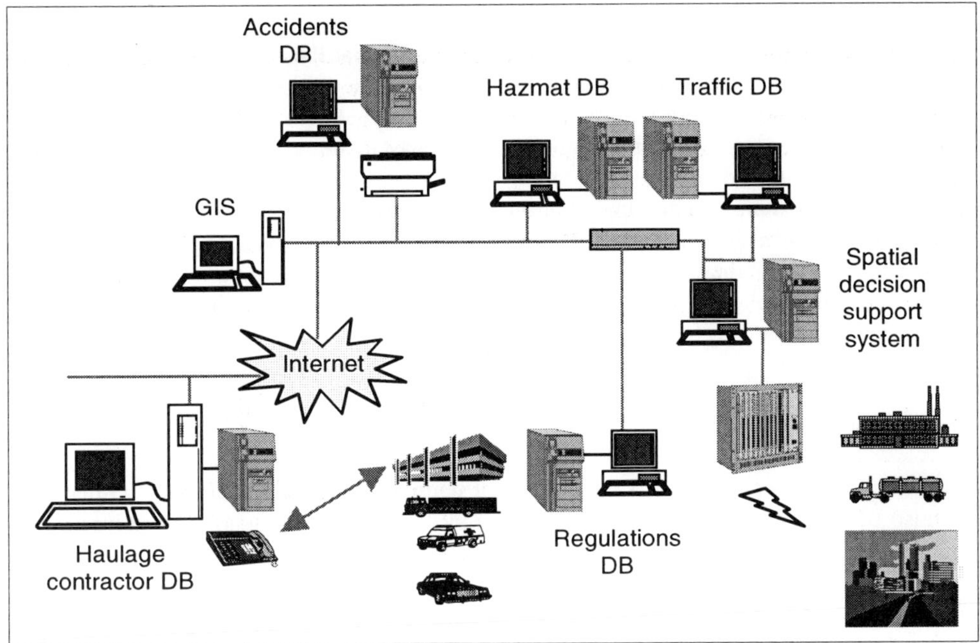

Figure 2.3 Global architecture of a system for the monitoring of hazardous material transportation (Boulmakoul and Laurini, 1998)

2.2.4 Monitoring river pollution

In river pollution monitoring, one deals with control along a tree-like graph (taking tributaries into account), or a more complex hydrologic graph (with canals). When pollution is detected, we have not only to determine its characteristics, but also its origin and the way to mitigate it. The key idea is to put sensors along rivers which measure some physical, chemical and biological parameters and which send these values to the control centre. These measurements are made periodically. When an emergency occurs, the control centre can modify the frequency of measurement in real time.

An interesting possibility is to connect each sensor to a cellular phone. In this case, data transmission is carried out by this device. For instance, the control centre can send a phone call to a sensor to modify its programming. So, by knowing the location of pollution (plants, etc.) and the characteristics of effluents they can emit in the river, thanks to information acquired by the sensors, the control centre can determine the exact location of the polluter. Figure 2.4 shows an example of such a control centre for flood monitoring in which the key element is a real time animated cartographic system based on information coming from sensors. In the same time operators can perform short term or long term simulations. In addition, GPS-equipped boats navigating along the river can also take measurements and send this information to the control centre.

Here, the decision-support system must be sophisticated enough to identify and locate the polluter without error, because on one hand pollutants can dissolve along the river, and on the other hand, one may face some pollutants which are not yet listed or which are very difficult to identify from sensor information. Indeed, sensors do not measure all possible parameters, but only a subset.

One of the non-robust aspects of the system is that the sensors which are located along the river can be damaged by animals or stolen. In case of floods, sensors can be washed away by the current. As a consequence, erroneous or biased information can reach the control centre. Similarly to the motorway example, some additional human control must be planned.

Some complementary information can be acquired by satellite and sent to the control centre. We should note that although images from SPOT satellites may be of interest in this case, they cannot help for real time decision-making: indeed, they are designed for time-delayed decisions. Existing image processing techniques are not robust enough and rapid enough for assisting real time decision making.

2.2.5 Major risk monitoring

The example of river pollution is a first case of natural or technological risks. Other risks can also be monitored by a telegeomonitoring system (under the name of cindynics, i.e. science of risks): let us mention, floods, landslides, mud or snow avalanches, earthquakes, hurricanes, volcanoes, vicinity of nuclear plants, and so on. Cindynics (from the Greek word kindynos which means danger) is a new field of knowlegde, dealing with all aspects of the problem (scientific, cultural, social and legal) as they affect individuals and organizations. The concept of hazard is the expression used by specialists but the research field described is much more open. This new field, exsentially interdisciplinary in character, is now known as the Sciences of Danger. The European Cindynics Institute was created in January 1990. The basic concepts emerged during the Symposium on Technological Risks, held in UNESCO in Paris, in 1987. See *http://www.cindynics.org.*

A global architecture for natural risks is given Figure 2.5, in which we can see sensors sending information to the control centre through a satellite system. For all these systems, fragility comes from sensors which can be damaged stolen, or taken away.

2.3 COMPARISON

After the description of several cases, some common characteristics can be distinguished, such as using GPS, data exchange, existence of a real time database, existence of fixed sensors, mobile or on-board computers, a control centre, a decision-support system,

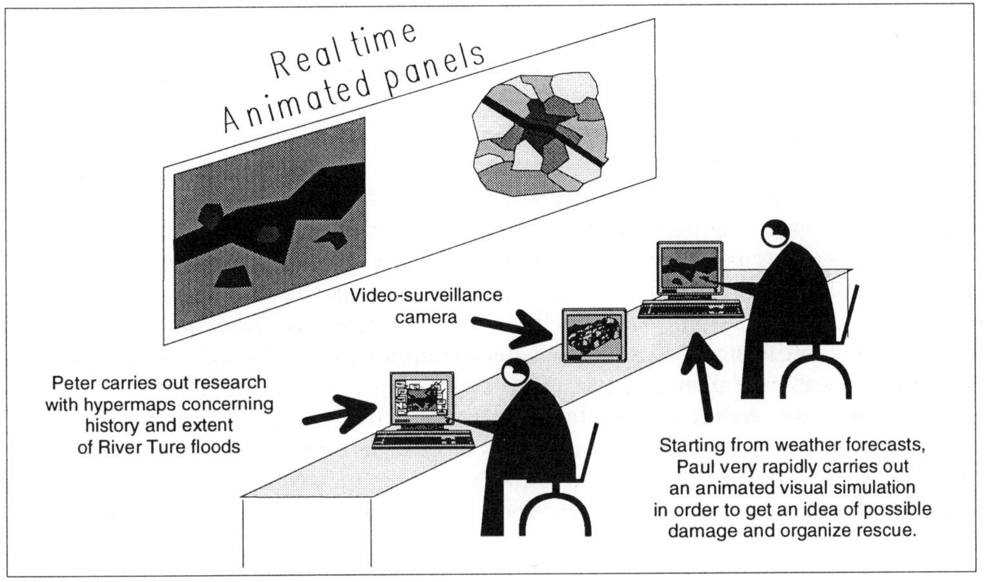

Figure 2.4 Example of a control centre for flood monitoring.

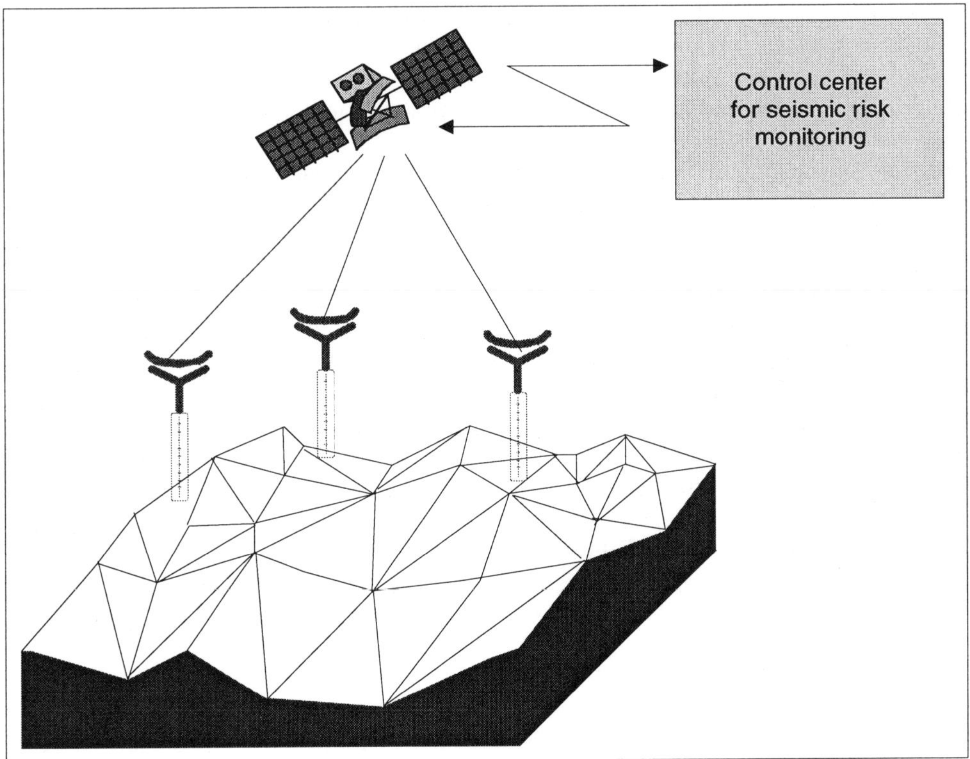

Figure 2.5 Architecture of a system for seismic risk.

anticipation by simulation, decision, animated cartography, etc. We can see that these common aspects are at functional or architectural levels.

2.3.1 Computer architecture

All the previous applications have one key component in common, namely the control centre, and some mobile components with embarked computers using differential GPS. Sometimes some fixed sensors are also present. Thus, we can distinguish passive vehicles which can emit and receive orders, and intelligent vehicles which can modify their path according to some local or global criteria. Similarly, one can imagine also intelligent sensors which can change their data acquisition strategy according to the environment.

Generally speaking, we can consider two types of sensors, fixed sensors and embarked sensors. Indeed, in addition to GPS receivers (which can be considered as special sensors), vehicles can possess cameras or other types of apparatus, for instance to measure external temperature. In later work, both sensors and embarked vehicles will be integrated into the concept of vehicles, and only the notion of isolated sensors will be used in this chapter.

The control centre and the other components exchange multimedia data from different sources by using primarily wireless telecommunications, perhaps with several satellites. All these pieces of information will be stored in a real time geographical database from which some other more elaborate information can be derived.

Another way to classify telegeomonitoring systems is to take the number of persons in manned vehicles, namely, one-person vehicles, or multi-person vehicles. When there is only one pilot, the program is much more complex than when there is a co-pilot. Indeed, when the pilot is alone, it is difficult for them both to drive the vehicle and to interact with the computer, whereas when there is a co-pilot, this person can be in charge of the computer. For one-person vehicles, an interesting solution is voice-based cartography replacing conventional graphic cartography by the emission of geographic information by means of a vocal device. The important consequence is that the one-person vehicle solution is presently too complex from a computing point of view.

Bearing all that in mind, it is possible to distinguish three architectures, centralised, co-operative and federated. We shall now examine them in detail.

2.3.1.1 Centralised architecture

The centralised architecture (Figure 2.6) shows the existence of a control centre where data converge, and which sends data and statements both to mobile vehicles and fixed sensors. We note that in this architecture, only the control centre has a global vision of the system at all times. Of course, a copy can be sent to the other components when necessary. The great weakness of this architecture is that in the case of a crash of the control centre, the telegeomonitoring system no longer functions.

When a new vehicle enters this system, it informs the control centre, which in return sends adequate information. Among initialisation operations, the control centre must update the local database so that the new vehicle can function efficiently. We note that presently this is the more common architecture. In order to palliate crashes, a solution is to create a mirror control centre; this solution is often taken for crisis management teams.

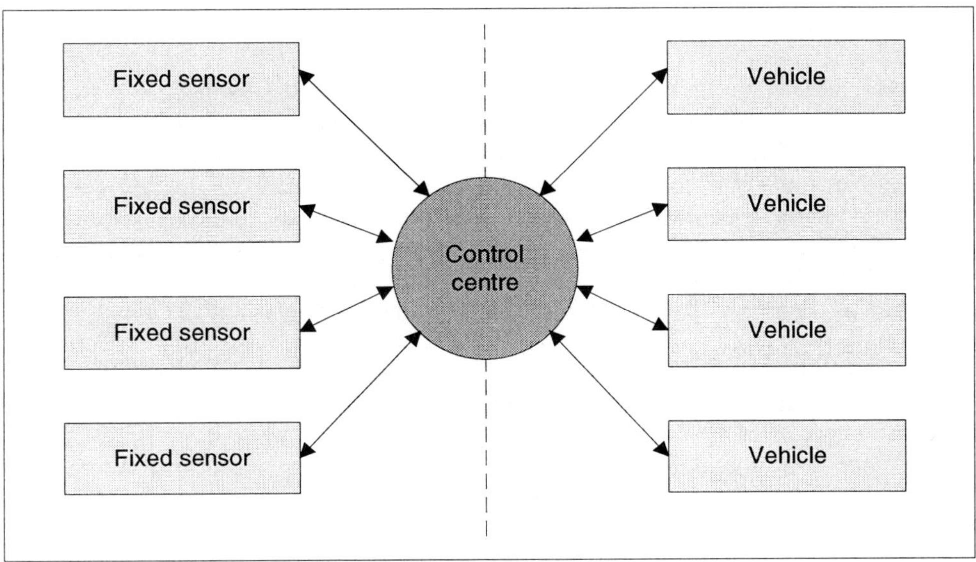

Figure 2.6 Centralised telegeomonitoring architecture

2.3.1.2 Co-operative architecture

In the co-operative architecture (Figure 2.7), that is to say without any central site, all fixed and mobile components exchange information between themselves. As a consequence all sites have a global vision of the context. When a new vehicle enters into the system, it informs all vehicles, which when necessary will send it information.

This architecture is much more robust to crashes than the centralised architecture. On the other hand, the quantity of information to be exchanged is much more important. Indeed, one of the priorities is the updating of all databases; if one copy is different from the others, some difficulties, sometimes drastic, can occur. In those cases, similar problems as those encountered in distributed database systems exist. Apparently, Date's rules can apply (Date, 1987; Burleson, 1994) concerning the autonomy of all databases within any grouping: one component database system must work even if the others are not functioning; in other words, there is no reliance on a central site, all sites are equal, and no one site has governing authority over another site

However, a more serious case is called the Byzantine situation in which one or several components are either defective or always sending erroneous messages, especially due to sensor drift (see for instance Ben Romdhane, 1996 or Simon, 1996 for more details). In this crucial case, the defective components must be detected as early as possible, and the messages must be corrected. Another problem is the authentification of messages received. Think for instance of the hijacking of a lorry transporting hazardous materials. In order not to attract attention from the control centre, the hijackers will send innocent messages. Of course, a soundly designed control centre must diagnose this situation.

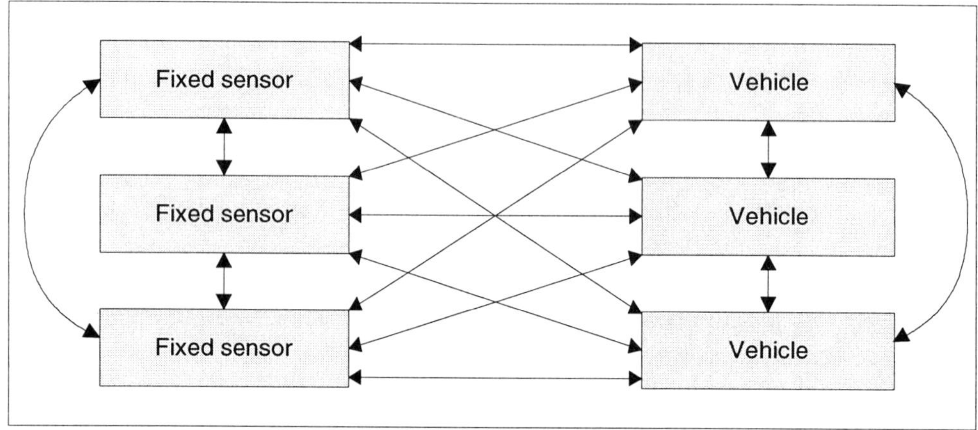

Figure 2.7 Co-operative architecture for telegeomonitoring.

2.3.1.3 Federated architecture

Some vehicles can be connected to several telegeomonitoring control centres. For instance a lorry transporting petrol must be in connection with its haulage contractor, the motorway system, and a more general system for hazmat surveillance. In this case, must the lorry have three distinct embarked computers, or a single one implying problems of interoperability between the three systems ?

As an example, Figure 2.8 depicts the case of a vehicle belonging to three telegeomonitoring systems (Figure 2.8a). From a general point of view, one telegeomonitoring system usually consists of several vehicles, and one vehicle can belong to several systems (Figure 2.8b). First, the vehicle (Figure 2.8c) belongs to system A, then to system B, and finally to C; then it exits from A, then from B, and so on. Therefore, we can see that, according to the number of systems and their nature, the kinds of interoperability problems will differ.

As a consequence, we will state that the belonging can be successive, for instance for aircraft which pass successively from one air traffic control tower to another one, or simultaneous, as in the example presented in the previous paragraph.

2.3.2 Functional aspects

From a functional point of view, we can state that these applications can be seen as variants of real-time spatial decision support systems, in which information pieces come from sensors and embarked components. In order to assist short term decision making, huge visual synoptics can be used, based on animated cartography. Or more exactly, animation is driven from the data received. For the long term, numeric simulations must be performed starting from data stored in the control centre, and some forecasting models. In this way, the consequences of decisions can be estimated.

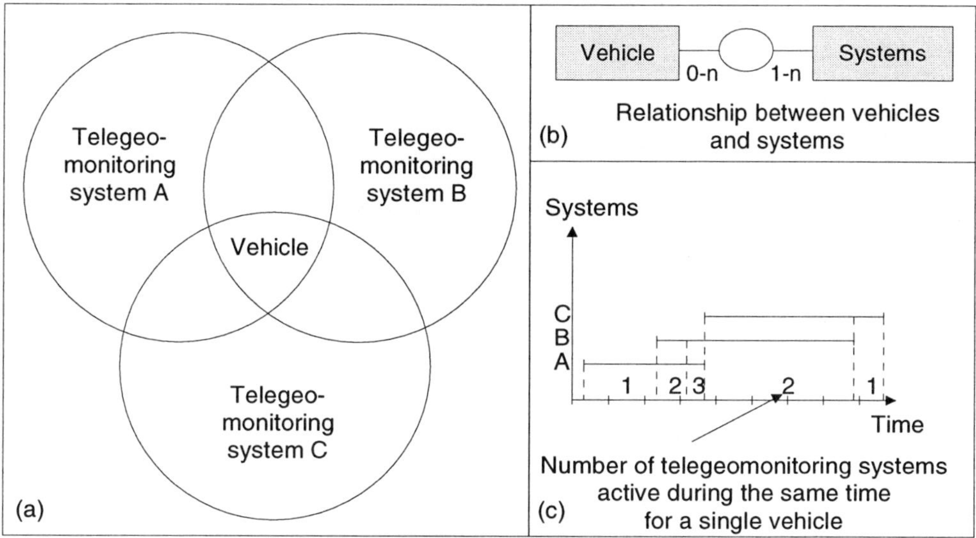

Figure 2.8 A single vehicle can belong to several telegeomonitoring systems. (a) ownership diagram. (b) relationship between vehicles and telegeomonitoring systems. (c) temporal sequence.

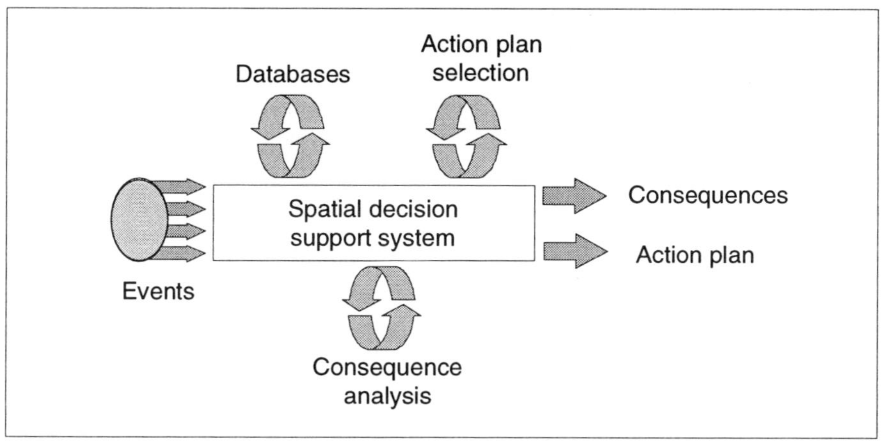

Figure 2.9 Environment of a real time decision support system (Tanzi, 1998)

An example of the functional architecture of such a real time decision making support system and of its environment in given Figure 2.9 (Tanzi, 1998). On the other hand Figure 2.10 details functional aspects of both central and embarked sites. A substantial literature exists on interactive decision support systems, but it is much more limited with respect to real time and spatial DSS; in addition, to our knowledge, little research has been done on cognition for real time systems.

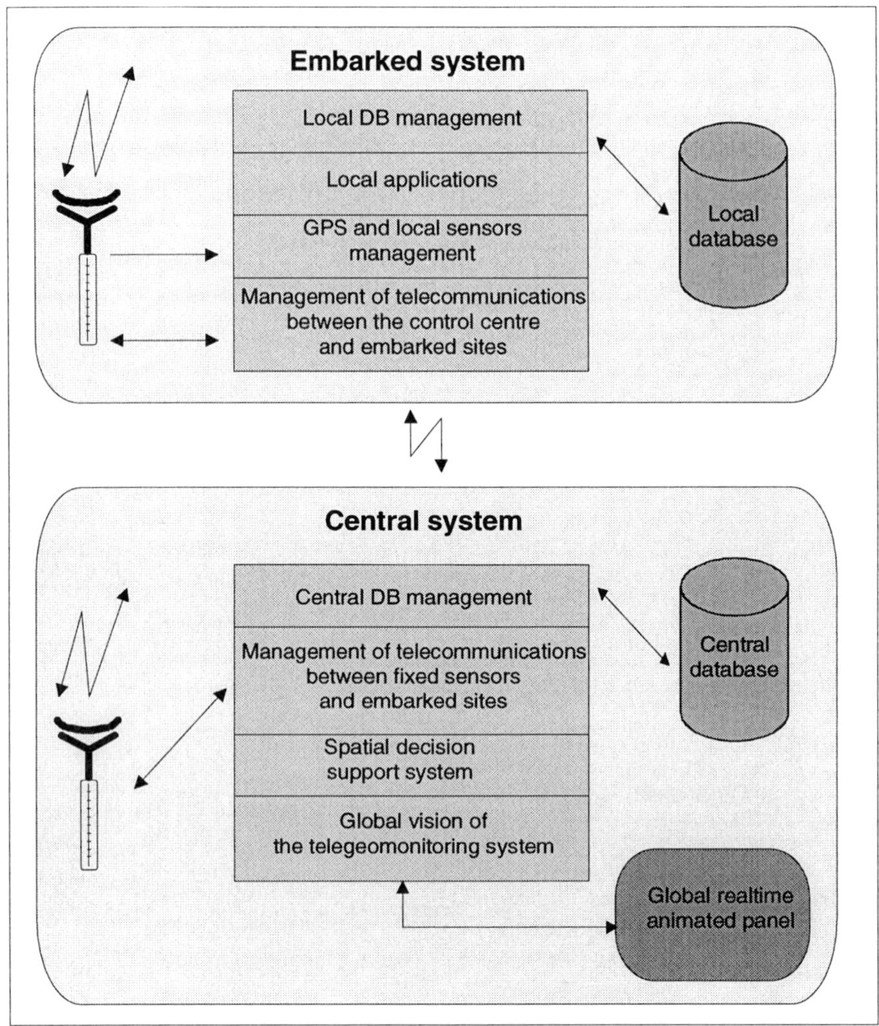

Figure 2.10 Functional details for central and embarked systems.

2.3.3 Comparison

Starting from these studies, a comparison can be made (Table 2.1) emphasising common issues and differences between several classes of applications, from a functional and an architectural point of view.

2.4 DEFINITION AND DIRECTIONS OF RESEARCH

Having presented and analysed some examples, in this section, we will try to give a definition of telegeomonitoring and some research directions.

2.4.1 Towards a definition of telegeomonitoring

Telegeomonitoring can be seen as an extension of GIS by adding some telecommunication aspects. In reality, we do think that it is more linked to real time spatial decision support systems. Now, we can define telegeomonitoring as 'a new discipline trying to design and structure geographical information systems functioning in real time with sensors and embarked or mobile components, exchanging data in order to allow short term and long term decision-making based on data regularly received in real time by any means of telecommunications'.

2.4.2 Research possibilities

Of course, we can state that these applications need first to carry out research separately in GIS (enhancing data structure performances for instance) and in telecommunications (increasing information flow rates, for instance). In addition, research must also be performed in multi-source data fusion, distributed computer systems, interoperability, and on real time cognitive aspects. But, we suggest that the most important research must be done when cross-fertilising the two "parents". Among the possibilities, we note:

- Architecture of telegeomonitoring systems, i.e. including several sites, embarked computers, sensors, by studying especially their characteristics, their advantages, their drawbacks, and so on;
- Architecture of the control centre on which data are converging from different multimedia sources and sensors distributed over space;
- Architecture of information systems with components mobile both in time and in space;
- Architecture of telegeomonitoring systems without any control centre; taking transmission errors and crashes into account must lead to the study Byzantine situations with geographical elements;
- Design of huge visual panels (synoptics), visual synthesis of real time spatial information, and real time generalisation;
- Real time fusion of geographical multi-source and multimedia data;
- The design of real time multimedia database systems, and especially spatio-temporal indexing;
- Interoperability of several telegeomonitoring systems;
- Cognition in real time decision-making.

Table 2.1 Common characteristics of some classes of telegeomonitoring systems

Activities		**Examples of applications**				
		Motorway traffic management	**Fleet management**	**Hazmat transportation**	**River pollution monitoring**	**Risk monitoring**
Computer Architecture	**Using of GPS**	Yes, on some vehicles	Yes, on all vehicles	Yes, on all vehicles	Possibly for some sensors	Possibly for some sensors
	Fixed sensors	Yes	Possibly	Possibly	Yes	Yes
	Embarked components	Yes	Yes	Yes	Generally no	Generally no
	Real time DB	Yes	Yes	Yes	Yes	Yes
	Data exchange	Yes, between vehicles, sensors and control centre	Yes, between vehicles, sensors and control centre	Yes, between vehicles, sensors and control centre	Yes, between all components	Yes, between all components
	Control centre	Yes	Yes	Yes	Yes	Yes
Functional Architecture	**Real time decision support system**	Yes	Yes	Yes	Yes	Yes
	Anticipation by simulation	Yes	Yes	Yes	Yes	Yes
	Animated cartography	Yes	Yes	Yes	Yes	Yes

2.5 CONCLUSION

The objective of this chapter has been only to give some founding elements of telegeomonitoring, to describe some classes of exemplary problems, to stress their similarities, and to launch some novel research directions. Obviously, other reflections must be launched in order to have a very satisfactory starting basis. One very important aspect is that the combination of GIS and telecommunications can be seen as an evolution of GIS, implying stronger connections between computer scientists and geographers. Many cases have not been examined in this chapter; for instance we might mention disaster management and the digital battlefield. See for instance the following URL's *http://www.ttcus.com/digbat/agenda.html* or http://www.ifp.uiuc.edu/nabhcs/reports/P3.html

2.6 ACKNOWLEDGEMENTS

The author wishes to thank all persons who have helped me in writing this chapter, and particularly Azedine Boulmakoul, Tullio Tanzi and Sylvie Servigne.

2.7 REFERENCES

Ben Romdhane, L. and Elayeb B., 1996, Using Beliefs for Faults Identification in the Byzantine Framework *Proceedings of the ISCA Conference on Parallel and Distributed Computing Systems*, edited by K. Yetongnon and S. Hariri, (Dijon: ISCA), pp. 567-574.

Boulmakoul, A. and Laurini, R., 1998, Système géographique environnemental pour la supervision des transports des matières dangereuses. Submitted to *Revue Internationale de Géomatique*.

Boulmakoul, A., Laurini, R., Tanzi, T., Elhebil, F., Zeitouni, K. and Aufaure, M. A., 1997a, *Un système d'information environnemental urbain pour la surveillance du transport des matières dangereuses de la ville de Mohammedia-Maroc. Conférence Européenne sur les Technologies de l'Information pour l'Environnement*, INRIA, Strasbourg, 10-12 Septembre 1997 Metropolis, pp. 187-196.

Boulmakoul, A., Zeitouni, K., Laurini, R. and Aufaure, M. A., 1997b, Spatial Decision Support System for Hazardous Materials Transportation Planning in the Mohammedia region. In *8th IFAC/IPIP/IFORS Symposium on Transportation* Systems'97. (Ghania: IFAC/IPIP/IFORS), pp. 611-616.

Date, C. J., 1987, *Twelve Rules for a Distributed Database. InfoDB*, **2**, 2 and 3, Summer/Fall

Burleson, D, K., 1994, *Managing Distributed Databases*. (Chichester: Wiley).

Laurini, R., 1998, La Télégéomatique, problématique et perspectives. Journées Cassini 1998, Marne La Vallée, 25-27 novembre 1998. *Revue Internationale de Géomatique*, **8**, 1-2, pp. 27-44.

Simon, E., 1996, *Distributed Information System, From Client-server to Distributed Multimedia*. (Columbus, OH: McGraw-Hill)

Tanzi, T., 1998, *Système spatial temps réel d'aide à la décision ; application aux risques autoroutiers*. Doctoral Thesis Ingénierie Informatique, (Lyon: INSA de Lyon).

Tanzi, T., Guiol, R., Laurini, R. and Servigne, S., 1997, Risk Rate Evaluation for Motorway Management. *Proceedings of the International Emergency Management Society Conference,* Copenhagen, June 10-13, 1997, edited by V. Andersen and V. Hansen, pp. 125-134. *Safety Sciences* **29**, pp.1-15.

Tanzi T., Laurini R., and Servigne S., 1998, Vers un système spatial temps réel d'aide à la décision, Journées SIGURA Octobre 1997. *Revue Internationale de Géomatique* **8**, 3 pp.33-46

Tanzi, T. and Servigne, S., 1998, A Real Time Information System for Motorway Safety Management. *Proceedings of the International Emergency Management Society Conference* (Washington DC: IEMS)

3

Models and queries in a spatio-temporal GIS

Baher El-Geresy and Christopher Jones

3.1 INTRODUCTION

Much interest has been evidenced lately in the combined handling of spatial and temporal information in large spatial databases. In GIS, as well as in other fields (Silva *et al.*, 1997), research has been accumulating on different aspects of spatio-temporal representation and reasoning (Stock, 1997). The combined handling of spatio-temporal information allows for more sophisticated application and utilisation of these systems. Developing a Temporal GIS (TGIS) leads to a system which is capable of tracing and analysing the changing states of study areas, storing historic geographic states and anticipating future states. A TGIS can ultimately be used to understand the processes causing geographic change and relate different processes to derive patterns in the data.

Central to the development of a TGIS is the definition of richer conceptual models and modelling constructs. Several approaches have been proposed in the literature for conceptual modelling in a TGIS. These have been previously classified according to the type of queries they are oriented to handle, viz. What, Where and When, corresponding to space, feature and event (Peuquet and Duan, 1995). Other classifications of these approaches were identified on the basis of the modelling tools utilised, for example, entity relationship diagrams, semantic or object-oriented models, etc. In this paper, an exhaustive review and classification of conceptual models for a TGIS is presented with the aim of understanding the different dimensions and complexity of the problem domain and identifying the gaps and links between the different views.

In section 2, the dimensions of the problem domain are identified. Possible query types in a spatio-temporal GIS are identified and discussed in section 3. Section 4 presents a framework with which conceptual modelling approaches for a TGIS can be categorised and studied. Models are classed as basic, composite and advanced. Discussion and a conclusion are given in section 5.

3.2 THE PROBLEM SPACE AND DATA SPACE

In what follows, the term *State* shall be used to denote the collection of spatial and non spatial properties of an object or a location at a specific point in time, and the term *Change* is the line connecting two different States of the same object.

In spatio-temporal applications of GIS, the main entities of concern are *States* of objects or features, their relations with space and time, and their inter-relations in space and time. Accordingly, the problem space of a TGIS can be modelled on three axes as shown in Figure 3.1(a).

1. A Spatial axis representing the location of object States,
2. A Temporal axis for time stamping object States and measuring the duration of Change between States.
3. A Semantic axis representing the classifications of objects or features.

The problem space defined by the three axes is infinite reflecting the infinite nature of space and time and all possible semantic classifications. For specific applications which are limited to considering specific object types, the problem space is reduced to a finite *Data Space*.

Each point in the data space represents a unique object *State* as shown in Figure 3.1(b). *Change* in the *States* of objects is represented by two or more points. In a rich data environment where States or Changes are monitored continuously, Change is represented by a connected line. Hence, the State ST_i of a location or an object in space can be represented by a tuple (O_i, S_i, T_i), where O_i is occupying a location S_i at time T_i. The incremental *Change* which an object undergoes in its spatial or non spatial properties can be represented by a vector which connects both its previous and current *States*.

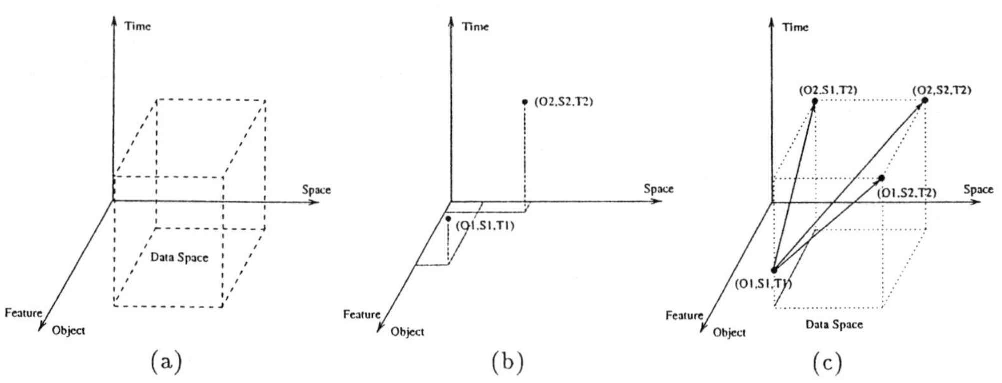

Figure 3.1 (a) Problem space and Data Space. (b) Object State defined as a point in the problem space. (c) Possible types of Change in object States.

Three possible types of Change can be distinguished, as shown in Figure 3.1(c). These can be defined as follows.

- Spatial Change: is the change in the spatial properties only of an object which transforms it from state $ST_1 = (O_1, S_1, T_1)$ to state $ST_2 = (O_1, S_2, T_2)$.
- Non Spatial Change: is the change in the non spatial properties only of an object which transforms it from state $ST_1 = (O_1, S_1, T_1)$ to state $ST_2 = (O_2, S_1, T_2)$.

- Total Change: is the change in both spatial and non spatial properties of an object which transforms it from state $ST_1 = (O_1, S_1, T_1)$ to state $ST_2 = (O_2, S_2, T_2)$.

3.3 A TYPOLOGY OF QUERIES IN A SPATIO-TEMPORAL GIS

Many points in the Data Space defined above may be void as they cannot have possible, valid values. This stems from the fact that features or objects may not exist at certain locations at specific times. Those void points can only be recognised when a query of the type, "Did object O_i occupy location S_j at time T_k?", posed to the spatio-temporal database fails to return a valid result. In this section, we attempt to identify the possible type of queries in a spatio-temporal database based on the type of search space they create.

3.3.1 Search Space for Spatio-Temporal Queries

A spatio-temporal query may include a reference to individual points on the different axes or to a range of points on those axes. A point on any one axis in the Data Space defines a plane which is parallel to the plane of the other two axes in that space. For example, Figure 3.2(a) shows how a point on the Space axis, representing the location S_1, defines a plane parallel to the Time-Object plane. Similarly, a range of values on any one axis defines a rectangular volume in that space. For example, Figure 3.2(b) shows the volume created by the time interval (t_1, t_2), and so on.

The plane and the volume defined in the above figure are representations of the confined search space of queries about S_1 in the plane and queries in the time interval (t_1, t_2) in the volume.

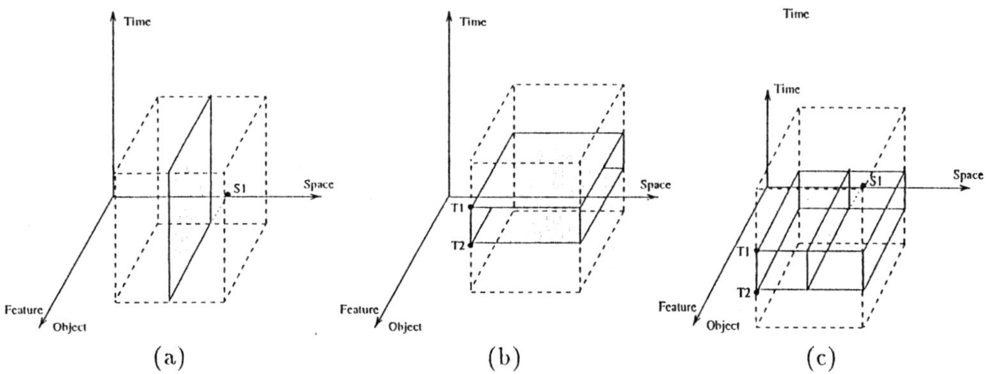

Figure 3.2 (a) Example of a Plane search space. (b) Example of a Volume search space. (c) Search space resulting from the intersection of planes and volumes.

More generally, the search space defined by a query to a spatio-temporal database is the result of the intersection of various planes and volumes defined by the elements of the query. For example, the search space of a query of the form "Find all the features which existed at location S_1 between times T_1 and T_2 " is defined by the plane resulting from the intersection of the plane defined at S_1 and the rectangular volume defined by T_1 and T_2 as shown in Figure 3.2(c).

Hence, spatio-temporal queries can be classified according to the dimension of their search spaces as follows[1]:

- **Point Search Space Queries (P_TSS):** where the search space is a point defined by the intersection of three planes. For example, the query "Did feature F_1 exist in location L_1 at time T_1 ?" defines a point search space as shown in Figure 3.3(a).
- **Line Search Space Queries (LSS):** where the search space is a line defined by the intersection of two different planes. Line queries can be further classified according to the axis to which the line is parallel.
    - **LSS on the Space Axis:** where the search space is a line defined by the intersection of Time and Feature planes. For example, the query "What are the locations of the features F_1 at time T_1 ?" is a Space-Line query as shown in Figure 3.3(b).
    - **LSS on the Feature Axis:** where the search space is a line defined by the intersection of Time and Space planes. For example, the query "What were the features that existed at location L_1 at time T_1 ?" is a Feature-Line query.
    - **LSS on the Time Axis:** where the search space is a line defined by the intersection of Feature and Space planes. For example, the query "When did feature F_1 exist at location L_1 ?" is a Time-Line query.
- **Plane Search Space Queries (P_NSS):** where the search space is a plane defined by the intersection of a plane and a volume in the data space. Plane queries can be further classified according to the axes plane to which the resulting plane is parallel.
    - **Space-Time P_NSS Queries:** where the search space is a plane defined by the intersection of Feature plane and a Time or a Space volume. The query "Where did feature F_1 exist between times T_1 and T_2 ?" is an example, as shown in Figure 3.3(c).
    - **Feature-Time P_NSS Queries:** where the search space is a plane defined by the intersection of a Space plane and a Time or a Feature volume. The query "What were the features that existed at location L_1 between times T_1 and T_2 ?" is an example.
    - **Space-Feature P_NSS Queries:** where the search space is a plane defined by the intersection of a Time plane and a Feature or a Space volume. For example, the query "Which features existed between locations L_1 and L_2 at time T_1 " is an example.
- **Volume Search Space Queries (VSS):** where the search space is a volume defined by the intersection of two different volumes. For example, the query "What were the

[1] Note the difference between the search space that the query defines and the result of the query.

features that existed between time points T_1 and T_2 in the space extension S_1 to S_2 is an example of a Volume query as shown in Figure 3.3(d).

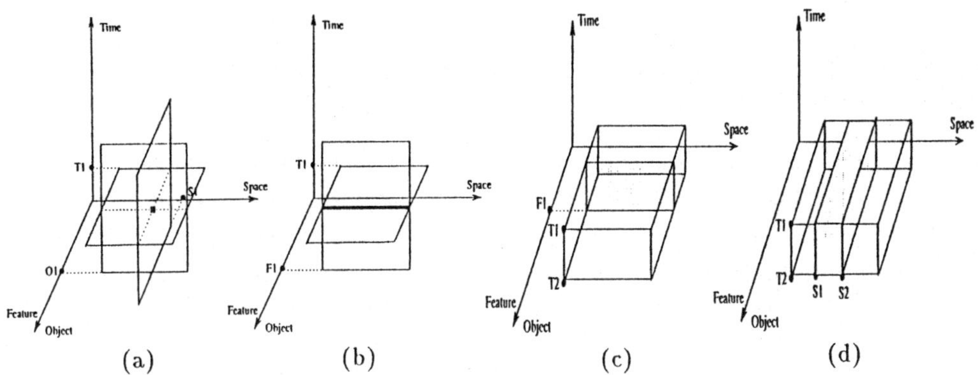

Figure 3.3 Examples of (a) P_TSS queries. (b) LSS queries. (c) P_NSS queries. (d) VSS queries.

3.3.2 Complex Queries

A complex query in a spatio-temporal GIS involves the derivation of objects or relations which are not defined explicitly. Those types of queries usually involve the definition of two or more search spaces which are studied in sequence. For example, the query "When were features F_1 and F_2 in close proximity anywhere in the space?" is a complex query. It involves several steps. First, a P_NSS is defined at F_1 and another at F_2 to identify all their instances. A space range is defined around the identified instances of F_1 to model its close proximity. Then the final search spaces are planes constructed by extending the space range around F_1 to intersect with the P_NSS at F_2. An illustrative diagram of these steps is shown in Figure 3.4, where instances of F_1 are shown as black dots and F_2 as white dots. F_1 and F_2 are in close proximity only at time T_2.

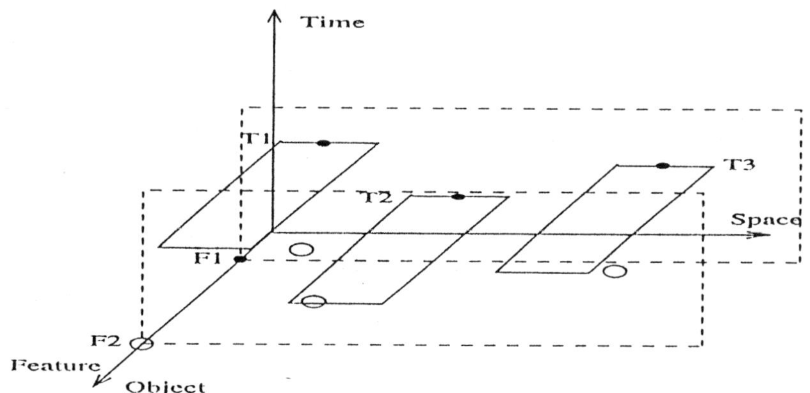

Figure 3.4 Example of a complex spatio-temporal queries.

3.4 CONCEPTUAL MODELLING IN A TEMPORAL GIS

Conceptual modelling is essentially a process of identifying semantic classification and relations between data elements. The process of classification is one which identifies distinguishing properties, relations and distinguishing operations for a certain group of entities. In general three possible types of relations can be classified between entities in a TGIS, namely, Spatial, Temporal, and Causal.

Levels of conceptual models may be distinguished by the semantic classifications used and types of relations that are explicitly defined.

In this paper, conceptual models for a TGIS are categorised by analysing their ability for representing and classifying entities in the Data Space and for representing different types of relations in that Space.

3.4.1 The Basic Models: Where, What and When

Basic conceptual models for a TGIS are built around the principal axes of the problem space: Space, Feature and Time.

3.4.1.1 Location-Based Models: The Where View

In this view, classifications are based on locations on the Space axis. A grid is used to divide up the space into locations. For each location, Changes are recorded in a list when they occur. This approach is illustrated by a set of n parallel Feature-Time planes in the data space, $\sum_{j=1}^{n} (O, S_j, T)$, as shown in Figure 3.5. No redundant data elements are stored in this model. Only the changing States are recorded and hence the model is compact. The model is most efficient for answering location-based queries, and is usually associated with raster data sets.

The main limitations are that object or feature-based queries are difficult to handle, and that Events are not explicitly recorded. An example of this model is given by Langran (1993).

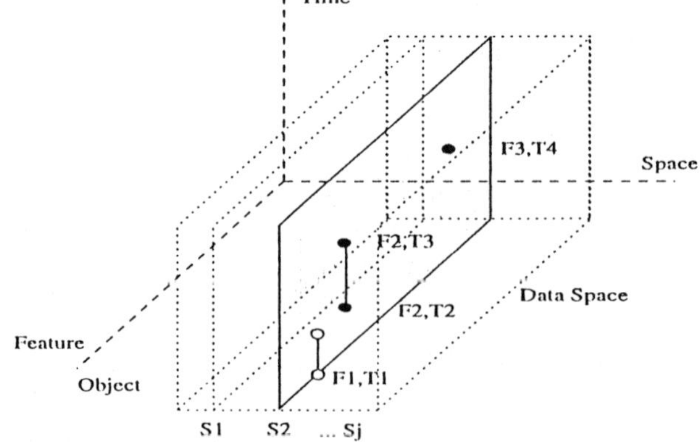

Figure 3.5 Location-based models. Information is recorded on locations in space, where S_2 first held F_1, then F_2 and finally F_3.

3.4.1.2 Object or Feature-based Models: The What View

In this view, classifications are based on geographic features or objects on the Feature axis. Changes are recorded by updating stored instances and reflecting the change of their spatial extent. For example, the extent of polygonal or linear geometries is changed incrementally over time.

This approach shares the same advantages of the location-based model, but from the view point of vector data sets and also shares the limitation of not recording Events explicitly. The feature-based approach is illustrated in Figure 3.6 as a set of n parallel Space-Time planes, $\sum_{j=1}^{n}(O_i, S, T)$, in the Data Space. This approach was first proposed by Langran (1993) and is the basis of the works in (Raffat *et al.*, 1994, Voigtmann *et al.*, 1996, Ramachandran *et al.*, 1996). Hazelton (1991) and Kemelis (1991) suggested extending the model of Langran by using an extended feature hierarchy of nodes, lines and polygons or surfaces within the context of a 4-dimensional space model. However, they preserved the main features of the original model.

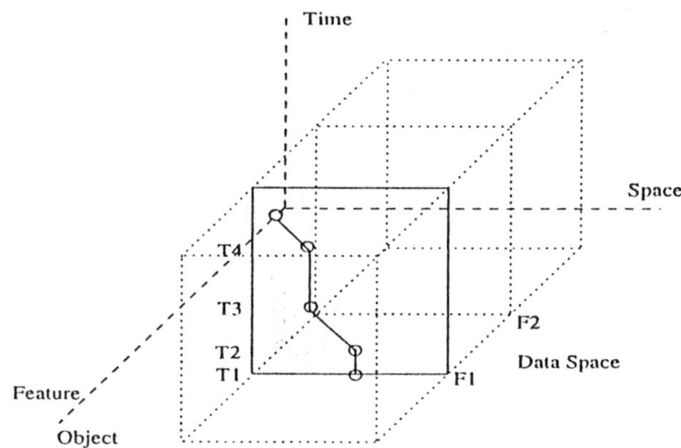

Figure 3.6 Object or feature-based view. Information is recorded on features and their locations at different times.

3.4.1.3 Time-based Models: The Snapshot View

In this view, classifications are based on the temporal axis, where snapshots of the State of the world at specific times are captured. Raster and vector data sets can be represented in this model. The main limitation here is the unavoidable redundancy in the data recorded where objects or locations do not change in a step-like fashion.

The approach is illustrated as a series of parallel Space-Object planes, $\sum_{k=1}^{m}(O, S, T_k)$, in the data space, as shown in Figure 3.7. This is the most common approach used in many approaches to this problem (Peuquet and Qian, 1995). As with the previous two approaches, queries about events in this model are complicated and are likely to require lengthy processing.

A State is the main entity type in all the above basic models. The main limitation is the inability to view the data as sets of events, that is, to represent the changes of different objects which makes it difficult to handle queries about temporally related events, for example, "which areas suffered a land-slide within one week of a heavy rainfall?".

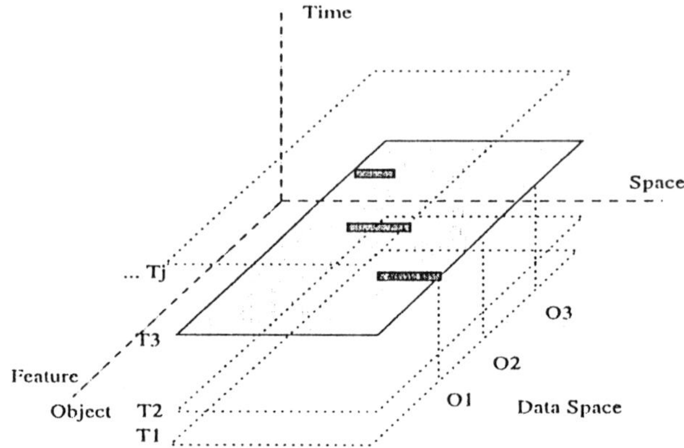

Figure 3.7 Time-based, or Snapshot models. Information is collected at time points on features in space.

3.4.1.4 Event-based Models: The When View

In this model, temporal relations between two successive States of objects or locations in space are defined explicitly, and Change is represented by an Event. Hence, an Event is defined as the line joining two States in the data space.

This model deals with more abstracted relations than the previous ones. It has the advantage of dealing equally with both location and object. Queries involving temporal relations between Changes can be handled efficiently. The model is illustrated graphically in Figure 3.8. In the figure, e_1 records the spatial change of F_1 and e_2 records the spatial change of e_2 , and e_1 overlaps e_2 .

The works of (Edwards *et al.*, 1993; Langran, 1993; Yuan, 1994; Peuquet and Duan, 1995; Worboys, 1995) fall into this category.

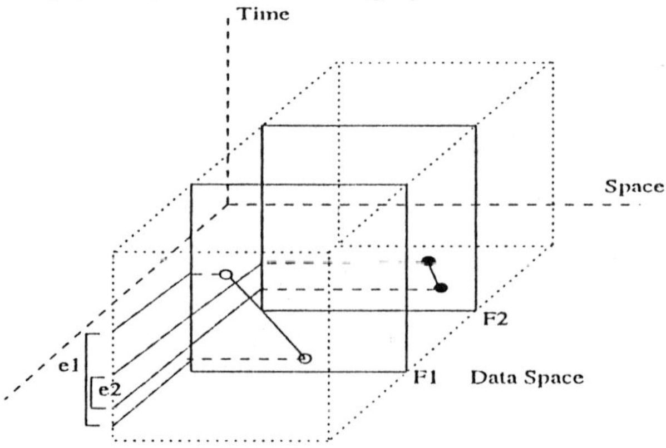

Figure 3.8 Event-based view. Relationships between events can be modelled, e.g. e1 overlaps e2.

3.4.1.5 Integrated Event Model

Events can refer to space locations or to objects and features. The TRIAD model presented in Peuquet and Qian (1996) uses pointers to link location, feature and time views. It stores successive changes in locations (as in the location-based view) which gives the full history of grid cells. However, it stores only two spatial delimiters of features representing their spatial extents at the time of their creation (or the time of recording their existence) and their current extent (or the extent at the end of existence).

This strategy is limiting as it assumes the history of spatial change of a feature to be dependent only on the comparison of their first and last instances without reference to change in-between. Also, the spatial extent of features is approximated by bounding rectangles for areal features and end points for linear features, which necessitates retrieving the history of enclosed grid cells inside the bounding frame to capture the full history of spatial change of the feature.

The TRIAD model can therefore be categorised as a Location-biased Event-based model. This integrated model retains the advantages of the Event-based model where locations and features are linked in an event-based view.

3.4.1.6 Space-Composite Models

In this model, intersection relations are defined explicitly between states of different objects at different times from the snapshots. Hence, the space is decomposed or reclassified as units of a coherent history. The approach was proposed by Langran (1993) where the method can be classified as a Space-Time composite. An illustrative figure of the model is shown in Figure 3.9. In 3.9(a) features F_1, F_2 and F_3 are shown at time slices T_1 and T_2. In 3.9(b), the time slices are overlaid, and the distribution of features in space is presented. The term $(O_1 \rightarrow O_2)$ denotes that the space was occupied by O_1 and then by O_2.

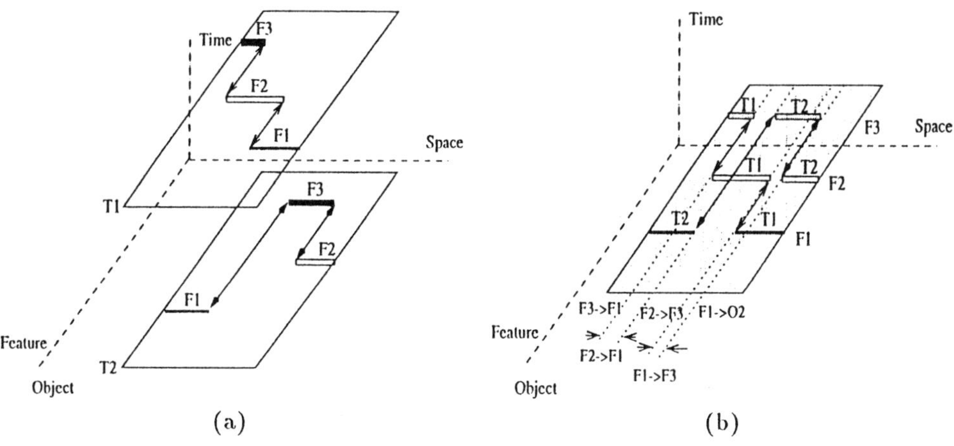

Figure 3.9 (a) Three features at different time slices. (b) Composite view stores the history of locations in space.

The main advantage of this model is the ability to answer queries about the feature history of a location. Using a join operation, the State of the world at a specific time can be rebuilt by searching the history list of each polygon to locate the attribute set that was current at the desired time slice. Hence, it combines the advantages of both the location-based and feature-based views. However, every time a new snapshot is added, an intersection operation must be carried out to reclassify the space and rebuild the history. This has the disadvantage of progressive fragmentation of objects into smaller ones and altering the identifiers of changed objects retroactively.

3.4.2 Advanced Models: How and Why

In all the previous views the main concern was to retrieve States and Changes based on location, object or feature type and temporal properties. A more advanced modelling exercise is to retrieve Changes based on their underlying processes and on their interaction. These type of models can be broadly classified into the How and Why views.

3.4.2.1 Process-oriented Models: The How View

In this approach, spatial relations between successive states of objects are defined explicitly and classified into specific processes. This is equivalent to defining a new axis in the Data Cube with Change, and not State as variable.

Three models in the literature can be classified as Process-oriented. Gagnon *et al.* (1992) presented taxonomies to define three types of Change: those involving one entity, two entities and *n* entities. The first, one-entity, group includes the change of position, size, shape and orientation. They noted that "spatial relationships can be modified only via modifications to the basic spatial variables of one or both of the spatial entities. Any change of spatial relationship between two spatial entities can be deduced from changes to the geometric data", that is, the change in spatial relationships can be brought about by changing the size, orientation, shape and proximity of one or both entities. In the case of two and *n* entities, two basic ways are defined to change spatial variables of the group: changing the basic spatial attributes of individual entities or changing the appearance (birth and disappearance) of individual entities.

Claramunt and Therialut (1995, 1996) proposed a conceptual process-oriented model where changes are classified into three categories. These are: a) evolution of a single entity, similar to what is proposed in Gagnon *et al.* (1992), b) functional relationships between entities and c) evolution of spatial structure involving several entities. However, the categories are not mutually exclusive, since any Change in the spatial properties of an object will change its neighbours, for example, transition of one object will cause, or will be coupled with transition, disappearance, contraction or deformation of another object.

Renolen (1997) classified six basic types of changes or processes of creation, alteration, cessation, reincarnation, merging/annexation and splitting/deduction. His types are a subset of the types classified by Claramunt and Theriault (1996), except for alteration which groups all possible spatial relations between object states. Cheng and Molenaar (1998) defined seven processes, namely, shift, appear, disappear, split, merge, expand, and shrink. Those processes are a subset of those defined by Claramunt *et al.* (1996).

3.4.2.2 Different How Views

If we consider the spatial change of an object O_i during a specific time interval Δt with a start state $(O_i)_s$ and end state $(O_i)_f$, the How question can be any of the following three views.

- **The Functional View**: Here, the bounding states, $(\Delta O_i)_s$ and $(\Delta O_i)_f$ are the only entities which may be considered, along with their spatial relations. The works in (Gagnon *et al.*, 1992; Claramunt and Theriault, 1995, 1996; Renolen, 1997; Cheng and Molenaar, 1998; Yuan, 1998) can be categorised under this view.
- **The Trend View:** Here, the change of object properties over time is the distinguishing factor. Different trends of change can be classified, for example, slow, fast, etc.
- **The Behavioural View:** Here Change is represented by the full spatio-temporal object. The work of Worboys (1992) can fall under this view. In his work, all the spatio-temporal objects either have a stable or an instantaneous change. Renolen (1997) distinguished four general classes of those types such as step-wise constant, gradually changing, discrete and continuous. The latter was further classified into uniform, smooth and irregular. This behavioural view can also be called a Pattern view, since it handles Change through an integrated pattern in time. In El-Geresy and Abdelmoty (1997), a unified framework for handling the representation and reasoning of both temporal and spatial relations was presented. The study and classification of Change, outside the context of data modelling, was carried out by Galton (1997), where continuous change in spatial regions was classified using three measures of separation of either boundary, interior or size. To represent the continuous change of spatial (or temporal) relations, Freksa (1991, 1992), in a seminal work coined conceptual neighbourhood as a means of expressing gradual change of relations. Egenhofer and Al-Taha (1992) used the intersection model to define "topological distance" factors to describe gradual change of relations. However, the need still exists for a comprehensive systematic study of Change as a distinct spatio-temporal object and the spatio-temporal relations between Changes.

3.4.2.3 Causal Models: The Why View

Causal relations are the third type of distinguishing relations in a temporal GIS. A specific temporal relation always exists between cause and effect. Cause always either precedes, or starts with, the start of its effect. Few models exist which addresses casual modelling in GIS. These are the works of Allen (1995) and Edwards *et al.* (1993). Allen differentiates between the effects caused by other events or an intentional agent (e.g. a person, an animal or an organisation). The uncertainty of the introduction of some attributes was also presented in his work.

Possible classifications of causal relations can be related to the temporal relations between cause and effect, for example, immediate, prolonged, delayed, etc. which can facilitate more sophisticated analysis.

3.5 CONCLUSIONS

In this paper, a typology of spatio-temporal queries is first presented. A critical review is then given of conceptual modelling approaches for a TGIS. The different approaches were classified according to their semantic expressiveness and the types of relations that can be defined explicitly. The following conclusions can be drawn.

- The taxonomy of models is extended to include How and Why views, in addition to the conventional, What, Where and When views.
- Process-based models can be further decomposed into functional, behavioural and trend models. The concepts of process and change as spatio-temporal objects need further comprehensive studies.
- A study of spatio-temporal objects and spatio-temporal relations is needed to understand non-discrete change.
- Few works have been reported on formal modelling of causal relations -a central issue in conceptual modelling for TGIS.

3.6 REFERENCES

Allen. E.,Edwards, G., and Bedard, Y., 1995, Qualitative causal modeling in temporal GIS. In *Spatial Information Theory: A Theoretical Basis for GIS, Proceedings of the International Conference COSIT'97*, (Berlin: Springer Verlag), pp. 397-412.

Cheng, T. and Molenaar, M., 1998, A process-oriented spatio-temporal data model to support physical environment modeling. In *Proceedings of the 8th International Symposium on Spatial Data Handling*, (Vancouver: IGU Commission of GIS), pp. 418-429.

Claramunt, C. and Theriault, M., 1995, Managing time in GIS: An event-oriented approach. In *Recent Advances on Temporal Databases*, (Berlin: Springer Verlag).

Claramunt, C. and Theriault, M, 1996, Towards semantics for modelling spatio-temporal processing within GIS. In *Proceedings of the 7th International Symposium on Spatial Data Handling*, Vol. 2, (Charleston: IGU Commission of GIS), pp. 27-43.

Egenhofer, M.J. and Al-Taha, K.K., 1992, Reasoning about gradual changes of topological relationships. *In Theories and Methods of Spatio-Temporal Reasoning in Geographic Space*, (Berlin: Springer Verlag), pp. 196-219.

Edwards, G., Gagnon, P., and Bedards, Y., 1993, Spatio-temporal topology and causal mechanisms in time integrated GIS: from conceptual model to implementation strategies. In *Proceedings of the Canadian Conference on GIS*, pp. 842-857.

El-Geresy B.A. and Abdelmoty A.I.. 1997, Order in space: a general formalism for spatial reasoning. *International Journal on Artificial Intelligence Tools*, **6**, pp. 423-450.

Freksa, C., 1991, Conceptual neighborhood and its role in temporal and spatial reasoning.. In *Decision Support Systems and Qualitative Reasoning*, (North Holland: Elsevier Science Publishers), pp. 181-187.

Freksa, C., 1992, Temporal reasoning based on semi-intervals. *Artificial Intelligence*, **54**, pp. 199-227.

Galton, A. 1994, Continuous change in spatial regions. In *Spatial Information Theory: A Theoretical Basis for GIS, Proceedings of the International Conference COSIT'97*, (Berlin: Springer Verlag), pp. 1-13.

Gagnon, P, Bedard, Y., and Edwards, G., 1992, Fundamentals of space and time and their integration into forestry geographic databases. In *Proceedings of IUFRO Conference on the Integration of Forest Information Open Space and Time*, (Canberra: Anutech Ply Ltd.), pp. 24-24.

Hazelton, N.W.J, 1991, *Integrating Time, Dynamic Modelling and Geographical Information Systems: Development of Four-Dimensional GIS*, PhD Thesis, (Melbourne: University of Melbourne).

Kelmelis, J, 1991, *Time and Space in Geographic Information: Toward a Four -Dimensional Spatio-Temporal Data Model*, PhD Thesis, (Pennsylvania: The Pennsylvania State University).

Langran, G., 1993, *Time in Geographic Information Systems*, (London: Taylor and Francis).

Peuquet D.J. and Duan N., 1995, An event-based spatiotemporal data model (estdm) for temporal analysis of geographical data. *International Journal of Geographic Information Systems*, **9**, pp. 7-24.

Peuquet, D. and Qian, L., 1996, An integrated database design for temporal GIS. In *Proceedings of the 7th International Symposium on Spatial Data Handling*, Vol. 2, (Charleston: IGU Commission of GIS), pp. 1- 11.

Renolen, A., 1997, Conceptual modelling and spatiotemporal information systems: how to model the real world, In *ScanGIS'97*.

Ramachandran, S., McLeod, F., and Dowers, S., 1996, Modelling temporal changes in a GIS using an object-oriented approach. In *Proceedings of the 7th International Symposium on Spatial Data Handling*, Vol. 2, (Charleston: IGU Commission of GIS), pp. 518-537.

Raafat, H., Yang, Z., and Gauthier, D., 1994, Relational spatial topologies for historical geographical information. *International Journal of Geographic Information Systems*, **8**, pp. 163-173.

Silva, F.L., Principe, J.C., and Almeida, L.B., 1997, *Spatiotemporal Models in Biological and Artificial Systems*, (Amsterdam: IOS Press).

Stock, O., editor, 1997, *Spatial and Temporal Reasoning*, (Amsterdam: IOS Press).

Story, P.A. and. Worboys, M.F. A design support environment for spatio-temporal database applications. In *Spatial Information Theory: A Theoretical Basis for GIS, Proceedings of the International Conference COSIT'97*, (Berlin: Springer Verlag), pp. 413-430.

Voigtmann, A., Becker, L. and Hinrichs, K.H., 1996, Temporal extensions for an object-oriented geo-data model. In *Proceedings of the 7th International Symposium on Spatial Data Handling*, Vol. 2, (Charleston: IGU Commission of GIS), pp. 25-41.

Worboys, M.F., 1992, Object-oriented models of spatiotemporal information. In *Proceedings of GIS/LIS'92*, Vol. 2, (Bethesda, MA: ASPRS), pp. 825-835.

Yuan, M., 1992, Wildfire conceptual modeling for building GIS space-time models. In *Proceedings of GIS/LIS'92*, Vol. 2, (Bethesda, MA: ASPRS), pp. 860-869.

Yuan M., 1998, Representing spatiotemporal processes to support knowledge discovery in GIS databases. In *Proceedings of the 8th International Symposium on Spatial Data Handling*, Vol. 2, (Charleston: IGU Commission of GIS), pp. 431-440.

4

A representation of relationships in temporal spaces

Christophe Claramunt and Bin Jiang

4.1 INTRODUCTION

Time is probably one of the most essential and paradoxical concepts that human beings face. Time is always present in our everyday life from the perception of events to the development of human thinking behaviours. However, time is still a difficult concept to describe and formalise as it has no obvious physical characteristics and properties. We can only establish a temporal statement from the observation or prevision of changes. The relationship between time and space is a consequence of the observation of changes as the perception of spatial alterations denotes the existence of time.

The representation of time within Geographical Information Systems (GIS) is still an important and expected development to make these systems more suited to the temporal analysis of real-world phenomena. Over the past years, the representation of spatio-temporal data has been extensively discussed by different research communities such as the Artificial Intelligence domain that provides a mathematical foundation to the representation of changes in space (Vieu, 1997); temporal database approaches that develop database models and query languages for the description and manipulation of spatio-temporal objects (Wu *et al.*, 1997) and studies oriented to the temporal extension of current spatial data models within GIS (Langran, 1992; Cheylan and Lardon, 1993; Peuquet, 1994; Frank, 1994; Worboys, 1994; Claramunt and Thériault, 1995 and 1996).

This chapter proposes a new reasoning and computational approach that integrates space and time within an integrated temporal space referential. The principles underlying temporal spaces are derived from time geography concepts, spatial and temporal reasoning formalisms. A set of minimal relationships and configurations in a temporal space are identified from the possible combinations of relationships in time and geographical space. Such a model allows the representation and computational study of independent trajectories in space and time. The algebra is illustrated with a case study that outlines some potential benefits of the temporal space model.

The remainder of this chapter is organised as follows. The next section briefly reviews current approaches in the combined representation of space and time. Then we introduce temporal and spatial relationships used for the development of our model. The following sections develop the concept of temporal space, propose a formal model for representing relationships in a temporal space, and illustrate the application of the temporal space model on a case study. The last two sections discuss some implications of our model and draw some perspectives.

4.2 TIME AND GIS

Despite its impenetrable nature, the understanding of time is still an important challenge in the comprehension of the evolution of Earth and the distribution of natural and anthropic phenomena. Until the 20th Century the absolutist (chronological) and relationist (topological relationships) conceptions of time were largely used and discussed by scientific communities. Then, the relativist revolution introduced a new relationship between time and motion that impacted the whole scientific community. Particularly, the relationship between motion, time and space was not henceforward representable using a constant and linear rapport. At present, many disciplines are implicitly integrating some of the outcomes of the relativity concept, cognitive sciences are yet discussing the influence of the environment on time perception. For example, an estimated duration is dependent on the degree of perceived changes or in other words the more static the environment the shorter the perceived duration (Zubek, 1969; Block, 1979; Glicksohn, 1996).

Nowadays the representation and understanding of time and motion are still an active research issue in many sciences related to the representation of real or abstracted phenomena. Particularly, the representation of time within GIS is an important research challenge. GIS is by nature a multi-disciplinary area that combines different sciences and techniques. Therefore, it is not surprising to notice that different research advances are contributing to the development of temporal GIS (TGIS). Over the past years, the representation of spatio-temporal data has been extensively discussed by different research communities. Various models have been proposed and examined, among others:

- Artificial Intelligence approaches that provide a mathematical foundation to describe the spatial changes of an individual region (see Vieu, 1997 for an overview). The objective of these models is the reduction of the representation of spatial changes to a minimal set of primitive concepts that can be used in spatial reasoning and computational implementations.
- Temporal database approaches that develop models and query languages for the description and manipulation of spatio-temporal objects (Wu *et al.*, 1997 provide a bibliography). The integration of the temporal dimension is generally considered as an extension of current database models and query languages. Such extensions integrate the time component as a representable and specific dimension and provide design patterns, temporal data types and operators that can be integrated at the data definition and manipulation levels.
- Temporal GIS research that aims at the extension of spatial data models towards the integration of time (Langran, 1992; Peuquet, 1994; Frank, 1994; Claramunt and Thériault, 1995). Current proposals extend the cartographical foundation of spatial data models towards the temporal dimension and attempt the identification of new spatio-temporal models for TGIS.

These different disciplines pursue a common objective: the representation and manipulation of the spatial changes of an individual region. They are complementary in terms of their achievements as they deliver formal, database or TGIS models, respectively.

Significant progress has been made in the qualitative description and manipulation of changes in space. Taxonomies of spatial changes (Claramunt and Thériault, 1995; Claramunt *et al.*, 1998; Horsnby and Egenhofer, 1998), and transitive changes of spatial relationships have been identified (Randell *et al.*, 1992; Egenhofer and Al-Taha, 1992; Galton, 1995). Representing individual changes in space (i.e., endogeneous changes) is a first step toward the integration of time within GIS. A complementary theoretical issue is the development of formal models for studying spatio-temporal patterns that involve the interaction of several regions in space and time (i.e., exogeneous changes). Interactions in space and time have been widely studied in time geography, individual trajectories are analysed and compared using an integrated time and space referential (Hägerstrand, 1967). The study of individual trajectories is particularly relevant for the analysis of diffusion mechanisms, crime patterns or the propagation of epidemics.

However, current spatio-temporal models still need some formal extensions oriented towards a qualitative and cross-comparative description of individual regions in both space and time, that is, the study of independent trajectories in space and time. The manipulation within GIS of individual trajectories in space and time still implies the development of formal operators that combine spatial and temporal properties. The objective of this research is the investigation and formalization of a set of minimal operators that support the cross-comparison of individual regions within an integrated space and time referential.

A formal foundation, i.e., a language and a set of minimal operators, is still required for the manipulation and comparison of regions in space and time. The research described in this chapter addresses the development of a formal language that can identify a set of basic operations between individual regions in space and time. Our objective is not to replace current spatial or temporal algebra with an unified language, but instead to present a complementary view of the possible spatial and temporal interactions of regions in space and time.

4.3 TEMPORAL AND SPATIAL RELATIONSHIPS

Representing time and space as an integrated referential implies the combination of temporal and spatial relationships within an integrated framework. This section introduces the temporal and spatial relationship principles used for the development of our model. Firstly, we introduce some basic temporal hypotheses. We assume time to be continuous, T is the set of measured times isomorphic to the set of real numbers and I is the set of time intervals. Let i be a time interval of I, $i = [t_1, t_2]$ where $t_1, t_2 \in T$ and $t_1 < t_2$. Relationships between these temporal intervals are defined using Allen's temporal operators that define mutually exclusive relationships between time intervals {*equals*, *before*, *meets*, *overlaps*, *during*, *starts*, *finishes*} and their inverses {*after*, *met*, *overlapped*, *contain*, *started*, *finished*}, respectively (does not apply for equal which is a symmetric operator) (Allen, 1984). They are defined in Figure 4.1. Begin(i) and End(i)

are temporal operators which provide respectively the beginning and ending instants of a time interval i of I.

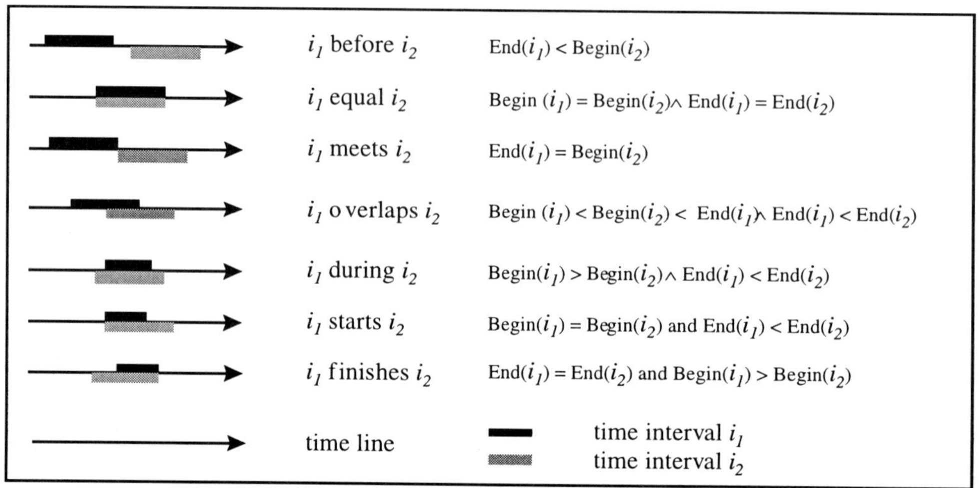

Figure 4.1 Temporal relationships

Similarly, we consider spatial relationships in a two-dimensional space. Different convergent languages have been proposed, using different primitive concepts and calculus, for the identification of exhaustive and pairwise disjoint spatial relationships (Pullar and Egenhofer, 1988; Egenhofer, 1991; Cui *et al.*, 1993; Clementini *et al.*, 1993). The basic relationships identified within these models are convergent. Let us consider the eight basic spatial relationships identified in two-dimensional spaces, i.e., {*equal, touch, in, contain, cover, covered, overlap, disjoint*} (Table 1). Let r_1 and r_2 be two regions in a two-dimensional space, r° denotes the interior of a region of space r, ∂r the boundary of a region of space r. Spatial relationships between r_1 and r_2 are expressed as follows (adapted from Clementini *et al.*, 1993).

Table 4.1 Spatial relationships between two regions in a two-dimensional space

r_1 *equal* r_2	⇔	$(r_1^\circ \cap r_2^\circ = r_1^\circ \cup r_2^\circ) \wedge (\partial r_1 \cap \partial r_2 = \partial r_1 \cup \partial r_2)$	
r_1 *touch* r_2	⇔	$(r_1^\circ \cap r_2^\circ = \varnothing) \wedge (\partial r_1 \cap \partial r_2 \neq \varnothing)$	
r_1 *in/contain* r_2	⇔	$(r_1 \cap r_2 = r_1) \wedge (\partial r_1 \cap \partial r_2 = \varnothing)$	
r_1 *covered/cover* r_2	⇔	$(r_1 \cap r_2 = r_1) \wedge (r_1 \neq r_2) \wedge (\partial r_1 \cap \partial r_2 \neq \varnothing)$	
r_1 *overlap* r_2	⇔	$(r_1 \cap r_2 \neq r_1) \wedge (r_1 \cap r_2 \neq r_2) \wedge (r_1^\circ \cap r_2^\circ \neq \varnothing)$	
r_1 *disjoint* r_2	⇔	$r_1 \cap r_2 = \varnothing$	

4.4 REPRESENTATION OF TEMPORAL SPACES

A combined representation of spatial and temporal relationships is a non straightforward task from both cognitive and formal points of view. Some of these difficulties are linked to the problem of perceiving and representing time in space, and formalising space with time. Spatial and temporal views provide an appropriate description of the relationships that apply in either the spatial or temporal dimensions, respectively. However, none of these views provides a significant understanding of binary relationships in both space and time. Firstly, spatial relationships identified in a spatially-oriented view are valid only during the intersection of the life spans of their operand regions. Secondly, temporal relationships identified by a temporally-oriented view do not provide any information about the spatial relationships of their operand regions. Temporal animation techniques can be used to provide a global view of geographical changes. However, the analysis of individual trajectories using a temporal animation is not suited for the evaluation of precise relationships and is limited to the potential of visual interpretation tasks which are irrelevant when the number of regions is high. Therefore, experimental representations still need to be explored in order to identify relationships in space and time between several individual regions.

We introduce a concept of temporal space that combines the temporal and spatial dimensions. Our model aims at the representation of the minimal combinations of temporal and spatial relationships between two regions in a temporal space. A region in a temporal space is defined as a region of space valid for a temporal interval. In order to construct a temporal space, binary relationships in two-dimensional spaces are mapped (not projected) towards corresponding binary relationships in a one-dimensional space. The second dimension of a temporal space is given by time. Therefore, relationships in a temporal space are derived from the combinations of the minimal spatial and temporal relationships. In order to maintain the properties of time, the temporal dimension of the temporal space is directed, i.e., a region cannot move backward in time. Moreover we assume that a region cannot be in two different locations in space at the same time.

Let us consider two regions in a temporal space, denoted $e_1(r_1, i_1)$ and $e_2(r_2, i_2)$, respectively. Let i° be the interior of an interval of time i, ∂i the boundary of an interval of time I, $i=[t_1, t_2]$, $i^\circ=]t_1, t_2[$, $\partial i=(t_1, t_2)$. Then the minimal set of relationships between two regions in a temporal space is deduced from the combinations of spatial relationships with the possible set-theoretic intersections between the interiors and boundaries of their temporal intervals (TR denotes a temporal relation, SR denotes a spatial relationship, TSR denotes a relation in a temporal space). Table 4.2 describes the possible combinations that lead to eight relationships in a temporal space by analogy to the relationships identified in a classic two-dimensional space (EQUAL for equals, TOUCH for touch, IN for in, CON for contain, CVR for cover, CVRD for covered, OVLP for overlap, DISJ for disjoint). The TSR relationships space qualify the semantics represented in a temporal space: two regions "intersect" in a temporal space if and only if they share a common region of space and a common instant of time, otherwise they are DISJoint. Such a representation provides an algebra that describes relationships using atomic operators. It presents the computational advantage of using existing spatial relationships and operations on temporal intervals.

Table 4.2 Relationships in a temporal space (TSR)

SR	$i_1^\circ \cap i_2^\circ$	$\partial i_1 \cap \partial i_2$	$i_1^\circ \cap \partial i_2$	$\partial i_1 \cap i_2^\circ$	TSR
any	∅	∅	∅	∅	DISJ
¬disjoint	∅	¬∅	∅	∅	TOUCH
disjoint	any	any	any	any	DISJ
touch	¬∅	any	any	any	TOUCH
in	¬∅	¬∅	∅	any	CVRD
	¬∅	any	¬∅	any	OVLP
	¬∅	∅	∅	¬∅	IN
contain	¬∅	¬∅	any	∅	CVR
	¬∅	any	any	¬∅	OVLP
	¬∅	∅	¬∅	∅	CON
cover	¬∅	any	any	∅	CVR
	¬∅	any	∅	¬∅	OVLP
covered	¬∅	any	∅	any	CVRD
	¬∅	∅	¬∅	any	OVLP
overlap	¬∅	any	any	any	OVLP
equal	¬∅	¬∅	∅	∅	EQUAL
	¬∅	∅	¬∅	¬∅	OVLP
	¬∅	any	∅	¬∅	CVRD
	¬∅	any	¬∅	∅	CVR

Table 4.3 summarises the resulting eight minimal and orthogonal TSR relationships in a temporal space obtained from the combination of minimal spatial and temporal relationships. Another interest of this classification is the identification of the different relationship configurations: we define a configuration as a triplet (SR, TR, TSR). Table 4.3 also represents the 71 configurations, 104 with their TR inverse (does not apply for the TR operation equals which is a symmetric one). Inverse temporal relationships lead to inverse TSR relationships in the temporal space, otherwise inverse TSR relationships in the temporal space are given.

The temporal space used for the definition of this algebra considers the spatial relationships line as a dimensional line and the temporal dimension as a second dimensional line. This two-dimensional representation of relationships in a temporal space integrates some visual similarities with the representation of spatial relationships. The main idea of that graphic representation is to project spatial relationships in the vertical spatial relationship axis (i.e., the notion of spatial distance is lost, however that is not a limitation as the spatial distance is a spatial measure available in the spatial dimension). TSR relationships have the computational advantage of being derived from existing spatial and temporal relationships so their implementation can be realised on top of existing TGIS models. Moreover, the different configurations can be represented with a visual language. As the spatial relationship dimension is qualitative only (i.e., the notion of distance between regions is not represented), the 71 configurations of relationships (104 with their TR inverses) in the temporal space are relatively complete in terms of their visual expression. The only minor variation is the possible stretch of the visual relationships in the temporal space (and not along the temporal dimension as the temporal line integrates the concept of temporal distance).

Table 4.3 Visual representation of TSR

TR / SR	equals	before/ after	meets/ met	overlaps/ overlapped	during/ contain	starts/ started	finishes/ finished
equals	EQUAL	DISJ	TOUCH	OVLP	CVRD/CVR	CVRD/CVR	CVRD/CVR
touch	TOUCH	DISJ	TOUCH	TOUCH	TOUCH	TOUCH	TOUCH
in	CVRD	DISJ	TOUCH	OVLP	IN/OVLP	CVRD/OVLP	CVRD/OVLP
contain	CVR	DISJ	TOUCH	OVLP	OVLP/CON	OVLP/CVR	OVLP/CVR
cover	CVR	DISJ	TOUCH	OVLP	OVLP/CVR	OVLP/CVR	OVLP/CVR
covered	CVRD	DISJ	TOUCH	OVLP	CVRD/OVLP	CVRD/OVLP	CVRD/OVLP
overlap	OVLP	DISJ	TOUCH	OVLP	OVLP	OVLP	OVLP
disjoint	DISJ	DISJ	DISJ	DISJ	DISJ	DISJ	DISJ

Relationships in a temporal space provide a set of low-level operations that can be reclassified according to application needs. For example a distinction can be made in terms of set operations. That leads to a distinction being made between DIS and other relationships in a temporal space (EQUAL, TOUCH, IN/CON, CVRD/CVR, OVLP) that can be reclassified to a CONNECT user-defined operation in a temporal space. Such an operation is of interest for identifying individuals in space whose trajectories intersect somewhere and sometime.

4.5 APPLICATION TO A CASE STUDY

In order to illustrate the potential of temporal spaces, let us consider a simplified example which presents five regions located in space (A, B, C, D, E) and valid for some interval of times. Two representations are possible with respect to the application of binary relationships in either time or space. Firstly, Figure 4.2 presents a spatially-oriented view in which spatial relationships between these regions are represented independently of the temporal dimension. Accordingly, Table 4.4 presents the binary spatial relationships (SR) that apply for these regions:

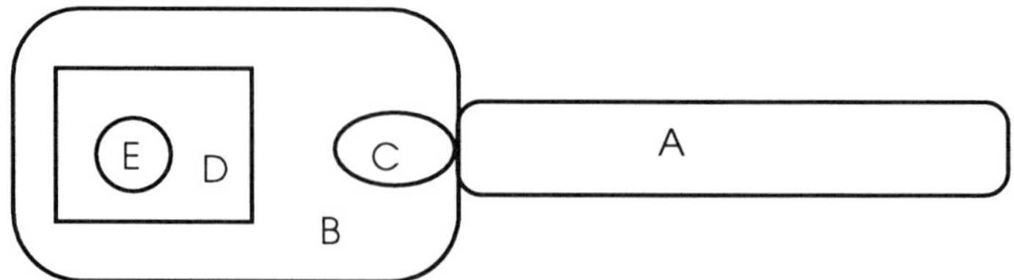

Figure 4.2. Spatially-oriented view (case study)

Table 4.4 Spatial relationships SR (case study)

SR	A	B	C	D	E
A	equals	touch	touch	disjoint	disjoint
B	touch	equals	cover	contain	contain
C	touch	covered	equals	disjoint	disjoint
D	disjoint	in	disjoint	equals	contain
E	disjoint	in	disjoint	in	equals

Similarly, Figure 4.3 presents the set of temporal relationships that hold between these regions independently of the spatial dimension; assuming that the regions of our example hold for some intervals of time.

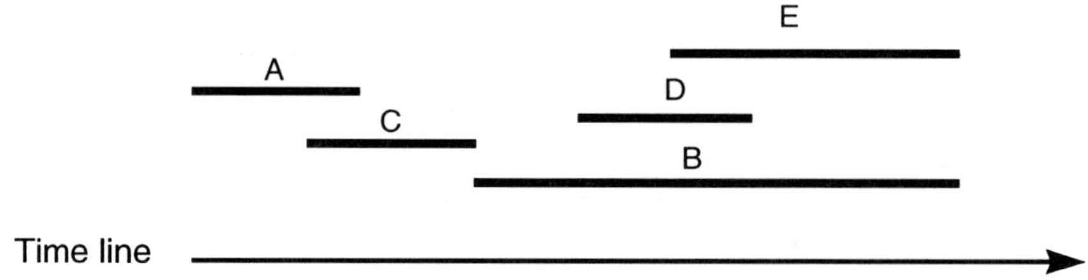

Figure 4.3 Temporally-oriented view (case study)

Temporal relationships between these regions can be expressed. Table 4.5 illustrates the temporal relationships (TR) between the regions of the above example.

Table 4.5 Temporal relationships TR (case study)

TR	A	B	C	D	E
A	equal	before	overlaps	before	before
B	after	equal	met	contains	finished
C	overlapped	meets	equal	before	before
D	after	during	after	equal	overlaps
E	after	finishes	after	overlapped	equal

In order to construct the representation of these regions in the temporal space, spatio-temporal relationships are mapped to their corresponding relationships in a one-dimensional space. The resulting spatial relationships are represented as follows (Figure 4.4).

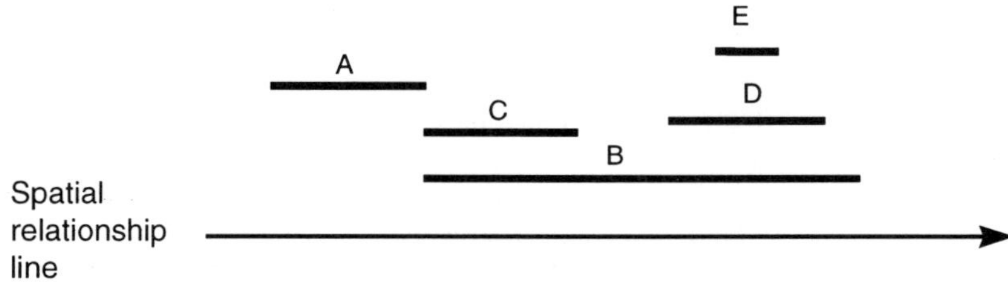

Figure 4.4 Spatially-oriented view (case study)

Therefore, a two-dimensional representation of relationships in a temporal space for our case study is proposed in Figure 4.5. This representation provides some nice properties as relationships in a temporal space integrate both the semantics of relationships in space and time, e.g., B contains D both in space and time, A touches C in space and overlaps C in time.

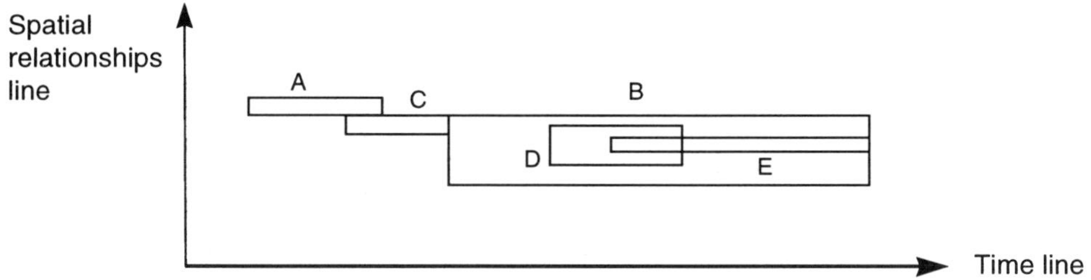

Figure 4.5 Visual representation of TSR (case study)

Accordingly, Table 4.6 summarises the relationships in the temporal space (TSR) that apply for our example:

Table 4.6 Relationships in the temporal space - TSR (case study)

TSR	A	B	C	D	E
A	EQUAL	DISJ	TOUCH	DISJ	DISJ
B	DISJ	EQUAL	TOUCH	CON	CVR
C	TOUCH	TOUCH	EQUAL	DISJ	DISJ
D	DISJ	IN	DISJ	EQUAL	OVLP
E	DISJ	CVRD	DISJ	OVLP	EQUAL

TSR relationships in a temporal space can be analysed in cascade. For example, A and E are DISJoint in a temporal space. However they are CONNECTed in cascade through C then B, and through C then E. Such a transitive property is fundamental for many applications such as epidemiological studies.

4.6 DISCUSSION

The concept of temporal space provides a new experimental approach to the representation of spatio-temporal relationships within GIS. A temporal space has the advantage of providing two levels of relationships: the corresponding 8 relationships commonly identified in two-dimensional spaces and the 71 configurations (104 with their TR inverses) directly derived from the combinations of temporal and spatio-temporal relationships. Temporal spaces do not replace the spatially-oriented or temporally-oriented views of space, instead they allow a different perspective and an integrated view that supports the identification of relationships in space and time. This algebra is suited to the analysis of distinct trajectories in space and time, and their effects in cascade. We can remark that the application of temporal space concepts to the life and motion process of an individual region is not relevant as temporal relationships between the successive regions that represent the evolution of an entity will be limited to meet/met, or even disjoint in the extreme case of a re-incarnation relationships. Then the range of resulting TSR relationships in a temporal space for these regions will be represented as TOUCH relationships (or DISJ in the case of a reincarnation process).

Qualitative representations of geographical spaces introduce the notion of conceptual neighbourhood to identify the direct transitions that can be made between spatial relationships (Freksa, 1991). Similarly, the temporal space model can be discussed with regard to the possible transitions of its TSR relationships. A set of axioms can be derived from the combination of the possible combination of spatial relationships identified in (Randell *et al.*, 1992; Galton, 1995), and from the study of the possible transitions of temporal intervals. The following diagram summarises the possible transitions of spatial relationships (adapted from Cohn *et al.*, 1998). It can be read as, for example, if two regions touch for an interval of time, and both continue to exist after the end of this interval of time, then the spatial relationships between these two regions for the immediate following interval of time are either touch, disjoint or overlap.

disjoint	←→ touch
touch	←→ overlap
overlap	←→ equal, cover, covered
cover	←→ contain
covered	←→ in
equal	←→ cover, covered, in, contain

Immediate transitions of temporal relationships between two regions in a temporal space can be analysed from the study of the possible relationships between the beginning and ending instants of the intervals of time of these regions as we consider the time line as continuous. For example, if the ending instants of the intervals of time of two regions in a temporal space are equal, and if these regions continue to exist after these intervals of time and the beginning instants of the immediate intervals of time of these two regions are equal, then the temporal relationships between these two regions for the immediately following intervals of time are either equal, starts or started. The possible transitions of temporal relationships between two regions in a temporal space are presented in Table 4.7.

Table 4.7 Transitions of temporal relationships betweentwo regions in a temporal space

End(i_1) < End(i_2)		Begin(i'_1) < Begin(i'_2)
before meets overlap during starts	→	before meets overlap contain finished
End(i_1) = End(i_2)		Begin(i'_1) = Begin(i'_2)
equal finishes finished	→	equal starts started
End(i_1) >End(i_2)		Begin(i'_1) > Begin(i'_2)
after met overlapped contain started	→	during after met overlapped finishes

The above table can be read as, for example, if the temporal interval of a region in a temporal space is before the temporal interval of a second region in the same temporal space, and if these two regions continue to exist after the end of this interval of time, then the temporal relationships between these two regions for the immediate following interval of time are either before, meets, overlap, contain or finished. The analysis of possible transitions of relations in a temporal space are also deductible from the possible transitions of their temporal and spatial relationships. Informally, a general axiom can be formulated: there is a possible transition between two TSR configurations in a temporal space (SR_1, TR_1, TSR_1), (SR_2, TR_2, TSR_2) if and only if there is both a possible transition between SR_1 and SR_2, and between TR_1 and TR_2. The application of conceptual neighbourhood principles is particularly interesting in situations in which the knowledge represented is incomplete. The evaluation of possible relationships between non well-defined locations in space and time can be derived and estimated from the analysis of preceding and posterior situations.

4.7 CONCLUSION

Many scientific and administrative studies require a combined representation of spatial and temporal relationships in order to analyse trajectories in space of time. For example, a spatial proximity of two regions in space can be relevant when these two regions exist for at least a common interval of time. Such requirements are important for epidemiological, crime pattern or spatial diffusion studies. This chapter experiments and proposes a concept of temporal space that provides a new approach to the representation of spatio-temporal data within GIS. Temporal spaces do not replace the spatially-oriented or temporally-oriented views of space. Instead they allow a different perspective and an integrated view that support the identification of relationships in both space and time.

From the identification of relationships in space and time, we propose a model that defines a set of minimal relationships within a temporal space that combines the spatial relationship and temporal dimensions. A set of basic relationships and configurations in a temporal space has been identified. As this model is based on an integration of relationships in space and time, its implementation is upward compatible with spatial and temporal models commonly identified within GIS. The proposed model is flexible enough to be applied and adapted to different application contexts. User-oriented relationships can be also defined using generalisation mechanisms. Further work implies the implementation of the formal algebra and the validation of its interest in real application contexts.

4.8 REFERENCES

Allen, J. F., 1984, Towards a general theory of actions and time. *Artificial Intelligence,* **23**, pp. 123-154.

Block, R. A., 1979, Time and consciousness. In *Aspects of Consciousness*, vol. 1: *Psychological Issues*, edited by Underwood, G. and Stevens, R. (London: Academic Press), pp. 179–217.

Cheylan, J. P. and Lardon, S., 1993, Toward a conceptual model for the analysis of spatio-temporal processes. In *Spatial Information Theory,* edited by Frank, A. U. and Campari, I. (Berlin: Springer-Verlag), pp. 158-176.

Claramunt, C. and Thériault, M., 1995, Managing time in GIS: an event-oriented approach. In *Recent Advances in Temporal Databases*, edited by Clifford, J. and Tuzhilin, A. (Berlin: Springer-Verlag), pp. 23-42.

Claramunt, C. and Thériault, M., 1996, Toward semantics for modelling spatio-temporal processes within GIS. In *Advances in GIS Research I*, edited by Kraak, M. J., and Molenaar, M. (London: Taylor & Francis), pp. 27-43.

Claramunt, C., Thériault, M. and Parent, C., 1998, A qualitative representation of evolving spatial entities in Two-dimensional Spaces. In *Innovations in GIS V*, edited by Carver, S. (London: Taylor & Francis), pp. 119-129.

Clementini, E., Di Felice, P. and Van Oosterom, O., 1993, A small set of topological relationships suitable for end-user interaction. In *Advances in Spatial Databases*, edited by Abel, D. J. and Ooi, B. C. (Singapore: Springer-Verlag), pp. 277-295.

Cohn, A. G., Gotts, N. M., Cui, Z., Randell, D. A., Bennet, B. and Gooday, J. M., 1998, exploiting temporal continuity in qualitative spatial calculi. In *Spatial and Temporal Reasoning in GIS*, edited by Egenhofer, M. J. and Golledge, R. G. (New York: Oxford University Press), pp. 5-24.

Cui, Z., Cohn, A. G. and Randell, D. A., 1993, Qualitative and topological relationships in spatial databases. In *Advances in Spatial Databases*, edited by Abel, D. J. and Ooi, B. C. (Singapore: Springer-Verlag), pp. 296-315.

Egenhofer, M., 1991, Reasoning about binary topological relations. In *Advances in Spatial Databases*, edited by Günther, O. and Schek, H.-J. (Berlin: Springer-Verlag), pp. 143-160.

Egenhofer, M. J. and Al-Taha, K. , 1992, Reasoning about gradual changes of topological relationships. In *Theories and Methods of Spatio-Temporal Reasoning in Geographic Space*, edited by Frank, A. U., Campari, I. and Formentini, U., (Berlin: Springer-Verlag), pp. 196-219.

Frank, A. U., 1994, Qualitative temporal reasoning in GIS - ordered time scales. In *Proceedings of the Sixth International Symposium on Spatial Data Handling Conference,* edited by Waugh, T. C. and Healey, R. C. (London: Taylor and Francis), pp. 410-430.

Freksa, C., 1991, Conceptual neighbourhood and its role in temporal and spatial Reasoning, *Technical Report FK1-146-91*, University of Munich, Institute of Information.

Galton, A., 1995, Towards a qualitative theory of movement. In *Spatial Information Theory: A Theoretical Basis for GIS*, edited by Frank, A. U. and Kuhn, W. (Berlin: Springer-Verlag), pp. 377-396.

Glicksohn, J., 1996, Entering trait and context into a cognitive-timer model for time estimation. *Journal of Environmental Psychology*, Academic Press, **16**, pp. 361-370.

Hägerstrand, T., 1967, *Innovation Diffusion as a Spatial Process* (Chicago: The University of Chicago Press).

Hornsby, K. and Egenhofer, M., 1997, Qualitative representation of change. In *Proceedings of the Conference on Spatial Information Theory COSIT'97*, edited by Frank, A. U. and Mark, D. (Berlin: Springer-Verlag), pp. 15-33.

Langran, G., 1992, *Time in Geographic Information System,* London, Taylor & Francis.

Peuquet, D. J., 1994, It's about Time: a conceptual framework for the representation of temporal dynamics in geographic information systems. *Annals of the Association of the American Geographers*, **84**, pp. 441-461.

Pullar, D. and Egenhofer, M. J., 1988, Towards formal definitions of spatial relationships among spatial objects. In *Proceedings of the 3rd International Symposium on Spatial Data Handling*, (Sydney: IGU), pp. 225-242.

Randell, D.A., Cui, Z. and Cohn, A.G., 1992, A spatial logic based on regions and connection. In *Proceedings of the 3rd International Conference on Knowledge Representation and Reasoning*, (Cambridge, MA), pp. 165-176.

Vieu, L., 1997, Spatial representation and reasoning in artificial intelligence. In *Spatial and Temporal Reasoning*, edited by Stock, O. (Dordrecht: Kluwer), pp. 5-41.

Worboys, M. F., 1994, A unified model of spatial and temporal information. *Computer Journal*, **37**, pp. 26-34.

Wu, Y., Jajodia, S. and Yang, X. S., 1997, *Temporal Bibliography Update*. George Mason University, Virginia, *http://isse.gmu.edu/~csis/tdb/bib97/bib97.html.*

Zubek, J. P., 1969, *Sensory Deprivation: Fifteen Years of Research* (New York: Appleton-Century-Crofts).

5

Multi-agent simulation: computational dynamics within GIS

Michael Batty and Bin Jiang

5.1 INTRODUCTION

As part of the long term quest to develop more disaggregate, temporally dynamic models of spatial behaviour, micro-simulation has evolved to the point where the actions of many individuals can be computed. These multi-agent systems/simulation (MAS) models are a consequence of much better micro data, more powerful and user-friendly computer environments, and the generally recognised need in spatial science to model temporal processes. In this chapter, we develop a series of multi-agent models which operate in cellular space. These demonstrate the well-known principle that local action can give rise to global pattern but also how such pattern emerges as a consequence of positive feedback and learned behaviour. We first summarise the way cellular representation is important in adding new process functionality to GIS, and the way this is effected through ideas from cellular automata (CA) modelling. We then outline the key ideas of multi-agent simulation. To set the scene, we discuss models based on agents whose purpose is to explore geographic space, and then illustrate these notions with three applications.

First we examine route finding in systems where distance and direction are largely unknown. The spatial properties of such systems need to be explored by agents rather than computed geometrically, and we show how we can implement the classic shortest route algorithm where the agents are used to 'discover' the route network. Second we look at spatial systems where gradients are important in directing movement and location such as river systems and watersheds. Finally we show how agents can detect the geometric properties of space, generating powerful results that are not possible using conventional geometry, and we illustrate these ideas by computing the visual fields or isovists associated with different viewpoints within the Tate Gallery. Our emphasis here will thus be on modelling how agents behave and can be made to behave in cellular systems where local action is the *modus operandi*, and where our focus in on the exploration of space-time rather than on the simulation of human behaviour *per se*. Our forays into MAS are all based on developing reactive agent models with minimal interaction.

5.2 CELL-BASED GIS, CELLULAR AUTOMATA (CA), AND MULTI-AGENT SYSTEMS AND SIMULATIONS (MAS)

In these models, the spatial framework within which agents behave are based on cell-based systems whose basic unit of representation is identical and regularly distributed such as those based on pixels, grids or any other regular tessellation of the plane. Typically spatial representation in such systems is based on a hierarchy of levels deriving from the cell at its most local level which is the basic unit of operation. Neighbourhoods are thence defined around each cell, usually formed from the cells at the 8 compass directions (where the cells are arranged in a grid), and this links such models formally to cellular automata where neighbourhoods are the basic structures within which action and decision take place. A higher level in which neighbourhoods (and of course the cells that comprise them) are arranged into zones is often defined as the focal level and this is akin to idea of the field which also appears in some CA models (Xie, 1996). Finally the top level is global where actions take place across the entire system (Gao et al., 1993).

CA models have been quite widely applied. Models based on urban and regional dynamics are significant (see White 1998 for a review) but the main problems have been twofold: developing spatial interaction in spaces wider than the neighbourhood itself and in enabling the model dynamics to take account of system-wide conservation constraints which are usually destroyed by CA transitions. A number of environmental models based on CA where local action is perhaps more significant than in urban systems, have been developed. The most promising areas involve wildfire propagation (Clarke and Olsen 1993), and ecosystem functioning based on predator-prey relationships (Camara et al. 1996). Burrough (1998) has recently suggested that temporal dynamics within GIS might be based on a tool kit comprised of the key elements of CA and cell-based GIS including map algebra.

Fully integrated systems incorporating GIS, CA, and map algebra have also been considered. However the basic problem with such models is that the concept of space is quite restrictive. Thinking of the elements of a geographical system as simply cells poses many difficulties especially as it is difficult to ascribe more than one set of transition rules to each cell and different mixes of states to the same cell. Such extensions can be implemented but only by disaggregating cells to more basic units, that is, to ever finer cells. To make progress without losing the intrinsic benefits of local operations across cells, what is required is a more active set of objects associated with those cells. Multi-agent simulations are being developed to meet this challenge.

Essentially agents are autonomous entities or objects which act independently of one another, although they may act in concert, depending upon various conditions which are displayed by other agents or the system in which they exist. Franklin and Graesser (1997) formalise the definition of an autonomous agent as "a system situated within and a part of an environment that senses that environment and acts on it, over time, in pursuit of its own agenda ...". Autonomous agents thus cover a wide variety of behaving objects from humans and other animals or plants to mobile robots, creatures from artificial life, and software agents. The key element is autonomy and in any system, there may be several different types of agent operating in quite different regimes.

Firstly, agents operate within environments to which they are uniquely adapted and just as there may be more than one type of agent in a simulation, there may be more than one kind of environment. In this context, the kinds of environments that we will deal with are, by their nature, 'distributed'. Spatial systems are extensive and in the form of cellular space, have the potential to represent highly decentralised behaviour. Indeed this is an intrinsic property of CA modelling. In fact there is an entire range of different kinds of environment from distributed to highly centralised (Ferber 1999). Secondly, the ways in which agents 'sense' and 'act' in their environment are central determinants of the behaviour which is to be modelled. Agents can be classified in many ways: a key distinction is between 'reactive' agents and 'cognitive' (or deliberative) agents, the difference being with respect to the conditions which drive their 'senses' and 'actions'. For example, a reactive agent is just as autonomous as a cognitive one but the central difference is that the reactive agent will behave entirely by reacting to its environment or to other agents. A cognitive agent may do the same but will also behave according to its own protocols.

This kind of agent technology is being slowly developed with geographic systems in mind. Rodrigues et al. (1996) employ spatial agents for geographic information processing. They have defined spatial agents as agents that make concepts computable for purposes of spatial simulation, decision making, and interface design. But agent-based simulation is perhaps most appropriate where local spatial operations are the focus and this suggests that physical, human, and software systems where movement and location is primarily structured in terms of neighbourhood action are the best candidates for application. In this sense, we can consider putting MAS together with cell-based and CA models. We illustrate the correspondence of these ideas in Figure 5.1 which shows that as we move from cell-based GIS to CA through to MAS, then our systems become ever more active in the temporally dynamic sense and ever more decentralised in terms of spatial decision-making.

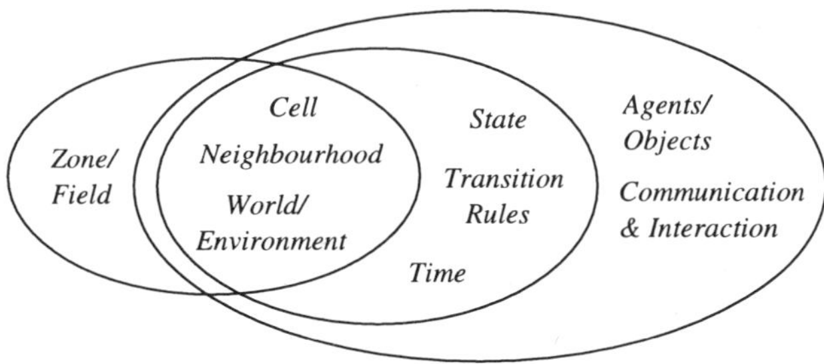

Figure 5.1 Relations between cell-based GIS, CA modelling, and MAS

There are now some significant applications of multi-agent models which are beginning to define the field, many of which are designed around appropriate "object-oriented" software. The *SWARM* project (Langton et al. 1995) is one of the most ambitious in that it is designed to serve as a generic platform for modelling and simulating complex behaviour in which agents generate artificial life. Based on this system, various projects are being developed such as the Los Alamos micro-simulation traffic model - *TRANSIMS* - as well as our own pedestrian model of Wolverhampton *STREETS* (Schelhorn et al., 1999). *SWARM* is not as easy to implement as might appear although attempts are being made to provide more user-friendly versions (Gulyás et al. 1999).

Less general but nevertheless quite highly structured dynamic models are being built with similar software: for example, Epstein and Axtell's (1996) *Sugarscape* model which attempts to simulate the evolution of an artificial society, and Batty, Xie and Sun's (1999) *DUEM* model of urban dynamics built around CA.

The software we use here is based on a massively parallel implementation of the graphics language *Logo*. *StarLogo*, as it is called by its originator (Resnick, 1994), is a language for CA modelling but with agents forming the critical reactive elements in the system. In effect, agents are able to roam across the space and sense the environment which is composed of cells. The key operations within the software enable agents to react to what is within their neighbourhood which in turn is based on information which is encoded within the cells forming their environment. Of course, agents can react to one another but in general, action at a distance is difficult (although not impossible) to represent within the software.

5.3 MOVEMENT IN CELLULAR SPACE: NAVIGATING BY THE STARS AND THE SUN

To develop multi-agent models that operate in cellular space, we must first introduce some formal notation. We will define each agent, or walker in our first example, by an identity k for a set of K agents (where $k = 1, 2, \ldots, K$) and a time t which is associated with when the agent acts in making some spatial decision involving movement, location, or interaction. Agents locate or move in cellular space which is defined according to the conventions of cellular automata modelling. Agents for example act locally in their immediate neighbourhood defined as the 8 adjacent cells (the Moore neighbourhood) surrounding their current location which is in turn given by the grid coordinates (x, y). We will use x and y to define distances within the grid but for purposes of reference we will also refer to coordinates x and y by their row and column identifiers i and j as x_i and y_j.

Motion in these models is always controlled by the angular heading $\theta_k(t)$ which is associated with walker k (and therefore by implication location ij). In each time period $[t \rightarrow t+1]$, a walker will move one cell step in its surrounding neighbourhood. Given a heading $\theta_k(t)$, an agent will walk to a new location with coordinates $x_i(t+1), y_j(t+1)$

$$x_i(t+1) = x_i(t) + \Delta x \quad \text{and} \quad y_j(t+1) = y_j(t) + \Delta y \tag{5.1}$$

$$\Delta x = r\cos\theta_k(t) \quad \text{and} \quad \Delta y = r\sin\theta_k(t) \quad . \tag{5.2}$$

As the step distance r is always assumed to be 1 which implies that agents always walk at the same speed, then equations (1) can be rewritten as

$$x_i(t+1) = x_i(t) + \cos\theta_k(t) \quad \text{and}$$
$$y_j(t+1) = y_j(t) + \sin\theta_k(t) \quad . \tag{5.3}$$

Motion is only possible within the Moore neighbourhood, and thus the heading $\theta_k(t)$ is approximated to one of the eight points of the compass around $x_i(t)$ and $y_j(t)$.

In many of our applications, movement or location takes place in space which is less than the dimensions of the entire cellular system. We show this by defining a constraints variable $C(x_i, y_j)$ as the state of the cellular system at each location x_i, y_j where

$$C(x_i, y_j) = \begin{cases} 1 & \text{if movement or location in } x_i, y_j \text{ is permitted} \\ 0, & \text{otherwise} \end{cases} \quad . \tag{5.4}$$

In our first example in this section, $C(x_i, y_j)$ will define a street or path system, with predominantly linear features which marks out the permissible routes along which movement might take place.

We now need to illustrate how movement and other actions based on equation (3) are affected by the constraints in equation (4). For this, we can prime the location coordinates as $x_i(t+1), y_j(t+1)$ to indicate that these are 'predicted' rather than actual locations. To check if movement to these predicted locations is feasible, we must make the following test:

$$\textbf{if } C(x_i, y_j) = 1, \textbf{ then } x_i'(t+1) = x_i(t+1) \text{ and } y_j'(t+1) = y_j(t+1)$$

and the move takes place. If $C(x_i, y_j) = 0$, then the move is not permissible and a procedure is activated to choose a new location which does meet the constraints. In short, this is the obstacle-avoidance algorithm. Typically in problems involving network systems, when motion reaches the point where it is not permissible, then this usually means that obstacles have been encountered.

There are many possibilities for avoiding obstacles but all these involve testing different, alternative directions to see what is possible. The simplest of these is to choose a feasible direction where $C(x_i, y_j) = 1$, for any of the remaining 7 compass points which have not been evaluated so far. It may be necessary to repeat the procedure until such a feasible direction is discovered although this will always be possible in the systems that we are dealing with here. The basis of this algorithm is as follows:

$$\begin{aligned} \textbf{if } C(x_i, y_j) = 0, \textbf{ then } \quad & \theta_k'(t) = \text{random}(\pi) \\ & x_i'(t+1) = x_i(t) + \cos\theta_k'(t), \text{ and} \\ & y_j'(t+1) = y_j(t) + \sin\theta_k'(t) \\ & \textbf{until } C(x_i', y_j') = 1, \quad . \end{aligned} \tag{5.5}$$

This is the simplest algorithm which enables agents to walk around obstacles but it destroys the original heading. If many walkers are launched from a single point in the system and given random headings to begin with, then in a system of connected streets, the steady state distribution of agents will mirror the density of the street pattern in that walkers will be distributed evenly (in statistical terms) per unit of street. Of course the configuration of the streets is important in this because the density of streets themselves, per unit of area, may well vary. Nevertheless this algorithm and the method of walking leads to uniformity in the probabilistic sense. If the heading $\theta_k(t)$ which is initially established is critical, then the algorithm in (5) above can be easily altered to reflect successive small changes ε to $\theta_k(t)$ which ultimately lead to a direction which is as near as possible to the original one but is then feasible.

Our thesis here is that multi-agent systems are useful not only for simulating spatial systems at the micro level but for actually exploring the properties of space that may not be accessible in any other way. Agents may well explore space as part of their behaviour but in this context we are focusing primarily on the use of agents to sense properties of systems that could not be measured or detected in any other way. Our first problem is one in which many walkers in the system need to find routes to different places which act as their destinations. The problem is to find 'feasible' but not necessarily shortest routes for these walkers from any and every point in the system to these destinations. The walkers know where their destinations are in global terms - in terms of their coordinates - but do not know how to get to those coordinates and thus need to explore the local street systems purposefully *en route* to these destinations. For each walker *k*, we can associate a destination X_k, Y_k which is uniquely specified in terms of the identity *k* of the walker. Agents can thus work out the distance from any location to their destination but do not know the route system and must thus make their way to the destination from their global knowledge of X_k, Y_k and from their ability to explore the street system locally avoiding obstacles. Then from any location x_i, y_j, the shortest distance to the destination as the crow flies for any agent *k* is

$$r_k(x_i, y_j) = \left[(x_i - X_k)^2 + (y_j - Y_k)^2\right]^{-\frac{1}{2}} \tag{5.6}$$

where the heading is then computed as

$$\theta_k(x_i, y_j) = \cos^{-1}\left\{\frac{(x_i - X_k)}{r_k(x_i, y_j)}\right\} \quad . \tag{5.7}$$

The model works as follows. Given any distribution of walkers, each walker fixes their heading from equations (6) and (7), and makes a first step which is checked using the obstacle-avoidance routine in equation (5). Because the heading is important, then walkers do not want to deviate from this heading too much. If they need to modify their heading to avoid an obstacle, they gradually modify their heading as $\theta_k^{''}(x_i, y_j) = \theta_k(x_i, y_j) + \varepsilon_k$ where ε_k is a very small increment to the angle. This is used in the algorithm in (5) and ensures that the correction to the global heading will be as small as possible. In essence, this model assumes that walkers navigate from their 'knowledge of the stars and the sun': they know the location of their destinations - the stars or sun at X_k, Y_k - but have to creep around the street system, mainly hugging the walls of the streets until they eventually reach their destinations. Much depends upon the local configuration because they might get stuck. It may be necessary to go further away from the destination at some point to actually get to it and thus the only way to do this is to give walkers some greater perturbation in their heading, letting them walk "the wrong way" for some little time. In fact, the original algorithm with random headings will ensure that most local loops are avoided. We have programmed several variants of the obstacle-avoidance algorithm for an hypothetical street system and we show a realisation of the path traced for one walker to one destination in Figure 5.2. Note how the path wanders around in small local loops on the street as it moves from origin to destination.

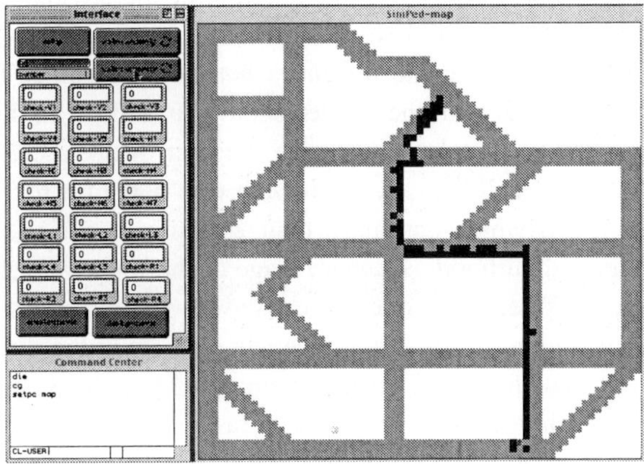

Figure 5.2 Approximating shortest routes by global navigation and local search

5.4 MORE PURPOSIVE NAVIGATION: SHORTEST ROUTES

The best developed navigation problem enables an exact calculation of the shortest routes between any origin and destination on a route system which is strongly connected in the graph-theoretic sense. We will digress a little and sketch the classic problem before we reformulate this in agent-based terms. We will call origins *m* and destinations *n*, and we will code the route network using the constraint variable $C(x_i, y_j)$ where the distance d_{mn} between any origin *m* and destination *n* is now computed as

$$d_{mn} = \left[(x_{im} - x_{in})^2 + (y_{jm} - y_{jn})^2\right]^{-\frac{1}{2}} \quad . \tag{5.8}$$

Note that $m, n \in N$ where *N* is the set of nodes and each node is associated with the coordinate pairs defined in equation (8). This problem is well-known and soluble using the Bellman-Dijkstra algorithm which was discovered 40 years ago (Bellman, Cooke, and Lockett, 1970). Starting from any origin node *O*, it is possible to compute all the shortest paths to the other *N-1* nodes from

$$f_{t+1,n} = \min_m \left\{ f_{t,m} + d_{mn} \right\} \tag{5.9}$$

where $f_{t+1,n}$ is the minimum distance so far at the $t+1^{th}$ step of the algorithm. The procedure begins with $f_{1,n} = d_{On}$, and essentially involves systematically tracing through all the routes of the network, keeping a running total of the minimum distances so far from node O to every other node, and finishing when all nodes and routes have been visited.

The usual problem is one where we know the nodes or junctions between the routes in the network and we know the distances between these nodes. However imagine a system in which we do not 'yet' know the location of the streets or the junctions and thus we have no idea how far it is between these unknown nodes. Thus the problem involves finding the junctions and the distances between the routes before the shortest routes can be found. However with the walker behaviour defined in the previous section, we can explore the space so that we can decide whether or not a cell in $C(x_i, y_j)$ is a node. We know that if the cell $C(x_i, y_j)=1$, then it is part of a route and we can then define a junction as a cell which has more than 2 positive cells in the eight cells that constitute its Moore neighbourhood. At every stage of the walk, we can thus decide if the cell is a node. As we walk between nodes (because we will always assume the origin O is a node from which all walking begins), then we can also count the distance travelled and in this way build up knowledge of local distances within the system. This kind of problem is not unusual where data is provided in raster form from satellite or where the system is so complex, that no graph theoretic definition of nodes and route distances has been made beforehand.

We can get a sense of how the model works from the following algorithm:

1 Start at the Origin O which is a node

↓

2 Find out how many exits there are from the node (s)

↓

3 Find out how many routes from these exits have already been traversed

↓

4 Generate as many new agents as there are routes to be traversed,
each having memory of the minimum distance to that node so far

↓

5 Set the heading of the relevant agent along its route

↓

6 Walk one step of the route at a time, count the distance travelled

↓

7 Check for the existence of a node

↓

8 If a node is encountered STOP; if not go back to 6

↓

9 Check all agents entering node; choose the agent with the minimum distance
so far, as the minimum distance for the node

↓

10 Check if all nodes have been visited and all routes traversed
If YES, STOP; if NO then go back to 2

Although the algorithm starts at node *O*, eventually the full set of nodes are encountered. A full matrix of distances from all origins to all destinations can be computed by running the algorithm *N* times.

We can explain how the algorithm works more effectively by tracing through an example. Imagine that we start at an origin node where there are three exit routes. We create three agents, give them a memory of zero distance, and then send them off along their respective routes, checking for the existence of junctions, and counting the increment of distance travelled one step at a time. When they encounter a node they stop, and we wait until all three have reached this state before we check the status of the problem. On this first pass through, we know that there are more routes to traverse from the three new nodes at the end of the three routes encountered (unless we have a triangular set of routes), so we take the shortest distances computed on these first three routes and find out how many new routes there are from the three nodes encountered. We dissolve the existing agents and from each of the three nodes, create as many agents as we need at each node so they can traverse the next set of routes. We give these agents memory of the shortest distance to date and off they go following the same procedure as previously. When they have all encountered nodes, we check to see if the routes from these new nodes have been traversed. If not we create new agents, and choose the shortest distance of any of the agents entering the nodes as the new memories of distance for the new agents. Note that at this point, we may have two or more agents entering a node from different routes (which will be the usual condition for networks with more than about 10 nodes), and thus we need to take the *minimum* distance associated with one of these agents. Eventually we will run out of nodes to visit and routes to traverse and at this point the minimum distances associated with the any of the agents at each node are the shortest routes from the origin to each of these nodes.

It is worth noting that at each step, we need to find out whether the pixel that the agent is on is a junction or not. We are also assuming that the route has been marked out by a single line of pixels, not by the full width of the street as shown in the examples. Then at each location x_i, y_j

$$\textbf{if} \sum_{\substack{i-1<i<i+1 \\ j-1<j<j+1}} C(x_i, y_j) > 3, \ \textbf{then} \ x_i, y_j \text{ is a junction} \quad . \tag{5.10}$$

We also need to compute the directions of the exits from the junction, check to see if the routes from these exits have been traversed, and give the agents who are created to leave the junction from these exits, the shortest distance to this junction computed so far. Note that agents communicate with one another when they have entered a node in that they pass their knowledge of distance travelled so far to the new agents who choose the minimum. Finally it is worth showing the typical outputs of a shortest route from a fixed origin to a fixed destination for the path system used in the previous section, and in Figure 5.3, we show the route traced out which is equivalent to that used in the global navigation model in Figure 5.2.

5.5 WATERSHED DYNAMICS: THE EVOLUTION OF RIVER SYSTEMS

We have seen how agent-based models can be used to track spatial properties associated with route and network systems which are highly constrained but equally effective procedures exist for cellular systems composed of surfaces. In this example, we associate agents with particles of moisture - rain droplets - falling on a landscape or indeed any phenomena which affects a surface such as pollution particles. In essence, when particles fall on a surface, they move across the surface; the usual assumption is that they move according to the principle of least effort, in the direction of the steepest gradients. Our model here simply traces the paths of these particles, demonstrating how local action/movement creates patterns that have a global morphology.

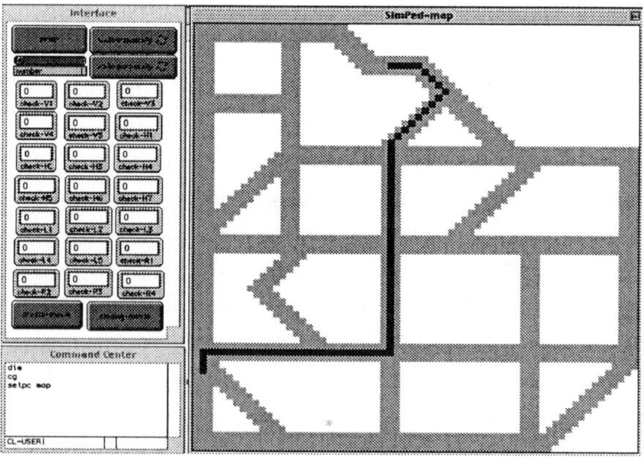

Figure 5.3: Agent-based implementation of the Bellman-Dijkstra algorithm

We first generate or observe a set of particles or agents at different locations x_i, y_j at initial time $t=1$. These agents might be randomly generated and as such act as sources for motion across the landscape. The terrain is represented by a height surface - a digital terrain model - which we represent as $S(x_i, y_j)$. The droplets of moisture now make their way across the landscape, 'walking' locally one step at a time according to the logic in equations (1) to (3). But in this case, the heading is not computed randomly or globally but in terms of the local gradient of the surface $S(x_i, y_j)$. The standard approximation to this gradient is given as

$$\text{grad}\,(x_i, y_j) \approx \Delta x \frac{\partial S(x_i, y_j)}{\partial x_i} + \Delta y \frac{\partial S(x_i, y_j)}{\partial y_j} \quad , \tag{5.11}$$

but in terms of the model implementation which is constrained to the Moore neighbourhood, then the minimum gradient and its appropriate direction are computed from

$$\min_{\{i\pm1,j\pm1\}} \left\{\text{grad}\ (x_{i\pm1}, y_{j\pm1})\right\} = \frac{\{S(x_i, y_j) - S(x_{i\pm1}, y_{j\pm1})\}}{\{(x_i - x_{i\pm1})^2 + (y_j - y_{j\pm1})^2\}^{-1/2}} \tag{5.12}$$

From the direction, we need to compute the appropriate heading, and then movement occurs according to the motion implied by equations (1) to (3).

From the initial configuration of agents, we trace out the paths formed by movement across the surface. These paths follow the steepest local gradients and essentially can be interpreted as continuing sources of water - springs if you like - which produce rivers across the surface. Of course this is not the way rivers are formed but simply give river-like patterns. As the distribution across the surface is random and takes no account of the fact that rivers are usually sourced on higher terrain, then the patterns produced are unrealistic. Moreover, when there is no exit from the landscape in terms of the terrain, lakes should form and the sinks shown in the examples below indicate where these lakes would be.

In Figures 5.4(a) and 5.4(b), we show two examples of these simulations of watershed dynamics in terms of terrain traces across two different landscapes. The first landscape in Figure 5.4(a) contains sources and sinks but no general drainage out of the region. In Figure 5.4(b), we show an example of the dendritic pattern which emerges when the general gradient of the terrain falls from north-east to south-west. What these examples both demonstrate is the classic conclusion that local action gives rise to global pattern, in this case fractal patterns which result from the action of movement which in essence is space-filling. However the terrain is somewhat artificial in both cases without the usual irregularity which leads to conventional stream patterns. To generate realistic phenomena, the environment, in this case the terrain, must be as realistic as the rules used to generate patterns across or within that terrain.

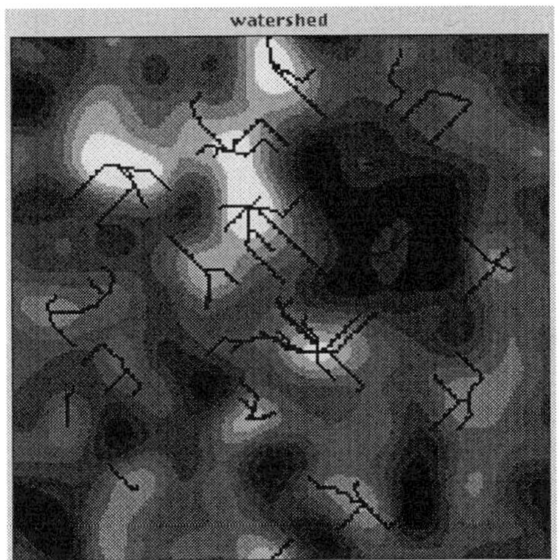

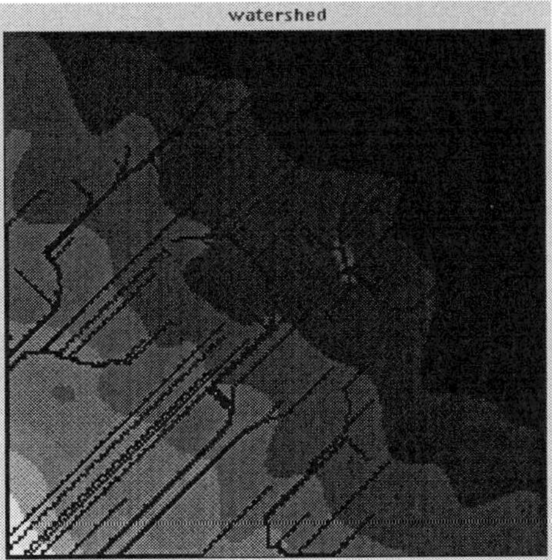

Figure 5.4 Watershed dynamics: (a) many sources and sinks (b) terrain with a more generally directed gradient

5.6 VISUAL FIELDS: A MORPHOLOGICAL EXPLORER

There are many problems where it is important to detect and measure the properties of a space which is visible from many viewpoints. These problems are intrinsic to architectural and landscape design but they are also significant in explaining human behaviour ranging from the routes that pedestrians might take through a town centre or large store to the imageability which is used by people to navigate around a city or landscape. Algorithms to compute lines of sight and viewsheds in the natural landscape have been quite widely applied in GIS based upon computational geometry but their equivalents in complex urban environments where geometric building blocks define and bound space are not well developed. Urban viewsheds have been called isovists by Benedikt (1979) who defines these strictly as all points which are visible from a fixed point or focus in space.

The isovist, urban viewshed, or visual field is essentially defined by the boundary which contains all the visible points. Within this boundary, we are also interested in the line of furthest distance from the given point or 'seed' as we will refer to it. The area of the field, the way in which visual fields overlap, and the way fields dominate other fields are all relevant properties that we will measure. There are however severe problems in defining such viewsheds due to the fact that their computation has traditionally been based on representing them as vectors and seeking geometrically exact boundaries. In short, their geometric computation for all possible configurations cannot be assured, at least not within reasonable computational limits.

In essence, what we will formulate here is a method for letting agents explore and thence define the properties of the visual field. We plant an agent on a seed and then program it to move in all directions from the seed, measuring distance as it moves from pixel to pixel; we terminate its walk once it reaches the boundary of the field which we will code into the environment by the constraint variable $C(x_i, y_j)$. The agent then returns to the seed and begins another walk in a different direction. We therefore need to choose a set of directions so that it covers the visual field and in this way we essentially let the agent explore the field. This method has similarities to ray tracing which is now widely used in problems of hidden line elimination in computer graphics. To measure all the properties of the space, we need to compute the isovists from every point or pixel in the system. This method in effect replaces geometric precision with pixel precision, that is replaces the vector representation of space with raster. It is effective only if there are enough agents and enough pixels to generate sufficiently precise measurements.

From any seed x_i, y_j, we define a ray or vista as a straight line $r_n(x_i, y_j)$ which will vary from 0 distance to a maximum $R_n(x_i, y_j)$. Note that the index n relates to the direction of the line. As these equations are the same for all seeds and all fields, to illustrate the computation we can drop the reference to the seed x_i, y_j. The computation of the area A of the field is

$$A = \int_0^{2\pi} \int_0^{R(n)} r(n)\, d\theta\, dr(n) \quad . \tag{5.13}$$

If $r_n = r$ and $R_n = R$, then it is easy to show that the area in equation (12) is πR^2. We do not know the function r_n but we can approximate $d\theta$ as $2\pi / N$ and (13) then becomes

$$A(N) \approx 2\pi \sum_n R_n \quad , \tag{5.14}$$

which is in fact approximately proportional to the total of the distances from the seed. In fact a better approximation and one that we use here is based on computing the triangle approximation to the segment between two distance lines R_n and R_{n+1} whose inclusive angle is $\theta = 2\pi / N$. Approximating equation (13) using these definitions leads to

$$A(N) \approx \frac{1}{2} \sin\left(\frac{2\pi}{N}\right) \sum_{n=1}^{N} R_n^{\,2} \quad , \tag{5.15}$$

which is the basis for the subsequent computations. With $R_n = R$, it is easy to show, using the appropriate series expansion of the sine, that (15) converges to πR^2 as $N \to \infty$.

The algorithm involves computing all visual fields simultaneously by creating an agent on each pixel. We then choose N so that we are sure that the angular variation $\theta = 2\pi / N$ is such that the triangle segment is always within one pixel, and then we move each agent in the direction θ until each agent hits a barrier which is the edge of a building given by $C(x_i, y_j) = 0$. As the agent moves, the distance is incremented as in previous examples and in this way the maximum distance in each field is identified as $R_n = \max_n R_{ij}$. The area is calculated from equation (15) which is in a form for cumulative computation.

These measurements are made independently of the procedure for computing the boundary of the visual field. The geometric limits of the field are identified by creating a very large number of agents on the relevant seed with random headings and letting these agents walk away from the seed until they reach the boundary defined by $C(x_i, y_j)$. As they walk, they identify whether the pixel they are on is in the field and in this way all the pixels in the field are identified. In fact the area can then be computed by counting the pixels. This process provides the set of pixels that mark out the boundary through a process that is akin to the flood-fill routine which is often now encoded in computer graphics hardware. Finally, we can identify various dominance relations associated with these isovists. For example, if we argue that a field dominates another, then the area of the field is greater than that of the other and if the two fields overlap at some point, we can rank fields according to this logic. We then argue that all the space associated with the most dominant visual field is associated with that field. We choose the next most dominant field whose seed is not within the first field and associate the space that is not in any field so far with that field. This process continues until all its space has been covered.

Our example is based on the Tate Gallery which is located on London's Millbank. This is a complex space composed of 45 distinct rooms but with only one entrance into the main Gallery building itself. The two key geometric measures - longest

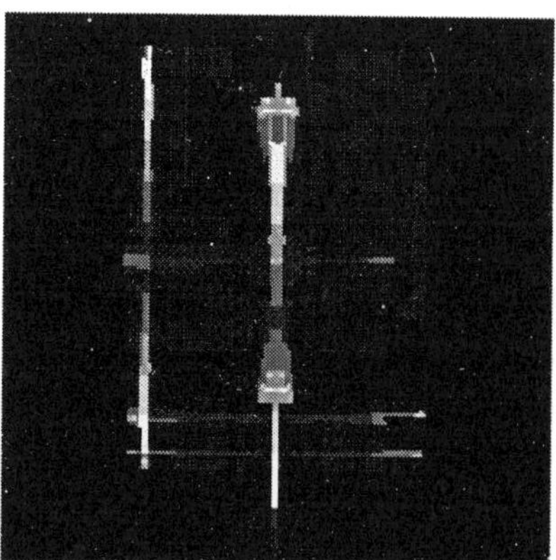

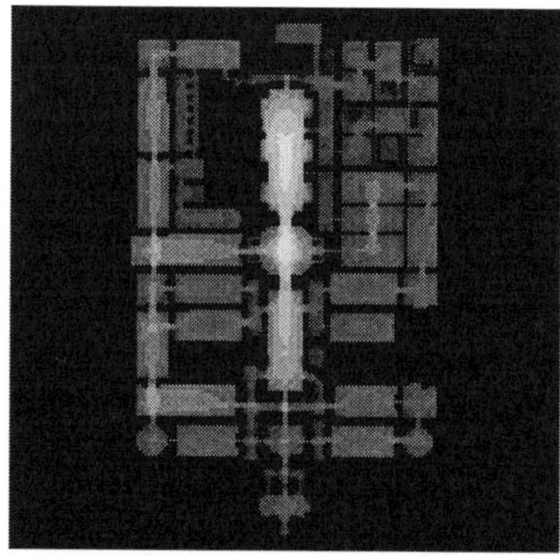

Figure 5.5 (a) How far one can see from each point/pixel
(b) The total area one can see from each point/pixel

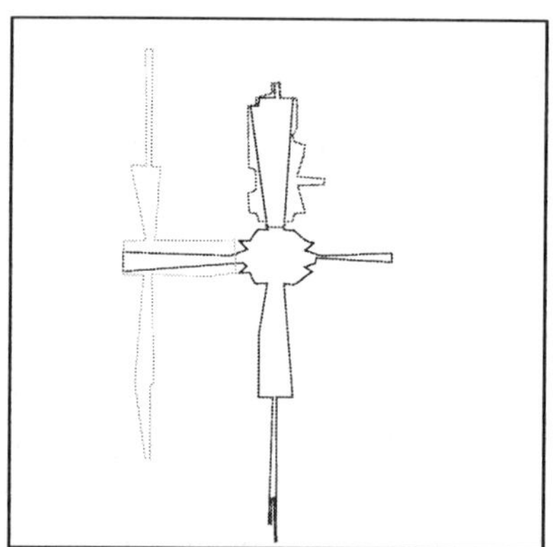

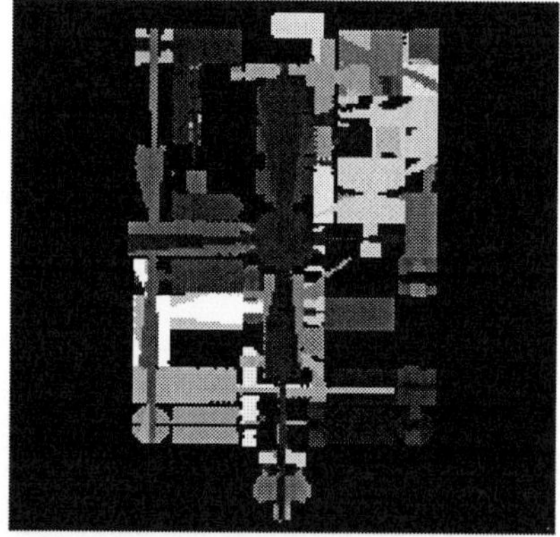

Figure 5.6 (a) The three most important isovists
(b) Fields associated with all the isovists that cover the gallery

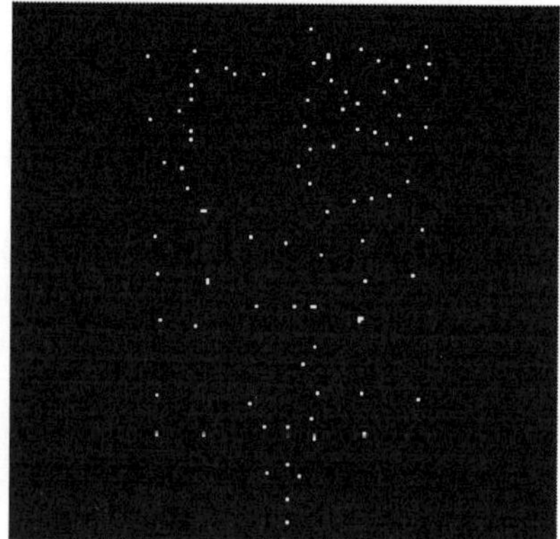

Figure 5.7 (a) The density of overlapping visual fields
(b) Seeds points for the 84 fields that cover the gallery

distances and areas viewed - are shown in Figures 5.5(a) and 5.5(b). The longest distances seen from every space within the Gallery show the obvious fact that you can see further along the axial lines of the building. The greatest areas seen also coincide with pivotal points in the space such as under the dome in the centre of the Gallery. The first three overlapping visual fields are shown in Figure 5.6(a) while the entire set of fields and what space they cover according to their dominance is shown in Figure 5.6(b). This is a complex mosaic and what is clear is that using our definition of dominance, some fields are disconnected, thus posing difficulties in interpretation from the individual standpoint. We also show the density of fields measured by the number that overlap each other in Figure 5.7(a). Finally in Figure 5.7(b), we show the 84 separate seed points which are associated with the most dominant fields that cover the space.

5.7 CONCLUSIONS: REAL AND FICTIONAL AGENTS

All our examples in this chapter have involved fictional agents in the sense that we have defined the agents to detect and measure various properties of space based on distance and area. In this sense, our models provide ways of exploring space in contrast to the wider class of agent-based models where the emphasis is upon simulating the spatial behaviour and decision-making processes of real agents. This distinction between real and fictional agents is similar to that between processes which simulate the actual mechanisms of behaviour in contrast to normative or idealised behaviour. For example, in our visibility explorer, the agents we created and the behaviour we gave them was fictional but only in the sense that this enabled agents to search the space in the most direct way. The same comments are true of the route finding applications. This shortest routes procedure is, in fact, one that would take far too long to operate were it to be used by real agents exploring space. The way real agents build up knowledge of shortest routes is by successive learning, usually through visiting places many times. The nearest application we have developed here to the simulation of real agent behaviour is the watershed dynamics model. Although rain particles are hardly agents in the behavioural sense, the use of the agent paradigm is very effective in demonstrating how such particles 'behave' physically in the landscape.

All the ideas in this chapter are demonstrations of processes and algorithms that we are using in a much larger and more obvious simulation problems of movement in town centres, in shopping malls, and in complex buildings such as museums and galleries. We are simulating actual movements in the Tate Gallery using a combination of route finding, obstacle-avoidance, and gradient search. Although we are concentrating on simulating real behaviour, we also pepper the model with various routines and actions that in which agents engage. These are clearly not real in the sense that they are designed so they can proceed or navigate efficiently but do not necessarily simulate what people actually do in real situations. Thus our simulations reflect a mix of real and fictional behaviour but nevertheless associated with real agents. This kind of geocomputation shows much promise as an effective way of building theories and models of spatial and temporal process into GIS.

5.8 ACKNOWLEDGEMENTS

The authors thank the UK Joint Information Systems Committee for support through their Joint Technology Applications Programme (JTAP) and the UK Office of Science and Technology for their support through the Technology Foresight Challenge Programme (EPSRC GR/L54950).

5.9 REFERENCES

Batty, M., Xie, Y., and Sun, Z., 1999, Modeling urban dynamics through GIS-based cellular automata, *Computers, Environments, and Urban Systems*, **23**, pp. 205-233.

Bellman, R., Cooke, K. L., and Lockett, J., 1970, *Algorithms, Graphs, and Computers*, (New York: Academic Press).

Benedikt, M. L., 1979, To take hold of space: isovists and isovist fields, *Environment and Planning B*, **6**, pp. 47-65.

Burrough, P. A., 1998, Dynamic modelling and geocomputation, in *Geocomputation: A Primer*, edited by Longley, P. A., Brookes, S. M., McDonnell, S., and MacMillan, B. (Chichester: Wiley), pp. 165-191.

Camara, A. S., Ferreira, F., and Castro P., 1996, Spatial simulation modelling, in *Spatial Analytical Perspectives on GIS*, edited by Fisher, M., Scholten, H. J., and Unwin, D. (London: Taylor and Francis), pp. 201-212.

Clarke, K. C., and Olsen, G., 1993, Refining a Cellular Automaton Model of Wildfire Propagation and Extinction, in *GIS and Environmental Modelling: Progress and Research Issues*, edited by Goodchild, M. F., et al. (Fort Collins: GIS World), pp. 333-338.

Epstein, J. and Axtell, R., 1996, *Growing Artificial Societies: Social Science from the Bottom-up*, (Cambridge, MA: Bradford Books, MIT Press).

Ferber, J., 1999, *Multi-Agent Systems: An Introduction to Distributed Artificial Intelligence*, (Reading, MA: Addison-Wesley).

Franklin, S. and Graesser, A., 1997, Is it an agent, or just a program?: A taxonomy for autonomous agents, in *Intelligent Agents III*, edited by Muller, J. P., Wooldridge, M. J., and Jennings, N. R. (Berlin: Springer-Verlag), pp. 21-35.

Gao, P., Zhan, C., and Menon, S., 1993, An overview of cell-based modelling within GIS, in *GIS and Environmental Modelling: Progress and Research Issues*, edited by Goodchild, M. F., et al. (Fort Collins: GIS World), pp. 325-331.

Gulyás, L., Kozsik T., Czabala, P., and Corliss, J. B. (1999) Telemodelling - overview of a system, *http://www.syslab.ceu.hu/telemodeling/index.html.*

Langton, C., Nelson, M., and Roger, B., 1995, The Swarm simulation system, (Santa Fe, NM: Santa Fe Institute), *http://www.santafe.edu/projects/swarm/*

Resnick, M., 1994, *Turtles, Termites, and Traffic Jams: Explorations in Massively Parallel Microworlds*, (Cambridge, MA: Bradford Books, MIT Press).

Rodrigues, A., Grueau, C., Raper, J., and Neves, N., 1996, Environmental planning using spatial agents, in *Innovations in GIS 5*, edited by Carver, S. (London: Taylor and Francis), pp. 108-118.

Schelhorn, T., O'Sullivan, D., Haklay, M. and Thurstain-Goodwin, M., 1999, STREETS: An agent-based pedestrian model, *6th International Conference on Computers in*

Urban Planning and Urban Management, (Venice, Italy: DAEST, University of Venice) *http://www.casa.ucl.ac.uk/~david/swarmStuff.html/*

White, R., 1998, Cities and Cellular Automata, *Discrete Dynamics in Nature and Society*, **2**, pp. 111-125.

Xie, Y., 1996, A generalized model for cellular urban dynamics, *Geographical Analysis*, **28**, pp. 350-373.

6

A parameterised urban cellular model combining spontaneous and self-organising growth

Fulong Wu

6.1 INTRODUCTION

Cellular Automata (CA) became a popular approach to heuristic urban simulation (Batty *et al.*, 1997; Couclelis, 1997; Batty, 1998). Nonetheless, many issues regarding its reliability as a modelling tool in the real world, such as model calibration, verification and validation remain unanswered. While it is interesting to observe, following experiments in physics and the life sciences, that uncoordinated local interactions can give rise to a structured global pattern, our capacity for identifying or devising such rules that can lead to a desired form is often questionable. This is particularly problematic in the domain of human geography where the rule space itself is complicated. Therefore, the aim of this paper is to identify whether it is possible to replicate a given pattern of land use change through CA simulation, and to propose a generic procedure which can be applied with modest modification to simulate urban land use changes. The method developed here is a parameterised approach which combines two different processes of urban growth - that of spontaneous growth which is independent of the evolving state of CA and that of self-organised growth which is directly controlled by the state of CA. This parameterised approach is appropriate because, firstly, factors affecting development such as access to roads can be considered and, secondly, the balance of these two processes is explicitly addressed. Bridging the gap between purely microscopic CA and conventional urban modelling, this hybrid approach is most appropriate for simulating urban land use changes.

Applying CA in geographical research often requires some modification and relaxation. In a strict sense, CA as discrete dynamic systems follow locally defined rules and these rules should be universally applied (Wolfram, 1984; Toffoli and Margolus, 1987). However, many geographical phenomena are not produced by purely local interaction in a universal way (Couclelis, 1997). While CA have wide appeal due to their simplicity, flexibility and transparency, there are some obstacles in application. Applying them to address a practical question often requires one to specify a set of transition rules. The definition of rules, however, relies on an intuitive understanding of the process. Building a plausible model requires the experience of model building. In fact, a quick glance at the growing literature on CA applications suggests that under the umbrella there are a variety of methods (White and Engelen, 1993; Batty and Xie, 1994; Portugali and

Benenson, 1995; Batty, 1998; Clarke and Gaydos, 1998; Wu, 1998a, 1998b). Unlike statistical methods such as multivariate regression, there is no fixed procedure for building CA models. The only common feature is perhaps the iteration of raster data with various sorts of microscopic mechanism.

Building a model on a heuristic basis is not wrong in itself. Unlike the physical process which is subject to a universal law, processes involving human and environment interaction such as urban growth are complicated and often ill-defined. However, it is exactly this level of complexity inherent in urban development that justifies a simulation approach, because it is often the only practical way to study these complex systems. Without actually building a model, often in the media of computer, it is hard to imagine their complex behaviour and associated spatial forms. The question is whether we can have confidence in this versatile approach?

As a flexible and open-structured procedure, urban growth simulation can adopt a variety of transition rules, even hypothetical ones. In essence, simulation is an analogy to the real world and thus simulation rules are often defined in terms of plausibility, that is, whether it fits the purpose of understanding the real world process. Nevertheless, if we want to develop CA as a widely used modelling approach, it would be useful to conceive a generic form of urban growth.

Basic to the CA simulation procedure proposed in this research is an argument that urban development is neither a pure locally-defined process described by classical CA such as *The Game of Life* nor is it a pure global process modelled in classical urban land use models such as the Lowry model. Urban growth is a process combining spontaneous and self-organising growth. The former represents the uneven distribution of development conditions. Regions are endorsed with different resources that form a basis for urban growth. This basis can be seen as a global structure which is relatively stable over a short period. In other words, such a structure is not changed by a single development. The global structure remains unchanged regardless of changes in the local sphere. The local changes, however, often witness fluctuations. Individual changes are likely to be affected by other changes nearby. The interactions at a neighbourhood, for example, often lead to the agglomeration of the same type of development. The question is therefore how to address the combined effect of global and local factors?

CA modelling would be an ideal approach to address this question because they are discrete dynamic systems which can be "solved" through GeoComputation. Conventional equilibrium urban models have enormous difficulty dealing with spatial detail - it is impossible to "solve" a complex set of partial differential equations describing space-time interactions at local levels. In the following section, a general procedure for simulating urban land use change is proposed, which combines spontaneous growth at the global level and self-organising growth at the local level. Then, an experiment involving simulating a hypothetical given land use pattern is conducted and its results are discussed. In conclusion, this experiment is used to address various neglected issues of CA simulation.

6.2 PROCEDURE FOR CA LAND USE SIMULATION

The procedure proposed here is a generic one (Figure 6.1). It generalises the common task of land use change simulation. Land use simulation is a process to replicate a given pattern from a number of known factors and extrapolate the trend into the future. The inputs are land use maps at time *t*1 and *t*2. Through the overlay of two land use maps, the pattern of land use change can be identified. This pattern is what we call the given pattern, or observation. The task of simulation is to replicate this given pattern by

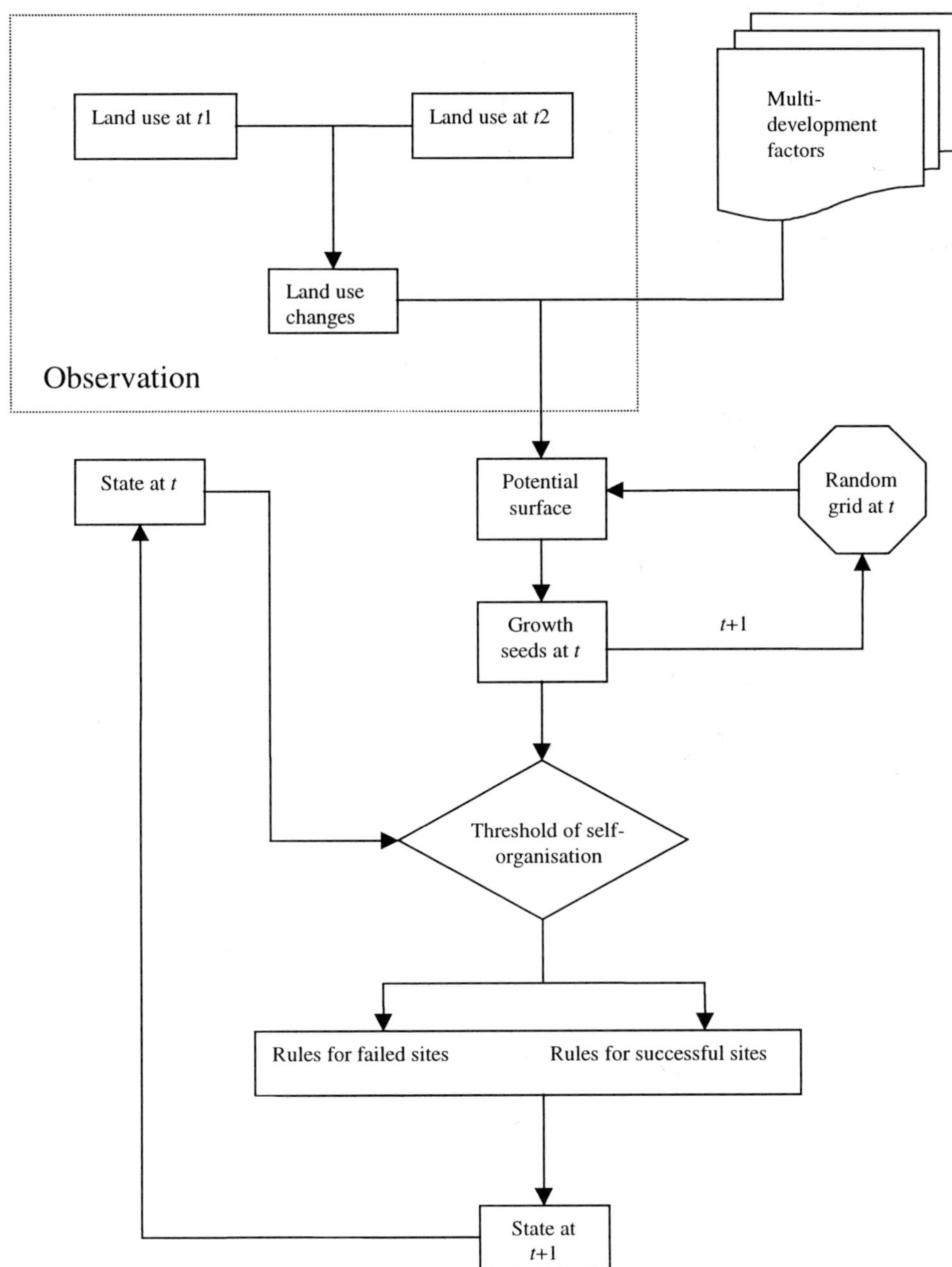

Figure 6.1 The general procedure of simulation of the combined spontaneous and self-organising growth

relating it to a series of development factors such as access to road networks and land suitability. The data required for this sort of analysis such as land use coverage, land suitability and infrastructure maps are usually stored in GIS. It is assumed that the factors affecting development pre-exist and are relatively stable during simulation. Therefore, these multi-development factors can be seen as global influences which often reflect uneven resource endorsement of regions.

The relationship between land use changes and multi-development factors can be identified though a variety of methods. Multicriteria evaluation is one such method. Most land development evaluation finally produces a map of development potentiality.

Conventional spatial analysis functions available in most GIS are very good for this type of analysis. But regression methods or other 'calibrated' approaches such as artificial neural networks (ANN) are preferable because the equation or the derived network are thought of as 'extracted' from observations. The result of this stage of analysis is a map showing the probability/potentiality of land use change.

The probability/potentiality map reflects the attractiveness of the development site. As development sites are pixels in CA, the map is in fact a raster surface. Spontaneous urban growth is referred to as land development following this attractiveness surface. From this step the model begins to use a CA approach. As discussed earlier, urban growth is conceived as a combined process of spontaneous and self-organising changes. Therefore, the contributions of static global factors and dynamic local factors need to be considered. The contribution of global factors is implemented by generating development seeds, reflecting spontaneous growth in response to the potential surface. As a stochastic simulation, seed generation can use a Monte Carlo method that compares a random number grid to the probability surface and thus produces the seeds. According to the Monte Carlo method, the exact location of a seed is subject to a stochastic process, but the frequency of seeding in an area conforms to the development probability of the area. However, it is necessary to control the overall-seeding rate if an appropriate level of global effect can be introduced. For a Monte Carlo simulation, a parameter - the seeding rate - is introduced to scale the calculated development potentiality/probability to a level which can best produce similar urban morphology. This can be achieved through a test run, that is, comparison of the random number grid with the scaled probability grid to see how many seeds are generated. By doing so, the number of seeds planted in one iteration can be controlled. For a deterministic simulation, seeds can be planted from the sorted potentiality index but it has been shown that such a method of introducing the effect of development factors only reflects a special case in site selection decision-making which assumes that perfect information exists across the whole development area (Wu and Webster, 1998).

The use of Monte Carlo simulation here lies in the proper derivation of development probability. Intuitively, the probability of development should follow a logistic or Poisson distribution which characterises the non-linear relationship between the score of development factors such as the distance to the city centre and the probability of a development event. Behind the use of a multinomial logit specification is the random utility theory (see Ben-Akiva and Lerman, 1985; for the application of multinomial logit model in land-use changes see McMillen, 1989). So far, unfortunately, most CA models have not specified the development probability / potentiality according to sound theory. The use of a statistical / calibrated method to describe the development probability is still a subject for future research.

The above procedure addresses explicitly the balance of spontaneous and self-organised growth through parameterisation. The balance is parameterised through the rate of spontaneous seeding in response to the global structure and the threshold of starting off a chain of self-organisation processes. As mentioned earlier the rate of seeding controls the total number of seeds at each iteration, which reflects the strength of globally determined growth. The growth seeds plus the developed cells at time t form the state from which a self-organising process may start off. At this point, another critical parameter - the threshold of self-organisation - is introduced.

The self-organisation in its broad sense means the process in which the state of a system is evolving towards a limited possibility of configuration through feedback between states. The growth of a city is a self-organisation process (Allen, 1997). Such interlock between states is realised through the feedback between existing development to the coming development in a local neighbourhood. Apparently, the developments existing in a place might attract further developments but this can also be conceived as congestion and hence they reduce the attractiveness of the area. Unlike a slope process in which the physical process such as connectivity controls the effect of self-organisation in a local kernel, the sphere of feedback in urban land use studies is often unknown. It is again, however, a complex issue - the mechanism is not subject to empirical studies, but rather is determined on the basis of intuition and common sense. Many models adopt heuristic specification of local feedback, for example a distance decay function (Batty and Xie, 1994), a predefined lookup table reflecting different interactions among land uses (White and Engelen, 1993), spatial interaction formulae (Portugali and Benenson, 1995) and weighted summation according to multicriteria evaluation (Wu, 1998a). More research, therefore, is needed in the future.

The difficulty in specifying the transition rule of land development does not, however, prohibit from building up models useful for a specific purpose. In fact, if the city is indeed a complex system and urban forms are qualitatively different from each other, we should adopt a new way of thinking about the result of simulation. CA models could reproduce urban morphologies similar to the one observed in the real world (White and Engelen, 1993), even apparently on the basis of heuristic rules in most CA models. How the model structure affects the robustness of the model is again a remaining issue.

The threshold of self-organisation specifies that self-organisation of land development starts only when the development intensity of a neighbourhood exceeds a threshold. A conventional local kernel operation is applied to measure the number of contiguous land uses around a cell. The kernel returns a value that is used to measure local development intensity. For a 3x3 kernel, the value 8 represents a state in which the neighbouring sites are fully developed and 0 for an isolated piece of developed land. The development intensity changes along with simulation and hence it should be standardised with respect to the maximum value at each iteration. The standardised development intensity is then compared with the threshold to determine whether a local change can happen. This, in turn, triggers respective rules that can be applied to successful and unsuccessful sites. For example, successful sites may be further subject to the examination of planning constraint. At the end of this process, the states of land use at time $t+1$ are determined.

6.3 EXPERIMENT

The experiment here is to apply the above general procedure to replicate the given pattern of urban land use through simulation. The main development factors of this simulation are the accessibility to the road network shown in Figure 6.2. The given pattern of urban system is shown in Figure 6.3. There is apparently some relationship between the pattern and the road network, especially the intersections of roads being attractive. Two factors are thought to be relevant to the growth of urban land: the distance to the ring road (d_1) and the distance to the cross road (d_2).

The pattern of land use is initiated as a grid of 100×100 pixels. The measure of distance is processed through conventional GIS analysis and exported together with the state of land use (being urban use or vacant land) to SPSS to calibrate a logistic regression model. To improve model performance, an interaction term is added to reflect the position relative to the ring road and cross road. The result of the logistic model is shown in Table 6.1.

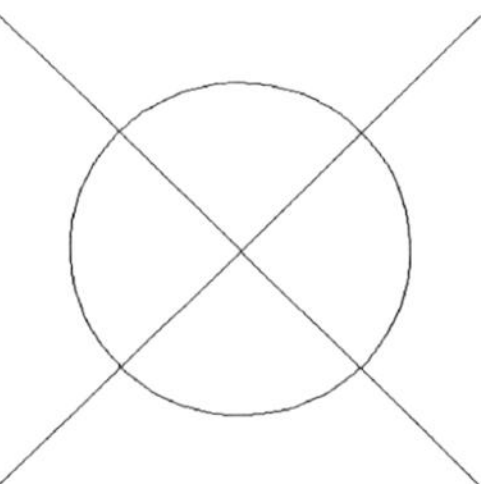

Figure 6.2 The road infrastructure in the experiment

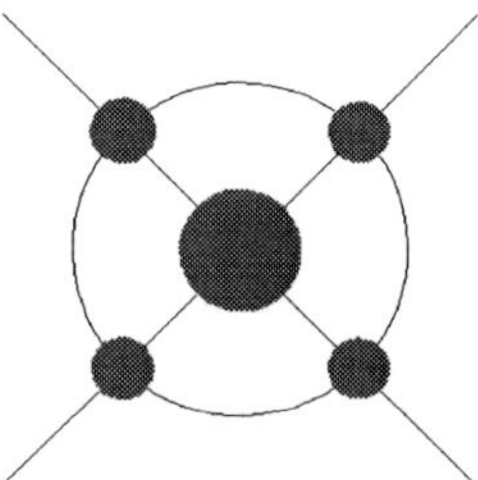

Figure 6.3 The pattern of urban system under simulation

Table 6.1 The logistic regression model to extract the rules of land use change

Variable	Coefficient	Standard error
The distance to the ring road (*d*1)	73.68	3.17
The distance to the cross road (*d*2)	104.8	3.85
*d*1**d*2	-228.27	7.87
a (constant)	-36.38	1.67
-2 Log Likelihood at convergence: 2739.46 (significant at 0.000)		

The model assumes the attractiveness of a site is a function of three attributes: the distance to the ring road d_1, the distance to the cross road d_2, and their interaction as d_1*d_2. According to the logistic model, the probability of a site being developed can be specified as:

$$p = \frac{\exp(a + b_1 d_1 + b_2 d_2 + b_3 d_1 d_2)}{1 + \exp(a + b_1 d_1 + b_2 d_2 + b_3 d_1 d_2)}, \tag{6.1}$$

where a is a constant, and b_1, b_2, b_3 are coefficients of the regression model.

Obviously, with the road network and the land use pattern given as Figure 6.2 and 6.3, the result of the regression is not surprising. The model effectively extracted the basic features of the land use distribution. This is then used to predict the probability of land use change in simulation. When the derived probability surface is compared with a random number grid (even distribution from 0 to 1), the site is tagged as a seed if the probability exceeds the random number. The total number of seeds is equivalent to the total number of observed sites as shown in the Figure 6.3. The rate of seeding is therefore used to scale down the probability:

$$p' = \alpha p \, , \tag{6.2}$$

where p' is the scaled probability, p is the probability derived logistic regression, and α is the rate of seeding.

The rate of seeding is, therefore, linked directly to the expected number of seeds. For example if the value is 0.001, the expected number of seeds will be 0.1% of the total observed number of urban sites in Figure 6.3.

Seeds are generated according to the Monte Carlo process. As more and more sites are developed along with simulation, the seeds may overlap with the developed sites and thus the impact on the simulation is nonlinear over time. This is further complicated by the stochastic effect together with a self-organisation mechanism discussed below.

For the measure of the strength of local growth, a conventional 3x3 kernel is used to count the number of contiguous land uses around the target site (Figure 6.4). At the beginning of simulation, there are few sites being developed. Therefore, to apply a threshold across iterations it is necessary to standardise the intensity of development which is defined as the proportion of developed sites to the total sites in a neighbourhood.

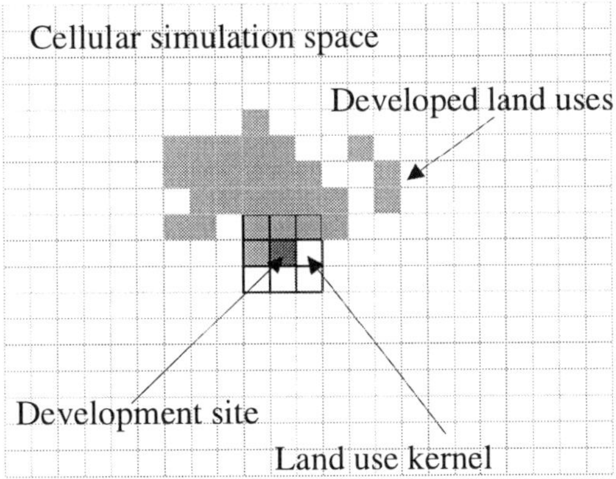

Figure 6.4 Cellular space of simulation and the land use kernel (in this case the neighbourhood kernel function returns a value of 4 for the counted number of developed neighbouring cells).

The standardised score is calculated as:

$$z_i = \frac{n_i - \min(n_i)}{\max(n_i) - \min(n_i)}, \tag{6.3}$$

where z_i is standardised score at location i, n_i is the counted number of developed sites, and max() and min() are the maximum and minimum number of developed sites in a neighbourhood at the iteration.

The standardised value means a relative position of the strength of local growth compared with other neighbourhoods. For example, the value of 0.6 means that the development site has 60% of the maximum intensity existing at the given iteration.

6.4 RESULTS

The experiment is quite computationally intensive. There is a need for automated running of the model. The simulation was programmed in the Arc Macro Language (AML) of ARC/INFO. The model varies systematically the parameters to evaluate their effects. Each combination of parameters runs for 50 iterations. The parameter space consists of the seeding rate varying from 0.001 to 0.005 and the threshold of self-organisation varying from 0.1 to 0.6. Each experiment varies in one step of 0.1 for the seeding rate and 0.001 for the threshold of self-organisation. In total, this consists of 30 experiments and 1500 iterations of all rules. The result was written into a log file, which was then processed. The simulation starts from the initial state in which all sites are vacant.

Figure 6.5 represents the results of these 30 experiments. The images are arranged according to the variation of parameters: the column standing for the rate of seeding and the row for the threshold.

From the figure, it is obvious that lower thresholds or higher seeding rates would lead to more developments. Simulations here did not constrain the total growth rate except that in a single iteration the number of land changes cannot exceed 25. With an increase in the seeding rate (i.e. from the left to the right column), the growth rate increases. With an increase in the threshold (i.e. from the top to the bottom row), the growth rate decreases.

Interestingly, the simulated land uses differ from each other significantly. With a high threshold, the growth mainly relied on the spontaneous growth. While this may produce a pattern similar to the surface of probability/potentiality, the sites are widely scattered and fragmented because of the lack of local growth to connect them together (see s54, s55, s64, and s65). In contrast, when a low threshold is adopted, self-organising growth quickly starts off but at the same time it follows a strong historically dependent path. In other words, the growth will be largely confined to the locations of the first few seeds. As the seeding process is a stochastic one, the structure developed is thus dependent upon the initial distribution of seeds. If a low seeding rate is adopted at the same time, which means that the spontaneous growth is weak, it is unlikely to develop a structure according to the probability/potentiality surface. This is because the seeds emerging at a later stage are not able to form enough strength to compete with the self-organising growth which entrapped the growth at the earlier seeds (see the urban patterns of s21 and s31).

Only with a proper balance of spontaneous and self-organising growth is it likely to produce a pattern similar to an intended pattern of growth. Through visual comparison, it can be found that the land use patterns s24 and s34 (where the seeding rate is 0.004 and the threshold is 0.2 or 0.3) are most similar to the pattern under simulation, despite some differences.

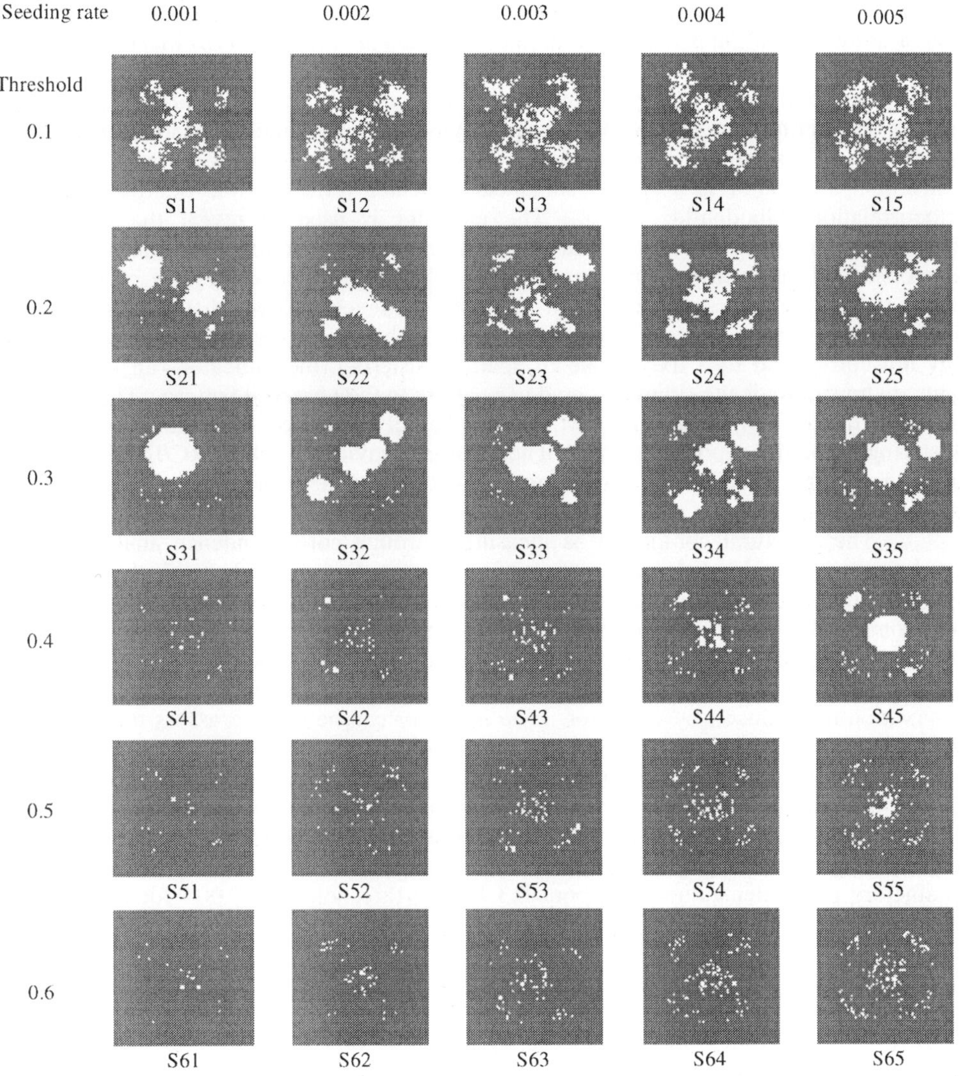

Figure 6.5 The simulated urban pattern with the systematic variation of parameters.

6.5 INDICATORS OF MODEL PERFORMANCE

Visual comparison is a straightforward way to find the parameters that can produce the desired simulation. However, to run a simulation in an automated way in the future, it is necessary to devise some indicators to measure the performance of each experiment and hence to refine the parameters through iteration.

The indicators of model performance, in the case of CA, can be divided into two types: those generic indicators which might reflect various facets of self-organisation processes and those performance indicators which serve as meaningful descriptions of socio-economic systems. The latter is largely dependent upon the domain of question and purpose of simulation and therefore cannot be addressed in a general way. However, the structural measure of land use distribution can be seen as a sort of performance indicator, which may be used to composite further indicators to describe different characteristics of land use distribution (for example, the extent of dispersal of suburban land).

Further measurement of the conformity between the simulated and intended land use patterns (observation) is analysed through some spatial statistics and structural indicators. Moran I is a spatial statistical indicator reflecting the degree of spatial autocorrelation (Goodchild, 1986). The indicator is used to reveal the pattern of clustering of the same type of use at adjacent cells and, therefore, the extent to which developed and undeveloped sites are mixed. The absolute concentration of land uses within a kernel generates a Moran I close to unity, while a more even distribution than can be expected by chance gives a value below zero. Although the absolute value of Moran I may not correspond to a fixed scale of spatial clustering, the indicator can be used to compare how close is the simulated land use pattern to the observed pattern. The function measures directly adjacent cells in the local kernel illustrated in Figure 6.4, that is, clustering at the finest possible scale. The measure provided by the ARC/INFO Moran I function does not describe the clustering of land uses at the level of spatial object or a structural level.

The structural conformity is measured through correspondence analysis. The simulation space is divided into 6 concentric rings and 8 sectors. In total, they produce 48 small areas. The total number of cells of developed land is counted in each area and then compared with the counted number in the given land use pattern. The process is again automated by an AML program that generates the final indicator to a log file. The counted numbers in these small areas are then cross-tabulated between simulation and observation to produce a χ^2 indicator. The indicator can be used to assess the deviation of simulation from the expected value of observation.

In summary, Moran I measures the extent of mixture of land uses or compactness of simulated urban pattern while χ^2 is a structural measure of the distribution of developed clusters. These indicators are plotted out in a three dimensional format in Figure 6.6 and 6.7. From the figures and together with visual comparison, it can be seen that the most satisfactory parameter values are about 0.3 for the threshold and 0.004 for the seeding rate.

6.6 CONCLUSION

This paper discussed the use of CA in urban morphological simulation and introduced a generic procedure that combines spontaneous and self-organising processes of urban growth to simulate land use changes. The experiment reported is aimed to test whether such a procedure can be applied to replicate a given pattern of land use change through simulation. The result of the simulation model suggests that it is possible to capture basic morphological features through simulation as long as the probability of land use change is extracted from the observation and appropriate parameters adopted. The procedure thus can be applied in the simulation of land use change in other contexts with some modest modification. The merit of such a generalised procedure is that it allows researchers to build up the experience of model building. This is particularly important considering that CA simulation is a versatile approach and that it takes a long time to understand the property of a novel procedure.

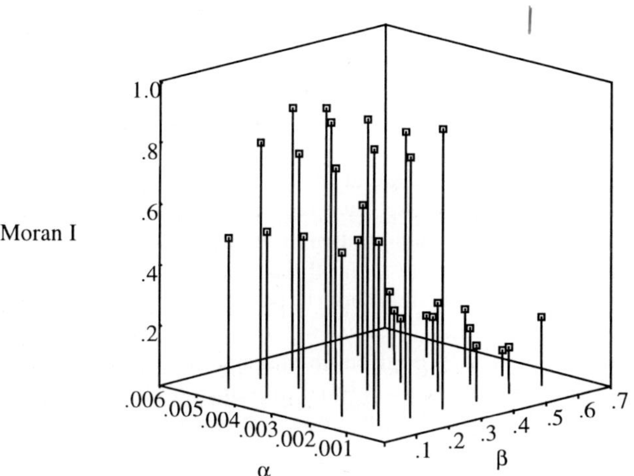

Figure 6.6 Measurement of the clustering of simulated developed land at the finest cell scale (note: Moran I of the given pattern of developed land is 0.925).

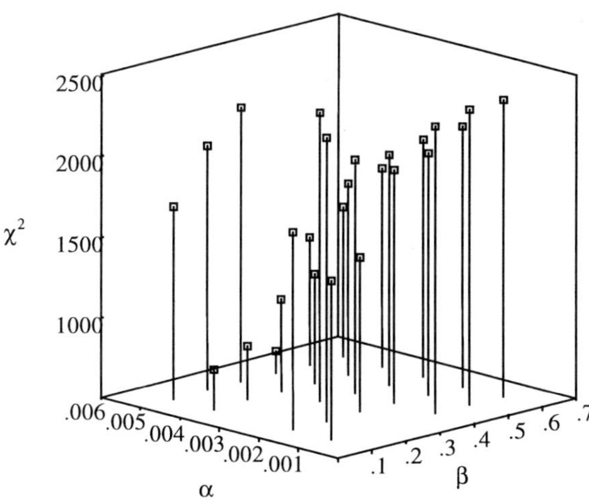

Figure 6.7 Measurement of the degree of structural fit by the χ^2 indicator

(note: χ^2 is a measure of the degree of deviance of simulation from observation, calculated as: $\chi^2 = \sum \frac{[f_s - f_o]^2}{f_o}$, where f_s is the number of simulated land development in an area and f_o is the observed number of the given pattern under simulation. The lower the value of χ^2, the better the simulation fits the observed pattern).

The result of simulation suggests that the actual urban growth is dependent upon the balance of global and local influences. Two parameters are introduced to reflect such influences. Respectively, one parameter controls the strength of global factors such as access to roads and the other reflects the strength built up from local interaction such as the proportion of land developed in a neighbourhood. For simplicity, it is assumed that global and local factors are separated, that is, not interchangeable within one cycle of simulation. But this can be relaxed through multi-stage simulations in which the result of the previous stage of simulation is seen as a global influence on the next-stage simulation.

Conventional models, solely built upon the global factors, are unlikely to reproduce the morphology of urban growth. It has been suggested that correct urban morphology can be reproduced through microscopic rules such as diffusion limited aggregation (DLA) (Batty and Longley, 1994). But the pure locally defined rules give no consideration to heterogeneous geographical conditions. A combined approach would be much more appropriate.

As mentioned earlier, the simulation of urban form is sensitive to parameters. While the importance of calibration is recognised, unfortunately, there is no calibration procedure that can be applied universally. CA models developed so far can be loosely classified into two types. The first type is generic and not aimed to reflect a particular decision-making process (e.g. Batty, 1998; Wu, 1998b; Webster and Wu, forthcoming). The second type addresses a practical urban development process (e.g. Clarke and Gaydos, 1998; Wu, 1998a). The former mainly aims to explore the relationship between form and process, that is, the different processes that lead to generically (qualitatively) different forms. The latter, however, bears a more empirical nature. Thus, model calibration is a critical issue. Calibration largely relies on repetitive runs of the same model with different combinations of parameters. However, this is computationally intensive, especially when there is a large set of parameters. Calibration should be further linked to validation and the measure of goodness-of-fit. Unless there is an acceptable measure of goodness-of-fit, it becomes meaningless to calibrate the model. As suggested, the validation of CA models is largely assessed on the basis of plausibility rather than goodness-of-fit. This is because the introduction of local interaction usually leads to the phenomenon of emergence and thus correlation measures such as *R* square are inappropriate (Batty and Xie, 1997; Wu, 1998b; Wu and Webster, 1998). It is therefore necessary to adopt a more structural measure. The tentative conclusion offered here is that, despite CA as a popular modelling method, there is a severe GeoComputation constraint, especially to deal with complex and ill-defined processes in which parameters cannot be clearly defined through observation and must be tested through repetitive experiments. Considering this constraint, a procedure that reduces the parameters to a few necessary ones may be useful.

6.7 REFERENCES

Allen, P., 1997, *Cities and Regions as Self-Organizing Systems: Models of Complexity*, (Amsterdam: Gordon and Breach Science).

Batty, M., 1998, Urban evolution on the desktop: simulation with the use of extended cellular automata. *Environment and Planning A*, **30**, pp. 1943-1967.

Batty, M., Couclelis, H. and Eichen, M., 1997, Urban systems as cellular automata. *Environment and Planning B*, **24**, pp. 159-164.

Batty, M. and Longley, P.A., 1994, *Fractal Cities*, (London : Academic Press).

Batty, M. and Xie, Y., 1994, From cells to cities. *Environment and Planning B*, **21**, pp. 531-548.

Batty, M. and Xie, Y., 1997, Possible urban automata. *Environment and Planning B*, **24**, pp. 175-192.

Ben-Akiva, M. and Lerman, S., 1985, *Discrete Choice Analysis: Theory and Application to Travel Demand*, (Cambridge, MA: MIT Press).

Clarke, K.C. and Gaydos, L.J., 1998, loose-coupling a cellular automaton model and GIS: long-term urban growth prediction for San Francisco and Washington/Baltimore. *International Journal of Geographical Information Science*, **12**, pp. 699-714.

Couclelis, H., 1997, From cellular automata to urban models: new principles for urban development and implementation. *Environment and Planning B*, **24**, pp. 165-174.

Goodchild, M. F., 1986, *Spatial Autocorrelation: Concepts and Techniques in Modern Geography 47*, (Norwich: Geo Books).

McMillen, D. P., 1989, An empirical model of urban fringe land use. *Land Economics*, **65**, pp. 138-145.

Portugali, J. and Benenson, I., 1995, Artificial planning experience by means of a heuristic cell-space model: simulating international migration in the urban process. *Environment and Planning A*, **27**, pp. 1647-1665.

Toffoli, T. and Margolus, N., 1987, *Cellular Automata Machines*, (Cambridge, MA: MIT press).

Webster, C.J. and Wu, F., Regulation, land use mix and urban performance, part 2: simulation. *Environment and Planning A* (forthcoming).

White, R. and Engelen, G., 1993, Cellular automata and fractal urban form: a cellular modelling approach to the evolution of urban land-use patterns. *Environment and Planning A*, **25**, pp. 1175-1189.

Wolfram, S., 1984, Universality and complexity in cellular automata. *Physica*, **10D**, pp. 1-35.

Wu, F., 1998a, SimLand: a prototype of simulate land conversion through the integrated GIS and CA with AHP-derived transition rules. *International Journal of Geographical Information Science*, **12**, pp. 63-82.

Wu, F., 1998b, An experiment on generic polycentricity of urban growth in a cellular automatic city. *Environment and Planning B*, **25**, pp. 731-752.

Wu, F. and Webster, C.J., 1998, Simulation of land development through the integration of cellular automata and multi-criteria evaluation. *Environment and Planning B*, **25**, pp. 103-126.

7

Testing space-time and more complex hyperspace geographical analysis tools

Ian Turton, Stan Openshaw, Chris Brunsdon, Andy Turner, and James Macgill

7.1 INTRODUCTION

This chapter presents the results of a novel experiment that seeks to compare the performance of several alternative exploratory geographical analysis methods. Simulated data sets containing different amounts of geographical, temporal, and attribute patterns are created and analysed using various Geographical Analysis Machines, commercial data mining software, smart geographical analysis tools, and artificial life based approaches.

The immense explosion in geographically referenced data occasioned by developments in IT, digital mapping, remote sensing and the global diffusion of GIS emphasises the importance of developing data driven inductive approaches to geographical analysis. For example, in the UK the Crime and Disorder Act (1998) places a statutory responsibility on the 464 District Authorities to perform crime pattern analysis. Geographically referenced crime data are not new: most Police Forces have, or are developing, on-line crime reporting systems with GIS or digital map based command and control computer systems. However, other than mapping, the idea and need to perform spatial analysis is relatively new and can be regarded as an activity that has been created as a by-product of GIS success. The problem for both researchers and developers is how best to try and meet these new emerging needs for useful geographical analysis. In the commercial sector, many large retail and financial organisations have been investing heavily in the development of data warehouses. They know that their future competitiveness may well depend on the uses they make of their data resources. Much of their data is or can be geographically referenced and may well contain business significant geographical patterns and relationships so tools are needed that can help find them.

Data mining is the business analysts' term for a particular form of interactive data analysis that employs at least one intuitive human expert and their available computing resources to explore and model patterns in data. It follows that a specific data mining application aims to generate information from data to help further create understanding of a particular problem or class of events via some kind of interactive exploratory data analysis using a software package of data mining techniques. IBM, the developers of Intelligent Miner define data mining as: '... the process of discovering valid, previously unknown, and ultimately comprehensible information from large stores of data. You can use the extracted information to form a prediction or classification model, or to identify similarities between database records. The resulting information can help you make more informed decisions.' The emphasis in this and other definitions is on the use of information extracted from databases to make informed business decisions.

7.2 WHAT IS SPECIAL ABOUT GEOGRAPHICAL DATA MINING?

It can be argued that conventional data mining tools can be usefully applied to mine GIS databases to extract pattern in the same way that conventional statistical methods can be applied to spatial data. There are some geoinformational data mining tasks that may be usefully performed by conventional data mining software. Table 7.1 outlines the range of tools that most data mining packages offer and many of these methods could be usefully applied to spatial data. For example, data reduction tools, such as multivariate classification, can be useful as a means of summarising the essential features of large spatial data sets; for instance, to create geodemographic classifications. Similarly, modelling tools such as neural networks and decision trees can be readily applied to some geographic problems. It can be argued that whilst these methods ignore all of the special features of geographical data (see table 7.2), they still 'work' to some degree but there are also many exploratory geographical analysis types of data mining task that seemingly cannot be performed. However, there is a major potential problem in that the use of conventional data mining tools implies acceptance of the key assumption that geographical data are the same as any other data, i.e. that is there is nothing special about geographical information, or indeed geographical analysis, that will prevent it being performed by conventional methods. These packages can only treat the X, Y co-ordinates as if they were merely two ordinary variables (such as age or income) and it is very likely that nothing useful will be achieved. There is no mechanism for handling location or spatial aggregation or for coping with spatial concepts or even mapping.

Table 7.1 Typical generic data mining functions

exploratory data analysis tools
linked map and graph displays
other visual data mining tools
most multivariate statistical methods
linear and logistic regression
classification
decision trees and regression trees
association rules detection
neural networks
memory based reasoning
sequence discovery

Openshaw (1999) argues that what is now needed are new types of data mining tools that can handle the special nature of spatial information and also capture the spirit and essence of geographicalness that a GIS minded data miner would expect to have available. There is a further problem that needs to be dealt with. If geographical data mining is simply equated with exploratory spatial analysis then maybe some will be misled into believing that this problem has already been solved. However, this overlooks the massive difference between exploring a manageable small data set with few variables and the need to perform the same process on massive databases (with two or three orders of magnitude more cases) and possibly high levels of multivariate complexity. A human being-based graphical exploration of spatial data just does not scale well. The bottleneck is not computational but is a result of limits on the speed and skills of the spatial analyst. There comes a point where adding extra dimensions to the analysis hyperspace overwhelms human abilities. Visualisation tools are useful but there are also limits as to what they can deliver and, in general, GIS databases often present too much complexity for such a simple minded (albeit technically sophisticated) approach.

This chapter describes and presents the results of a series of empirical experiments concerned with evaluating the abilities of a range of data mining tools in detecting patterns in synthetic spatial data. The patterns can be purely spatial, or temporal, or space-time, or space-time-attribute based. Synthetic data sets are used so that the 'true' results are known allowing the abilities of the various methods to be investigated. Openshaw (1994, 1995) identifies seven possible interactions of the tri-space that characterises GIS databases (Table 7.2). The need is for tools relevant to GIS, which can successfully search for patterns in some or all of these hyperspaces. The problem is that these hyperspaces interact to create or hide patterns. For example, in a database with X, Y for space, T for time, and C type of event, different events may well contain different geographical patterns. The same event may also exhibit different geographical patterns for different time periods. The essence of the problem is that the time-event type interactions cannot be studied separately or sequentially, as it may be that the strongest patterns are found only when certain time periods and event types are analysed together. Currently the best available methods tend to ignore totally these interactions and would study the data either together as a single data set or else rely on *a priori* research design decisions that effectively strangle the data before it can speak, albeit unintentionally.

Section 3 outlines two synthetic data generators. Section 4 outlines the five methods that are applied. Section 5 presents the results, and ideas for further research are contained in the conclusion.

Table 7.2 Trispace characteristics of a GIS database

Datatype	Nature of Data Interaction
1.	spatial data
2.	time data
3.	multiple attribute data
4.	geography and time data
5.	time and multiple attribute data
6.	geography and multiple attribute data
7.	geography, time, and multiple attribute data

7.3 TWO SYNTHETIC DATA GENERATORS

7.3.1 Objectives

A major perceived difficulty is the current lack of significant practical success stories in real world geographical analysis applications that can be used to demonstrate successful projects and hence function as exemplars. For instance, in spatial epidemiology there are very few documented successes whereby exploratory geographical analysis tools have been used to correctly identify hitherto unknown disease clusters. Indeed many researchers are fairly pessimistic about the prospects; see Alexander and Boyle (1996); although this seems to reflect a reluctance to perform exploratory analysis and a preference for testing hypotheses. As a result there are currently no generally available pattern benchmark data sets that can be used to measure the performance of old or new methods of exploratory geographical analysis.

The strategy adopted here is to create synthetic data sets with varying degrees of pattern and then assess the success of a selection of methods in analysing these data. This need becomes even more critical if the patterns being concealed in these data sets are not just localised spatial clusters of varying intensity but also include space-time interactions and more complex structures. The data sets are available on the WWW (*http://www.ccg.leeds.ac.uk/-smart/data/index.html*) for others to test out their favourite methods.

The study region was defined as Yorkshire and Humberside as this yielded a sufficiently large data set. The data relates to 10,430 Census Enumeration Districts (EDs) for which persons were used as the population at risk factor. Each census ED had a corrected 100 metre grid-reference attached to it. For the purposes of this exercise 1,000 events were to be generated from a total population at risk of 4,820,129 persons. Figure 7.1 shows one of the data sets.

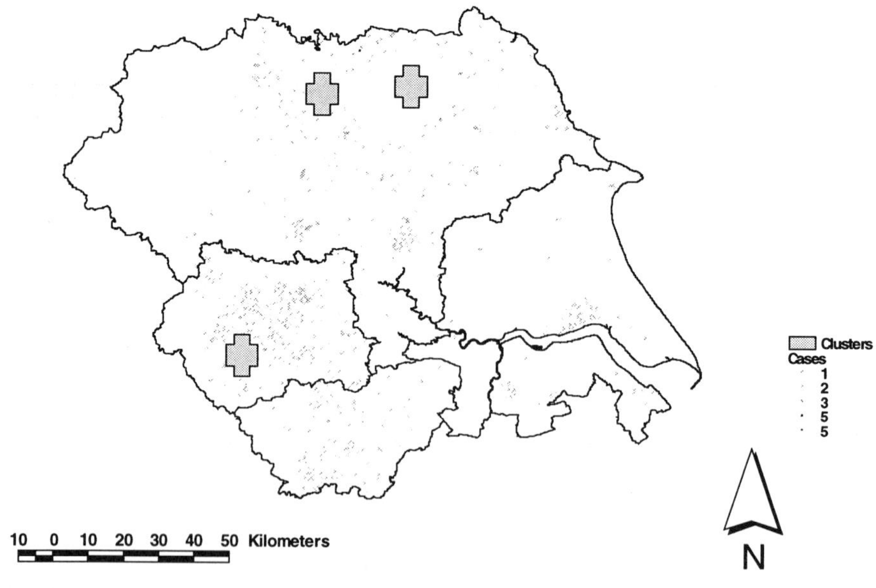

Figure 7.1 One of the test data sets showing the locations of the three clusters

7.3.2 Synthetic data generator 1

The first data generator used created 10 synthetic data sets that displayed varying degrees of spatial and temporal clustering; from purely random to 25% clustered. The spatial clustering algorithm is that described in Alexander et al (1996). Each synthetic data set had a different degree of clustering and often different parent locations. The random cases were selected by multinomial allocation with multinomial probabilities proportional to the population at risk (i.e. total persons). This approximates an inhomogeneous Poisson process, the intensity of which depends on the distribution of the population. The non-random events were allocated as follows:

1. identify parent locations
2. cases are allocated to clusters using an inhomogeneous Poisson process that reflects a Gaussian risk function. There is a distance cut-off so that 90% of the cases will occur within 10 km of their parent.
3. A time clustering component was added by assigning each random spatially clustered event a uniformly distributed time period between 1 and 255. Those cases which are part of the spatial clusters are assigned a uniformly distributed random period limited to time periods between 50 and 60.

This produced 10 data sets with varying degrees of spatial patterns but with a common space-time component. There was no attempt to generate attribute interactions and the attribute field was randomly set to 1 or 2 with equal frequency. Table 7.3 describes the characteristics of these data.

Table 7.3 Characteristics of the synthetic data sets 1

Data Set	Clusters	%Clustered	Size of Clusters
1	3	20	132 48 20
2	1	20	200
3	0	0.0	0
4	4	15	123 1 1 25
5	2	10	87 13
6	2	5	45 5
7	2	2.5	24 1
8	1	2.5	25
9	1	1.0	10
10	2	1.0	8 2
11	0	0.0	0

7.3.3 Synthetic data generator 2

The second data generator is more ambitious and included the full range of trispace interactions described in Table 7.4. The true results were only revealed after the analysis task had been completed. The method is based loosely on the 'raised incidence' model. Consider first the space-only case; under a null hypothesis of no clustering other than that due to variation in underlying population at risk, the probability of incidence in any place would be proportional to the population at risk in that place. If we denote the probability of incidence at a point (x,y) by $p(x,y)$, then this simple model is:

$$p_1(x, y) = kh(x, y) \tag{7.1}$$

where $h(x,y)$ is the density of population at risk at (x,y), and k is a constant of proportionality chosen to ensure that $p(x,y)$ integrates to unity. Clearly, this model is of little use if one wishes to consider clustered data. However, the model can be modified, so that the likelihood of incidents occurring around some places exceeds that expected due to population density alone. One way of achieving this is by multiplying (7.1) by a spatial kernel function centered around some point (x_l, y_l), with a bandwidth b. For example one could have

$$p_2(x, y) = kh(x, y)\exp\left[\left((x - x_1)^2 + (y - y_1)^2\right)/2b^2\right] \tag{7.2}$$

Here the likelihood of a point occurring at location (x,y) depends not only on the population at risk, but also on the closeness to some 'hot point' (x_l, y_l). Points will cluster spatially around (x_l, y_l). The rôle played by the bandwidth, b, is to control the 'tightness' of the cluster. Low values of b will produce more concentrated clusters. Note finally that although (7.2) described a clustered process, in reality it is likely that some cases will be part of a cluster while others will be general 'background' cases. Indeed, if (7.2) were the only generating process, identification of clusters would be a trivial point-plotting exercise. For this reason, the final model will be a mixture of both processes:

$$p_3(x, y) = kah(x, y)\exp\left[\left((x - x_1)^2 + (y - y_1)^2\right)/2b^2\right] + (1 - a)h(x, y) \tag{7.3}$$

The newly introduced parameter a can be interpreted as the proportion of the data which is part of the cluster.

7.3.3.1 Computational issues

The next issue to be addressed is how one can generate random points from the distribution p_3 above. To do this in practice, two issues must be considered. Firstly, the data apply to discrete spatial units (census enumeration districts), and not to continuous space. Secondly, it is not immediately clear how one can simulate random numbers in the distributional form given above. To resolve the first problem, instead of randomly generating a real-number pair (x,y), we generate an index to the enumeration district. Call this index i. Since there are 10,430 enumeration districts, i is a random integer between 1 and 10,430. Of course, different values of i will have different probabilities of selection. For example, in the simple model (7.1), the probability of selection for a given i is proportional to the population at risk resident in the corresponding enumeration district. For the 'hot point' model (ii), the probability of selection for a given i is proportional to the population at risk multiplied by the 'kernel factor' $\exp[((x-x_l)^2+(y-y_l)^2)/2b^2]$. Here (x,y) for each enumeration district is taken as the zonal centroid.

Having re-specified the models for discrete data, the next problem is that of actually generating the data. For models (7.1) and (7.2) this can be achieved using rejection sampling, as set out below:

Step 1 For each enumeration district, compute a number proportional to the probability of selection. Store these in array X.
Step 2 Compute M, the maximum value in array X.
Step 3 Generate a uniformly random integer in the range 1...10400. Call this J.
Step 4 Generate a uniform continuous number in the range 0...M. Call this U.
Step 5 If U < X(J) then return j as the selected index. Otherwise repeat from step 3.

(NB. It is not necessary to compute the normalising constant, k in this algorithm.) Thus, we now have a method for generating cases for models (7.1) and (7.2) - the next stage is to generate the mixture model (7.3). This is relatively simple once there is a method for generating models (7.1) and (7.2). If one wishes to condition on a value of a (say 0.2), and a given sample size *n* (say 1000), then draw a selections from model (7.1), and *n(1-a)* selections from model (7.2), and merge these. For the example values suggested above, one would draw 800 model (7.1) selections, and 200 model (7.2) selections.

Table 7.4 Characteristics of the synthetic data sets 2

Data Set	Clusters	%Clustered	Cluster Radius km	Type of Clustering
1	3	30	5	Space
2	3	30	20	Space
3	3	60	5	Space
4	3	60	20	Space
5	3	30	5	Space-Time
6	3	30	20	Space-Time
7	3	60	5	Space-Time
8	3	60	20	Space-Time
9	3	30	5	Space-Attribute
10	3	30	20	Space-Attribute
11	3	60	5	Space-Attribute
12	3	60	20	Space-Attribute
13	3	30	5	Space-Time –Attribute
14	3	30	20	Space-Time-Attribute
15	3	60	5	Space-Time-Attribute
16	3	60	20	Space-Time-Attribute

7.3.3.2 Adding Time and Attribute Interaction

The above section sets out a method for drawing spatially clustered data, but does not consider time or attribute information. The aim here is to detect interactions between space, time and attributes in clusters. Generating random time or attribute data in itself is straightforward, given the methods described in section 2.4. For example, suppose we wished to generate a day in the range 1...365 in addition to the spatial information. Assuming initially that there were time clusters, but that these were independent of any spatial clustering, one could use the mixture method set out in section 3.3 to generate a data set. One could choose a 'hot day' and a bandwidth, similar to the 'hot point', and have some cases clustered around this point in time (with probability *a*), whilst others occurred uniformly throughout the year (with probability *(1-a)*). To introduce a degree of interaction here, one could link the probability of the observation coming from the clustered model to the spatial location. For example, suppose the probability *a* was a kernel function of (x_1, y_1), with a maximum height of 0.75 - this implies that incidents occurring near to the 'hot point' are more likely to also cluster in time - around the 'hot day'. When the spatial location of an observation lies exactly on the 'hot point', then there is a 0.75 chance that the observation will also be clustered in time. Finally, one can extend the method to incorporate attributes in the same way. Initially this is carried out only for a single, dichotomous attribute taking the values 0 and 1. In the unclustered version, the respective probabilities are c_0 and c_1 for the values to be selected. However, around the point (x_1, y_1) these probabilities may be adjusted using a kernel model. All of these interactions are used in this study.

7.4 GEOGRAPHICAL ANALYSIS TOOLS

7.4.1 GAM/K

GAM/K is designed to detect localised spatial clustering without knowing either or at what scales to look for patterns. Location and scale can both interact in that clustering may occur at different scales on different parts of the map. The algorithm is extensively described elsewhere; see Openshaw (1998), Openshaw et al (1987, 1999a, b). However, there is no provision in GAM/K for detecting space-time or attribute specific clustering other than the ability of the end-user to segment the data into appropriate subsets. It is not expected to do so well on the more complex data sets because here no attempts are made to subset the data. This may undervalue its potential in the real world because not all end-users would be that ignorant of their data or the patterns they suspect it may well contain.

7.4.2 GAM/K-T

This is a recently developed version of GAM/K in which permutations of time period are also examined. The search is now focused on identifying where to look for localised clusters, at which spatial scales, and in which time periods. All three now interact. Tests indicate it will detect space, space-time, or time only patterns although there are now multiple output maps necessitating the development of special results viewers (see Figure 7.2). GAM/K-T, in common with GAM/K, employs an exhaustive search strategy. The additional computational load of handling time was reduced by algorithmic design improvements and by data sparsity, which dramatically reduces the maximum number of time permutations that need to be searched.

7.4.3 MAPEX

MAPEXplorer (MAPEX) seeks to perform a GAM/K and GAM/K-T function via a smart search based on a genetic algorithm. The original method described in Openshaw and Perree (1996) only performed spatial searches. This has now been extended to handle space-time patterns. The hope is that MAPEX will perform well on the space and space-time pattern detection tasks but it will be unable to handle (except by accident) attribute clustering. Although once again in a real-world application the end-user is unlikely to always be totally ignorant of the data.

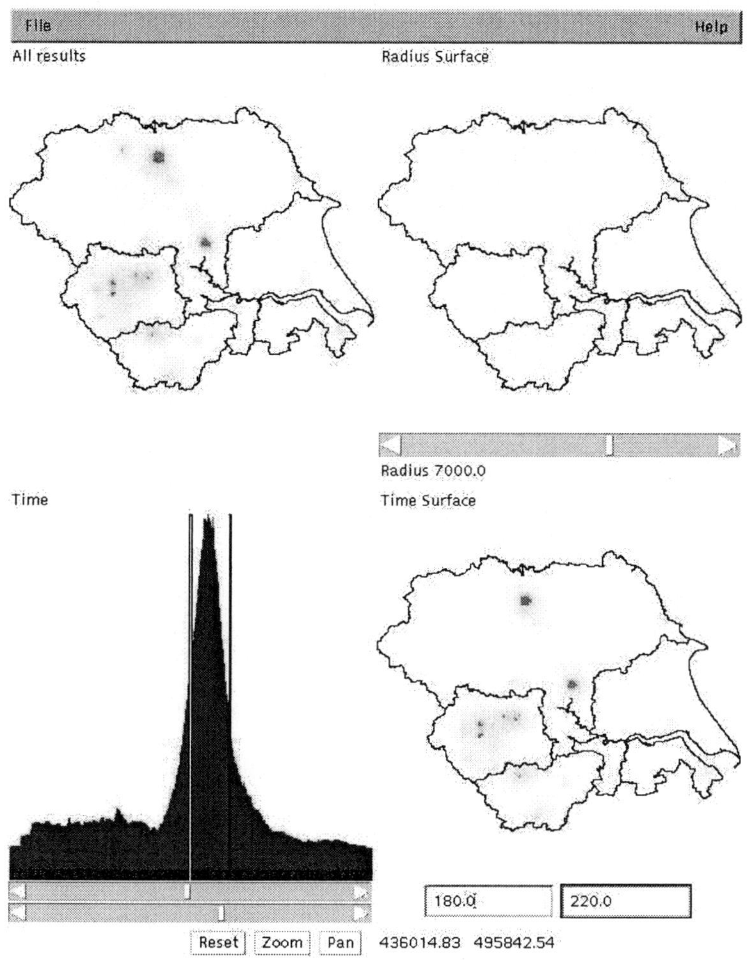

Figure 7.2 The results viewer developed to investigate the results of this project. The top left window shows the total results surface, the top right shows the surface created by circles of a given radius, the lower half shows the time periods of the circles and the surface created by the selected time periods.

7.4.4 GDM/1

Geographical Data Miner (GDM/1) is another recent development. Its design is described in Openshaw (1999). GDM is a development of MAPEX to handle event characteristics and include GIS coverage linkages as performed by GEM (Openshaw, 1998; Openshaw and Turton, 1999a; 1999b). The search now has to handle where to look for localised clustering, at which spatial scales, at which time periods, and identify what event characteristics define the patterns. All the spaces interact to create a highly complex search hyperspace. It is emphasised that this not an uncommon problem in GIS where it is increasingly common for the data to contain some, or all, of the following: map location information, time, event attributes, location attributes. Location attributes are based on

PRIFYSGOL CAERDYDD ✱ CARDIFF UNIVERSITY ✱

typical GIS coverage details; i.e. geology, rainfall, etc. that apply to all the data. Event attributes only apply to the cases (events) which are being analysed. Here time and type are both event attributes that do not exist for locations in the database, which are not events. The data being analysed consist of two types: (1) X, Y for census ED and population of census ED and (2) X, Y for event, time, and type. Note that the type (1) population data could also be indexed by time but here it is constant. The design purpose in GDM/1 was to extend the MAPEX approach to handle space only, space-time space-event type, time-event type, and space-time-event type interactions. Note that a spatial dimension is always present which makes sense because of the geographical purpose of the method. The interaction effects are handled via a series of AND and OR implicit operators. Another way of thinking about GDM is as an intelligent query generator. The GDM searches the universe of all possible relevant queries (given the available data) to suggest the most promising ones (as measured by some pattern detecting statistic). GDM too needs a sophisticated viewer in order to make sense of the results.

7.4.5 Commercial Data Mining Tools

Data mining tools need to be guided by users who understand the problem, the data and the general nature of the analytical methods involved. The methods now used in data mining are usually extensions and generalisations of methods known for decades. Despite their high cost, there is little prospect of finding any explicitly geographical analysis tools. Many will not even draw maps. The synthetic data were pre-processed to become a complete flat file for each census ED; thus X,Y, population, 0-1 event, time, type. Each event is represented as either a 0 or 1 count (some records may be duplicated). Where the event count is zero the time and type values were generated randomly with a uniform distribution between relevant limits (1-365, 1-2). Three different strategies were used:

- compute event rates for each ED and attempt to predict the rates given population, time, type, X, and Y; co-ordinates using neural networks, rule generators and decision tree type data mining tools;
- classify the data and then display incidence rates for each group in the classification using available classification tools such as Kohonen nets and K-means,
- train a neural net to model a version of the data that was purely random and then apply to the synthetic data to identify the largest residuals which would provide evidence of clustering.

The greatest problems here are the learning curves of using the two different data mining packages, numerous operational problems with them, and the large number of alternatives that could be investigated under each of the three headings. It was not possible to investigate hybrid combinations of different tools, such as the use of decision trees or other classifiers to select between alternative data mining methods for a given data set. In general, none of the three direct approaches listed above worked at all well. The problems are intensified by the absence of any mapping capability and other difficulties concerning visualization of the results. It is possible that with more complex multiple category data there will be more opportunities for data mining packages to find interesting relationships as the relative importance of the spatial variables may reduce. Certainly at present with these synthetic test data sets the two data mining systems appeared to struggle. It is important to note that these tentative, preliminary views are limited to the application of the basic data mining tools: the systems do provide opportunities for developing more innovative analysis procedures.

7.4.6 Flocking hunting agents

The flock engine utilises a number of independent agents (called geoBoids) that explore the spatial database foraging for clusters. They communicate and direct their movements co-operatively by borrowing behavioural traits that are based on flocking motion (Reynolds 1987). Each member of the flock is able to evaluate and broadcast its performance. Based on this, other flock members can choose to steer towards well-performing geoBoids to assist them, or to steer to avoid poorly performing boids. In addition, each geoBoid changes its own behaviour based on its performance. For example, a poorly performing boid will speed up in order to find a more interesting area. Likewise, a well-performing geoBoid will slow down to investigate an interesting region more carefully. Finally, if a boid finds an empty section of space with zero population, it masks off that area blocking other boids from entering it.

For the Time-Space-Attribute analysis a genetic algorithm based optimiser was built and retrofitted to the flock-based search. Interaction between the flock of agents and the GA is, at this time, minimal: essentially, the Flock suggests geographic regions of interest, and the GA tries to find the best combination of attribute and time span to describe any clustering found there. As its stands, there is no 'memory' within the GA to carry information between each suggestion from the flock. In the future, each agent in the flock would carry around its own mini-GA allowing the search to take advantage of the inter-agent communication. Macgill (1998) goes into more detail about the flock algorithm.

7.5 RESULTS

The results are summarised in Table 7.5. The GAM/K worked surprisingly well and found most of the clusters in both data sets despite the clever time-attribute interactions in dataset 2. GAM/K revealed cluster locations located fairly rapidly. The results are easy to interpret and the technology is fairly mature and well established. It may not work so well if the data contain more complex structure although the two data sets are by no means simple. It struggled most with dataset 2 cluster 2. This was because the cluster centre is a rural location and the results were pulled towards and mixed with the nearest large town, which had a weakening effect.

The GAM/K-T results were similar to GAM/K except that it correctly identified the time clustering in dataset 2. It struggled with dataset 1 time clusters (as did all the time sensitive methods) leading to the suggestion that the data are incorrect in that the intended 50 to 60 time cluster was randomised over the complete range.

The MAPEX software performed quite well. It easily found spatial clustering and space-time clustering in two of the three dataset 2 clusters. It was interesting that this method was able to find the locations of clusters without any awareness of the interactions with the attribute. This was unexpected and may well suggest that the attribute interactions were insufficiently subtle. However, it failed totally to find cluster 1 indicating a likely software bug in the coding of the co-ordinate ranges. On dataset 1 there was a good performance in detecting the spatial clustering but almost total failure to identify the correct time period.

The GDM/1 results were potentially promising but reflect the early stage in its development. It worked quite well but was a little unreliable and required large amounts of processor time. It also displayed the same generic fault as did MAPEX when processing data 2 clusters. However, it correctly spotted that the attribute information for dataset 1 was random (by ignoring it) but it seemed to experience the same difficulties in identifying th e time period, again supporting the data error hypothesis.

The flocking results are currently only available for dataset 2. The flock worked well in detecting purely spatial clustering. The poorest results relate to those data sets where the cluster radius size was large (20 km). Here the boids became very susceptible to artefacts from the background population that tended lead the flock away from the real cluster centers. For the smaller radius clusters the performance was good in that it often found two of the clusters with strong indications of the third (slightly offset to the nearest population center). The majority of the development of the GA side of the method had gone into the detection of attribute interactions and the system correctly identified all but two of the attribute, space-attribute, time-attribute effects. It also identified time only clustering but more work will be needed on the time-space interactions of the system. With closer integration between the GA and the flock agents the performance of the system should improve even further.

Finally, there are no Data Mining results because none of the methods that were applied worked particularly well. It was also very difficult to interpret the outputs in a manner equivalent to that used for the other methods. There are also some other problems in that the interpretation of whether or not a synthetic data cluster was found was subjective. The identification and measurement of false positives is also fraught with difficulty because of probabilistic uncertainties as to what is a "real" error. Maybe a more deterministic synthetic data generator needs to be developed so that the deviations in space-time-attribute interactions can be assessed directly. There is clearly scope for considerable additional research in these and related aspects.

7.6 CONCLUSIONS

The results are very interesting and clearly require further study and investigation. They demonstrate that there are now fairly reliable methods capable of detecting spatial clustering and space-time clustering. Their ability to also analyse space-time-attribute and space-attribute interactions is less well understood. One surprise that in retrospect should have been obvious was the apparent failure of the commercial data mining packages that were investigated although their testing is continuing. Another surprise was the need to develop data result viewers that would allow the end-user to interpret the results. The extension of exploratory geographical analysis into more complex hyperspaces generates orders of magnitude more results to investigate. The design objective of creating an intelligent human-machine partnership may require re-thinking. As it stands there is considerable subjectivity in "interpreting" the results. Maybe the old GAM/K approach of doing this interpretation automatically or of adding a results filter before human beings start to visualise them might well be a useful subsequent development. Meanwhile we would hope that other interested researchers will use our data sets to test out and develop new methods of geographical analysis.

7.7 ACKNOWLEDGEMENTS

Census data used here were purchased by ESRC/JISC for the academic sector and are Crown Copyright. The ESRC funded much of the GAM/K-T, MAPEX, and GDM1 research via grant No. R237260. The assistance of Capital Bank plc in providing the data mining evaluation is gratefully acknowledged.

	Dataset			GAMK	GAMKT		MAPEX		FLOCK			GDM1		
	C	T	A	C	C	T	C	T	C	T	A	C	T	A
S1	3	Y	N	1	1	Y	3	Y	--	--	--	3	Y	N
S2	1	Y	N	1	1	N	1	Y	--	--	--	1	N	N
S3	0	N	N	0	0	N	0	N	--	--	--	0	N	N
S4	4(2)[1]	Y	N	2	2	N	2	N	--	--	--	2	N	N
S5	2	Y	N	2	1	N	2	N	--	--	--	2	N	N
S6	2	Y	N	2	1	Y	2	Y	--	--	--	2	Y	N
S7	2	Y	N	1	1	N	1	N	--	--	--	0	N	N
S8	1	Y	N	1	1	N	1	N	--	--	--	0	N	N
S9	1	Y	N	0	0	N	1	N	--	--	--	0	N	N
S10	2	Y	N	1	2	Y	0	N	--	--	--	1	Y	N
C1	3	N	N	2+?[2]	1	N	2	N	2+?[2]	N	N	2	N	N
C2	3	N	N	2+?[2]	1	N	2	N	1+?[2]	N	N	2	N	N
C3	3	N	N	2+?[2]	2	N	2	N	3	N	N	2	N	N
C4	3	N	N	2+?[2]	1+?[2]	N	2	N	2+?[2]	N	N	2	N	N
C5	3	Y	N	2+?[2]	2+?[2]	Y	2	N	3	Y	N	2	Y	N
C6	3	Y	N	2+?[2]	2+?[2]	Y	2	N	1	Y	N	2	Y	N
C7	3	Y	N	2+?[2]	2+?[2]	Y	2	N	3	Y	N	-	-	-
C8	3	Y	N	2+?[2]	2+?[2]	Y	2	N	1	N	N	-	-	-
C9	3	N	Y	2+?[2]	2+?[2]	N	2	Y	2+?[2]	N	N	-	-	-
C10	3	N	Y	2+?[2]	2+?[2]	N	2	Y	1+?[2]	N	S	-	-	-
C11	3	N	Y	2+?[2]	2+?[2]	N	2	Y	2+?[2]	N	S	-	-	-
C12	3	N	Y	2+?[2]	2+?[2]	N	2	Y	2	N	S	-	-	-
C13	3	Y	Y	2+?[2]	2+?[2]	Y	2	Y	3	Y	T	-	-	-
C14	3	Y	Y	2+?[2]	2+?[2]	Y	2	Y	2+?[2]	Y	T	-	-	-
C15	3	Y	Y	2+?[2]	2+?[2]	Y	2	Y	2+?[2]	Y	ST	-	-	-
C16	3	Y	Y	2+?[2]	2+?[2]	Y	2	Y	2+?[2]	Y	ST	-	-	-

Table 7.5 Summary of results

Notes: C = number of clusters; T = Time clustering (Y/N); A = Attribute interaction (Yes/Space/Time);
1 two of the clusters in this data set are very small, all the methods missed these two clusters;
2 one of the clusters was very rural and these methods centred the detected cluster over a nearby town.

7.8 REFERENCES

Alexander, F. E., and Boyle, P., 1996, (eds) *Methods for Investigating localised Clustering of Disease*, (Lyon: IARC Scientific Publications)

Alexander, F. E., Williams, J., Maisonneuve, P. and Boyle, P, 1996, The simulated data-sets, in *Methods for Investigating localised Clustering of Disease*, edited by Alexander, F. E. and Boyle, P. (Lyon: IARC Scientific Publications), pp. 21-27

Besag, J. ,Newell, J., 1991, The detection of clusters in rare disease, *Journal of the Royal Statistical Society A*, **154**, 143-155

Dobson, J. E., 1983, Automated geography, *Professional Geographer* **35**, 135-143

Macgill, J., and Openshaw, S., 1998, The use of flocks to drive a Geographic Analysis Machine., Proceedings of the Third International Conference on GeoComputation (Bristol: GeoComputation)

Openshaw, S., Charlton, M E., Wymer, C. and Craft, A., 1987, A Mark I Geographical Analysis Machine for the automated analysis of point data sets, *International Journal of Geographical Information Systems* **1**, 335-358.

Openshaw, S. and Craft, A. 1991, Using the Geographical Analysis Machine to search for evidence of clusters and clustering in childhood leukaemia and non-Hodgkin lymphomas in Britain. In: Draper, G., ed., *The Geographical Epidemiology of Childhood Leukaemia and Non-Hodgkin Lymphoma in Great Britain 1966-83*, (London: HMSO), pp. 109-122

Openshaw, S., Fischer, M M., 1995, A framework for research on spatial analysis relevant to geo-statistical information systems in Europe, *Geographical Systems* **2**, 325-337

Openshaw, S. and Perrée, T., 1996, User centred intelligent spatial analysis of point data, in *Innovations in GIS 3* edited by Parker, D. (London: Taylor and Francis), pp.119-134

Openshaw, S., 1994, Two exploratory space-time attribute pattern analysers relevant to GIS in *GIS and Spatial Analysis* edited by Fotheringham, S. and Rogerson, P. (London: Taylor and Francis), pp. 83-104

Openshaw, S., 1995 Developing automated and smart spatial pattern exploration tools for geographical information systems applications, *The Statistician* **44**, pp. 3-16

Openshaw, S. and Openshaw, C., 1997, *Artificial Intelligence in Geography*. (Chichester: Wiley)

Openshaw, S., 1998, Building automated Geographical Analysis and Exploration Machines in *Geocomputation: A primer* edited by Longley, P. A., Brooks, S. M., Mcdonnell, R. and Macmillan, B. (Chichester: Wiley), pp. 95-115

Openshaw, S. and Turton, I., 1999a An introduction to High Performance Computing and the Art of Parallel Programming: for geographers, social scientists, and engineers. (London: Routledge), forthcoming

Openshaw, S. and Turton, I., 1999b, Using a Geographical Explanations Machine to Analyse Spatial Factors relating to Primary School Performance (forthcoming)

Openshaw, S., Turton, I, Macgill, J., and Davy, J., 1999b, Putting the Geographical Analysis Machine on the Internet In *Innovations in GIS 6* edited by Gittings, B. (London: Taylor and Francis)

Openshaw, S. 1999, Geographical Data Mining: key design issues, In Proceedings of the fourth International Conference on GeoComputation (Fredericksburg: GeoComputation)

Openshaw, S., Turton, I. and Macgill, J, 1999a, Using the Geographical Analysis Machine to analyse limiting long term illness, *Geographical and Environmental Modelling* **3**, pp. 83-99

Reynolds, C W., 1987, Flocks, herds, and schools: a distributional behavioural model, *Computer Graphics* **21**, 25-34

PART II

Zonation and Generalization

8

Automated zone design in GIS

David Martin

8.1 INTRODUCTION

The concept and effects of the modifiable areal unit problem (Openshaw and Taylor, 1981; Openshaw, 1984) have been well documented in quantitative geography (Fotheringham and Wong, 1991; Amrhein, 1995). Fundamentally, if areal units are imposed onto a discrete geographical distribution for the purpose of data aggregation, then the resulting areal values will be conditional on the locations of the boundaries. This situation frequently arises, for example, when administrative boundaries are imposed onto the distribution of a human population. The modifiable areal unit problem comprises two separate issues, known as the scale and aggregation problems: aggregated values will vary according to the scale of aggregation, and at any given scale, will vary according to the particular boundary configuration chosen. Even the simplest real-world areal aggregation schemes present a massive number of possible alternative configurations. In conventional choropleth (shaded area) mapping, the problem is one of map interpretation (Monmonier, 1996), but in GIS applications, the implications are more complex, due to the multiple and unpredictable uses to which the area data may be put. Recognition that the problem exists leads directly to the possibility of adapting the design of areal units to provide the best configuration for any particular application. Defining and obtaining the best configuration becomes computationally intensive as the number of areas rises above very small numbers, and this chapter presents a range of research which deals with automated zone design in a GIS context. Despite being both geographical and computational, zone design applications do not fit neatly within the definition of geocomputation expressed by Longley (1998), for example, who identifies 'research-led applications which emphasize process over form, dynamics over statics, and interaction over passive response'. Nevertheless, these procedures are clearly in keeping with Openshaw and Alvanides' (1999) view of geocomputation as 'the application of a large-scale computationally intensive approach to the problems of physical and human geography' (p. 270).

This chapter presents a brief review of the emergence of zone design concepts and their implementation in a GIS environment, and moves on to look firstly at the techniques and secondly at applications of zone design research with particular reference to the application of zone design tools to 2001 UK Census output geography.

8.2 ZONE DESIGN IN A GIS ENVIRONMENT

8.2.1 Introduction to the problem

The creation of geographical zoning schemes with some particular purpose in mind is a very practical application, which is not well served by standard GIS tools. However, the idea of using a computer to evaluate many possible configurations of a set of areal units in order to create a configuration which maximizes the value of some objective function is a relatively simple one. A computational solution to the zone design problem was first proposed by Openshaw (1977), using a simple geographical framework built up from regular grid squares. Possible design criteria (Openshaw, 1978) might include equality of population size, equality of geographical area, minimization of boundary length, maximisation of social homogeneity etc. In many cases, these will conflict with one another, such that improvement in one will result in deterioration of others, and an 'optimal' solution is therefore one which achieves the desired degree of balance between the competing design criteria.

In order to set these techniques in context, a simple zone design problem is considered here with reference to Figure 8.1. Figure 8.1(a) shows a set of geographical building blocks A-E, each of which has an associated population value, totalling 300. Supposing the objective is to create two output areas of equal population size, ie. 150 each, we could measure the success of any proposed solution by summing the differences between the populations of each output area and the target value of 150. An initial configuration with two output areas is shown in Figure 18.1(b), which results in output areas with populations of 110 and 190. The total population error is therefore 80 (40 for each of the two output areas). Exchanging building block D between the output areas, as illustrated in Figure 18.1(c) improves the solution such that the total population error is now only 40 (20 for each output area). A final exchange of building block B between the output areas achieves the desired solution in Figure 18.1(d), with a total population error of 0. In this case, there is only one objective function, namely the sum of output area population sizes measured from the target size, and it has been possible to achieve an exact solution. In real world situations there are often a variety of competing objectives, and perfect solution of any one of them may be impossible. When the number of building blocks becomes large, manual evaluation of all possible combinations is intractable.

8.2.2 Computational approaches

It is not necessary to have a GIS in order to implement automated solutions to the zone design problem. Mehrotra *et al.* (1998) and Horn (1995) describe approaches couched in graph theoretical terms. They treat the problem as a constrained graph partitioning task, in which each building block is a node in a graph, and output areas are built by clustering adjacent nodes. Horn (1995) notes that exhaustive search procedures are useful only in relatively small problems, due to the heavy computational requirements. Automated approaches to the zone design problem require that all the necessary information be available in digital form. In early work, one of the most difficult issues was the availability and management of the building block boundary data and the corresponding contiguity matrix, but these are aspects of the task with which contemporary GIS software can be of considerable assistance. The contiguity matrix defines which building blocks are adjacent to which, and this is necessary in order to determine which moves are possible while retaining contiguous output areas. For example, in Figure 18.1(c), it is necessary to know the building block membership of the output areas in 18.1(b) and to determine that building block D may be swapped without breaking the contiguity of the donor area. The populations of each building block and formative output area must also be known. An automated approach to the zone design problem was therefore not an applicable technique until the advent of widely available digital boundaries and attributes for small areas, and the associated emergence of GIS software with data structures able to handle the necessary adjacency information. In the GIS context, it becomes possible to consider the problem in its broader context, dealing with boundary placement at the micro-scale in addition to the topology and grouping of the areal units.

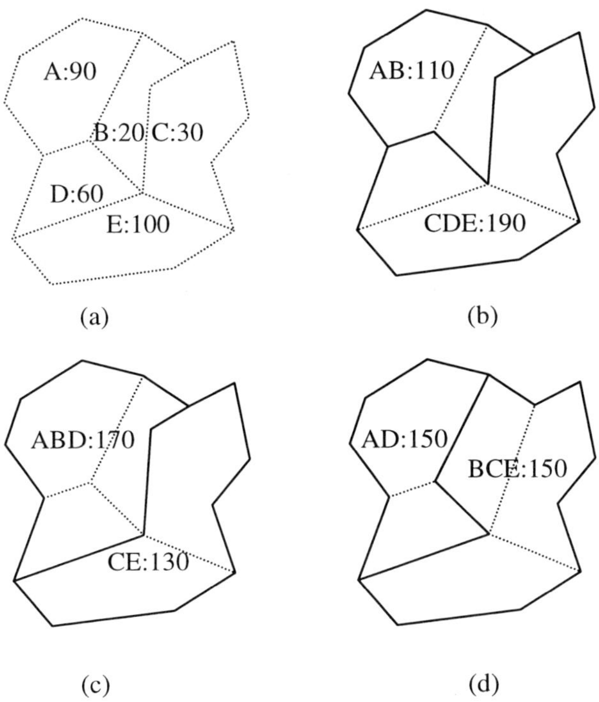

Figure 8.1 Illustrating the zone design problem

Openshaw and Rao (1995) examine several alternative computational approaches to the problem in a GIS environment. The simplest, termed the automated zoning procedure (AZP) is based on Openshaw's original (1977) work. It proceeds from an initially random aggregation of building blocks into output areas, by the evaluation of moving randomly chosen building blocks between adjacent output areas, an idea introduced in relation to Figure 8.1 above. If a move results in an improvement in the objective function, it is retained, otherwise further moves are evaluated until an improvement can be obtained. A problem with the simple AZP is the possibility of its becoming trapped in local sub-optima, although for practical purposes this may usually be overcome by restarting the algorithm from a number of different random initial configurations.

One variant of AZP is to incorporate simulated annealing (Aarts and Korst, 1989), which draws on analogies with the physical properties of materials. This is a relatively simple modification to the AZP in which the 'temperature' is gradually reduced through the duration of the zoning process. In the early stages, when the temperature is high, moves may be permitted which actually reduce the value of the overall objective function, but the probability of allowing such moves decreases as the zoning environment gradually 'cools' and the solution becomes less malleable. The simulated annealing process results in a much slower solution than AZP, but can produce more satisfactory results. The second alternative considered by Openshaw and Rao is to use a tabu search heuristic. In this case, a tabu period is established, during which it is not possible to reverse moves which have already been made (ie. they become 'tabu'). As with simulated annealing this may sometimes result in the adoption of moves which result in a deterioration of the objective function, but allow the algorithm to escape from local sub-optima. As a final approach, they consider the possibility of operationalizing AZP in a parallel computing environment. AZP is not directly amenable to parallel implementation because the algorithm focuses on sequential evaluation of swaps and their impact on the objective function. It would be possible to parallelize this approach by working on separate sub-regions simultaneously, providing the swaps being evaluated did not interfere with one another, and providing the objective function could be broken down into independently evaluated and modified components for each sub-region.

This work is implemented in the Zone Design System (ZDES) which comprises a Fortran program, integrated with the Arc/Info GIS using AML routines, available from *http://www.geog.leeds.ac.uk/pgrads/s.alvanides/zdes3.html*. AZP uses Arc/Info's internal functions for the construction of the contiguity table, and for the presentation of the analysis to the user, while the zone design routines themselves are embedded in the external program. The zone design code is not specifically dependent on Arc/Info, but could be adapted to any GIS software offering a suitable macro programming environment and spatial data structure.

8.3 APPLICATIONS

Application of automated zone design procedures requires suitable building blocks to be available with associated data values. Flexible output area creation requires that the building blocks be small, and it is only since 1991 in the UK that there have been nationally available digital boundaries for enumeration districts (EDs), the smallest geographical units for which census data are available. Strictly, the issue of boundary placement is part of the overall zone design problem, as will be illustrated below, but most researchers in the mid-1990s have been limited to using pre-existing building blocks. The usual choice has been to use census EDs, although they are not ideal for this purpose, themselves being the result of manual zone design processes which lead to wide variations in geographical area and population size. Discussion of current developments towards the 2001 UK Census indicates how the generation of new, smaller, building blocks is actually being driven in part by the desire for an automated solution to the zone design problem.

8.3.1 Practical applications

Haining *et al.* (1994) illustrate the use of a zone design methodology for epidemiological analysis by the combination of EDs from the 1991 census. They are particularly concerned that the areal units used to link health and census data should contain large enough populations to ensure that apparent incidence rates are reliable, and that they should be homogeneous with regard to major socioeconomic characteristics. In this context, the use of individual EDs would create severe small number problems. Their approach is based on classification of the non-spatial census characteristics of the EDs, using a method based on information theory which minimizes information loss resulting from the classification (Johnston and Semple, 1983). This analogy between aspatial classification and geographical regionalization problems was explored as long ago as Grigg (1965). In Haining *et al.*'s (1994) study, 1159 EDs in Sheffield are grouped on the basis of a series of deprivation-related variables into different numbers of statistical classes and the resulting information loss observed. The classes are then mapped, and boundaries which are internal to each class dissolved. Additional mergers are then conducted according to two rules. The 'island' criterion states that a region of one class entirely surrounded by another region of another class will be absorbed into the surrounding region, and the 'like neighbours' criterion states that small regions could be merged into large neighbouring regions if they had similar classifications. This approach avoids the creation of regions with very small populations but still results in wide variations in population sizes from 4000-23000. Their work is further developed in Haining *et al.* (1998), in which the utility of region-building as an exploratory spatial data analysis tool is discussed, as a means of creating regions which provide a satisfactory basis for statistical analysis. Homogeneity, equality and compactness criteria may be combined by the application of weights within the overall objective function. Their system, SAGE, is another example of integration with the Arc/Info GIS, and is available at ftp://hippo.shef.ac.uk/pub/uni/academic/D-H/g/sage/sagehtm/sage.htm. As with ZDES, noted above, the GIS environment is used to provide the necessary geographical information, while the zone design algorithms reside in the specialist software tools.

Openshaw and Rao (1995) also present applications based on the aggregation of 1991 Census EDs, and illustrate their work with reference to the study of ethnicity, unemployment and car ownsership in Merseyside. They devise zoning schemes in which the zones are of approximately equal size, to assist in the study of strongly clustered variables like ethnicity, and others which maximise the correlations between the variables. These applications, as with the Sheffield studies noted above, are concerned more with spatial analytical questions than with practical zone design problems with some particular purpose in mind. Openshaw and Alvanides (1999) explore these ideas further, extending the ethnicity data to the national scale, and demonstrating how a simple spatial interaction model may be used in place of a more conventional objective function. A notable feature of these optimal solutions is that they can display highly irregular boundaries unless some form of shape control is retained. In the analytical context, the zone design tools serve primarily to illustrate the sensitivity of mapped patterns and observed relationships to the particular zonal configuration chosen. While demonstrating powerfully the effects of the modifiable areal unit problem, they neither propose specific solutions nor inform any particular operational context.

Mehrotra *et al.* (1998) address zone design in the rather more applied context of political districting, one of the oldest areas in which the problem has been understood, and where the deliberate manipulation of zone boundaries in order to achieve a particular outcome is popularly termed gerrymandering. They are concerned with the creation of districts in the USA from counties as building blocks, with equality of population and compact geographical shape as constraints. Horn (1995) also considers political districting, in the Australian context, but also notes workload equalization applications in administration, service and sales territory planning, noting that although absolutely optimal solutions may not be achieved, zone design tools can be intuitively plausible and useful for such tasks. It would appear that in each of these application domains, one of the key difficulties becomes that of choosing appropriate objective functions.

8.3.2 2001 Census output geography

In 2001, it is proposed to separate the sets of geographical areas used for the collection and output of census data in England and Wales. The design of a census output geography presents a massive zone design task, and is the largest application challenge to date for the kinds of automated zone design tools discussed here. In recent decades, the small areas for which UK Census data have been published have been the EDs used for data collection. The design of EDs has been dominated by the considerations of data collection, and they consequently vary widely in shape and size, and were not available in digital form prior to the 1991 census. With the advent of GIS use for census geography management (HM Government, 1999), it is proposed that output geography be designed with reference to the requirements of census data users.

The design considerations for census output areas are complex, in that it is only possible to publish data for a single small-area geography, due to the requirements of data confidentiality. Within the selected geographical framework, no data are published if the population of an areal unit falls below certain thresholds. If aggregated data were published for two sets of areal units, which differed only by a very small amount, it would theoretically be possible to overlay the two geographies in order to construct small geographical areas whose populations fell below the threshold. This is known as the differencing problem, and has been addressed in detail in the case of the UK Census by Duke-Williams and Rees (1998). If only one geography is therefore to be produced for census output, it is important to all users that the areal units chosen are the 'best' possible. Coombes (1995) identified a number of desirable characteristics of the areas making up a census geography, including: that they be as small as possible, consistently defined, coterminous with significant physical and administrative features and available in digital form. It is not conceivable that the traditional manual approach to census area design could produce such a geographical base. There is thus a very strong case for the use of automated zone design procedures in order to create the 2001 census output geography.

Initial experiments with the construction of an output geography for 2001 by automated means are reported in Martin (1997), and the development of these into a working prototype in Martin (1998). In this application, where the output areas are themselves to be the smallest geographical units for which census data will be published, it is necessary to provide a further, lower level, set of geographical units for use as building blocks. The only smaller areas covering the UK are unit postcodes, typically containing around 15 households compared with 200 in a census ED, but these units do not have formally defined boundaries, and this therefore becomes an additional component of the zone design task. Use of the unit postcodes as building blocks has the added advantage that the resulting zoning scheme will automatically have a strong relationship with the postcode geography which is used for the georeferencing of a wide range of non-census socioeconomic data collection. Census output area boundaries must be constrained within higher level statutory boundaries, such as wards, for which it is a basic requirement of the census that precise counts be provided. The zone design task is thus one of repeated small problems (building output areas within statutory areas) rather than one enormous combinatorial problem (building output areas within the entire country).

In this major practical application, the overall design process can be divided into three stages, as indicated in Table 8.1, of which the need to generate building block boundaries becomes the first stage. This is primarily a GIS task concerned with the generation, intersection and overlay of digital boundary data. Stages two and three are part of the automated zone design procedure, stage two comprising 'hard' constraints which must always be met in order to achieve an acceptable solution, and stage three comprising 'soft' constraints which must be traded off and compromised, and for which there will not be a perfect solution available.

Table 8.1 Three stages in the 2001 census output area design system

Design stage	Nature of design constraints	Examples
Stage One	Boundary placement	Boundaries should follow real-world topographic features Boundaries should nest within higher level statutory boundaries
Stage Two	Application of 'hard' zoning constraints	All output areas should comprise contiguous building blocks No island areas are allowed All output areas should be above the chosen population threshold
Stage Three	Application of 'soft' zoning constraints	Equality of population size Compactness of output area shape Internal social homogeneity of output areas

Stage one is the creation of boundaries for the unit postcodes (and part unit postcodes where they are split by statutory boundaries). Individual postcoded property locations are recorded as point coordinate references to 0.1 m precision in the ADDRESS-POINT product produced by Ordnance Survey, Britain's national mapping agency. Thiessen polygons are created around each of these property locations, and the boundaries between polygons sharing a common postcode are then dissolved. The resulting postcode polygon set is then intersected with a series of additional features which are to be taken into account in the postcode boundary creation, such as road centrelines. This operation is performed for each statutory area at a time, to ensure that postcode and part postcode polygons are created which aggregate neatly to statutory areas, and which observe identifiable physical features wherever possible.

The extent to which postcode boundaries should follow road centrelines and other physical features is an important design issue because it has implications for the detailed form of the final output area geography, affecting both the contiguity matrix (which postcode is adjacent to which), and the smoothness of the output area boundaries.

Output areas constructed from building blocks with irregular and jagged shapes will themselves display these characteristics, which are unacceptable to many users of census data. Conversely, integration with more than one or two ancillary feature sets becomes complex in different ways. Use of (for example) road centrelines and existing ED boundaries may help to create postcode building blocks with smoother edges, but introduces many sliver polygons where the additional data layers both represent the same geographical features (such as roads) which have been digitized differently. Although not conventionally considered part of the zone design task, these first stage considerations are particularly influential over the appearance of the final zoning scheme, and its acceptability to potential users. The reason that these have not been given such prominence in earlier applications is that it has usually been necessary to accept the building blocks as given, although a national digital data infrastructure and widespread GIS functionality will make these issues more important in the future.

Once unit postcode building blocks have been constructed, a contiguity matrix is assembled for the postcode polygons within each statutory area. The requirement for contiguity and the avoidance of island output areas (areas completely surrounded by another output area) are absolute requirements of stage two, directly testable from boundary topology. Within the Census Offices, population counts and other socioeconomic data required for homogeneity assessment can be tabulated for each of these units, although these data are not available to external researchers due to the confidentiality constraints noted above. The population size of each output area is tested to ensure that no areas fall below the confidentiality threshold, providing the third essential requirement of a valid solution at stage two.

For the third stage evaluation, various statistical measures of the current configuration are computed, which can be combined to obtain the value of the objective function at each iteration. Squared difference from the target population size has been minimized in order to achieve uniformity of population. Different shape statistics have been used, including minimization of boundary length, and minimization of the distance between the building block centroids and the overall centroid of the output area, both of which favour smooth compact shapes which tend towards circularity over elongated and irregular shapes. Maximization of the proportion of the households falling into the dominant tenure class has been used as a measure of social homogeneity, although more sophisticated measures based on levels of intra-area correlation (Tranmer and Steel, 1998) are currently being considered. The rationale behind these measures is that social neighbourhoods should be recognized by census output area boundaries. One of the most difficult issues is the weighting and combination of these different objective functions, which is essentially arbitrary. None of the measures is directly comparable, thus when assessing building block swaps it is difficult to determine the relative importance of an improvement in population size compared to a deterioration in shape statistics (for example). This remains one of the most important research issues concerning automated zone design.

These ideas are taken further by Openshaw *et al.* (1998) who consider the use of individual address locations as the building blocks for census output area creation. They propose Thiessen polygons constructed around each individual address location as the building blocks. The difficulty with this approach is that the resultant output areas may bear no clear relationship with the underlying built form, with properties on different streets being grouped together, even when they are separated by long distances and many intervening properties on the street network. It may be possible to achieve better results when measured by the objective functions at stages two and three, but only at the cost of sacrificing boundary placement constraints at stage one. They use geographical compactness as the primary design criterion with all other objectives as inequality constraints.

8.4 CONCLUSION

Automated zone design procedures offer an automated solution to a family of complex real world problems which cannot be solved realistically by manual methods. Although the basic algorithms required have been developed for some years, it is only with the combination of high computational power and readily adaptable GIS data structures that it has become practical to consider these tools for operational use. GIS development has also been important in so far as it has fostered the development of a digital geographical data infrastructure which provides the necessary building block data. The choice of design criteria is very important, and is an essentially subjective process, as is the method adopted for their weighting and combination, and these remain some of the most important issues for further research. Openshaw and Rao (1995) stress the importance of educating potential users of the importance of zoning schemes in affecting statistical analysis, and of the possible danger of zoning 'anarchy' if such computational tools are used indiscriminately. The review given here has shown how the most applied work to date has tended to concern political redistricting and other applications have been concerned primarily with the investigation of the statistical analytical implications of alternative zoning schemes. The design of output areas for the 2001 UK census is a most interesting application because it undoubtedly represents a major step up in the application and use of geocomputational tools which will affect the lives of a far wider community of data users, and which brings to fruition aspects of computational geography originally proposed over 20 years ago.

8.5 ACKNOWLEDGEMENT

The author is particularly grateful to staff in the Census Division at the Office for National Statistics (ONS) in Titchfield UK for their assistance with various aspects of the work reported here. All views and opinions remain those of the author, and do not necessarily reflect the policies of ONS.

8.6 REFERENCES

Amrhein, C.G., 1995, Searching for the elusive aggregation effect – evidence from statistical simulations. *Environment and Planning A*, **27**, pp. 105-119.

Aarts, E. H. L. and Korst, J., 1989, *Simulated Annealing and Boltzmann Machines: a Stochastic Approach to Combinatorial Optimization and Neural Computing* (Chichester: Wiley).

Coombes, M., 1995, Dealing with census geography: principles, practices and possibilities. In *Census Users' Handbook*, edited by Openshaw, S. (Cambridge: GeoInformation International), pp. 111-132.

Duke-Williams, O. and Rees, P., 1998, Can Census Offices publish statistics for more than one small area geography? An analysis of the differencing problem in statistical disclosure. *International Journal of Geographical Information Science*, **12**, pp. 579-605.

Fotheringham, A.S. and Wong, D.W.S., 1991, The modifiable areal unit problem in multivariate statistical analysis. *Environment and Planning A*, **23**, pp. 1025-1044

Grigg, D., 1965, The logic of regional systems. *Annals of the Association of American Geographers*, **55**, pp. 465-491.

Haining, R., Wise, S. and Blake, M., 1994, Constructing regions for small-area analysis – material deprivation and colorectal cancer. *Journal of Public Health Medicine*, **16**, pp. 429-438.

Haining, R., Wise, S. and Ma, J., 1998, Exploratory spatial data analysis in a geographic information system environment. *Journal of the Royal Statistical Society D*, **47**, pp. 457-469.

HM Government, 1999, *The 2001 Census of Population* Cm 4253, (London: The Stationery Office).

Horn, M.E.T., 1995, Solution techniques for large regional partitioning problems *Geographical Analysis*, **27**, pp. 230-248.

Johnston, R. and Semple, K., 1983, *Classification using information statistics*. Concepts and Techniques in Modern Geography, **37**, (Norwich: Geo Books).

Longley, P.A., 1998, Foundations. In *Geocomputation: a primer*, edited by Longley, P.A., Brookes, S.M., McDonnell, R. and Macmillan, B., (Chichester: Wiley), pp. 3-15.

Martin, D., 1997, From enumeration districts to output areas: experiments in the automated creation of a census output geography. *Population Trends*, **88**, pp. 36-42.

Martin, D., 1998, 2001 Census output areas: from concept to prototype. *Population Trends*, **94**, pp. 19-24.

Mehrotra, A., Johnson, E.L. and Nemhauser, G.L., 1998, An optimization based heuristic for political districting. *Management Science*, **44**, pp. 1100-1114.

Monmonier, M., 1996, *How to lie with maps*. 2nd Edition, (Chicago: University of Chicago Press).

Openshaw, S., 1977, A geographical solution to scale and aggregation problems in region-building, partitioning and spatial modelling. *Transactions of the Institute of British Geographers* NS, **2**, pp. 459-472.

Openshaw, S., 1978, An empirical study of some zone design criteria. *Environment and Planning A*, **10**, 781-194.

Openshaw, S., 1984, *The modifiable areal unit problem.* Concepts and Techniques in Modern Geography, **38**, (Norwich: Geo Books).

Openshaw, S., Alvanides, S. and Whalley, S., 1998, Some further experiments with designing output areas for the 2001 UK Census. In *The 2001 Census: What do we really, really want?*, edited by Rees, P., Working Paper 98/7 (Leeds: School of Geography, University of Leeds).

Openshaw, S. and Alvanides S. 1999 Applying geocomputation to the analysis of spatial distributions. In *Geographical Information Systems Principles, Techniques, Management and Applications*, edited by Longley, P. A., Goodchild, M. F., Maguire, D. J. and Rhind, D. W., (Chichester: Wiley), pp. 267-282.

Openshaw, S. and Rao, L., 1995, Algorithms for reengineering 1991 Census geography. *Environment and Planning A*, **27**, pp. 425-46.

Openshaw, S. and Taylor, P., 1981, The modifiable areal unit problem. In *Quantitative Geography: A British View*, edited by Wrigley, N. and Bennett, R. J., (London: Routledge and Kegan Paul), pp. 60-69.

Tranmer, M. and Steel, D., 1998, Using census data to investigate the causes of the ecological fallacy, *Environment and Planning A*, **30**, pp. 817-831.

9

Designing zoning systems for flow data

Seraphim Alvanides, Stan Openshaw and Oliver Duke-Williams

9.1 INTRODUCTION

It is now accepted wisdom that the geography used to report census and other spatially aggregated data affects the results to some degree. The modifiable areal unit problem (MAUP) is a fact of life and it is important to discover how best to manage it rather than just ignore it and hope (Openshaw, 1996). One way of managing the MAUP is by designing appropriate small areas for data collection and dissemination. A move in this direction is the reporting of the 2001 UK Census of Population for census output areas based on a carefully and explicitly designed zoning system which will replace the census Enumeration Districts (EDs) as the most detailed spatial unit (Martin, 1998). However, the output areas designed for reporting household and person based data are inherently unsuitable for reporting flow data (journey to work and migration tables). These flow data types also require special data structures in order for the flows to be handled efficiently. In addition, developments in georeferencing technologies create the prospect of increasing levels of geographical precision.

Unfortunately, the explosion in spatial flow data is considerably greater than for any other type because of their matrix nature. If N is the number of zones for a given area of interest and K the number of variables, then flow data are dimensioned $N^2 \times K$ compared with $N \times K$ for other census data. As a result the computational time and modelling effort in analysing flow data are substantial even though very high levels of sparsity are often involved. On the other hand, more than for other types of data, spatial aggregation can serve an extremely useful purpose in reducing vast data volumes to a manageable and meaningful level. The problem is how to perform this type of flow data spatial aggregation so as to minimise the loss of useful information and to avoid undue distortion due to scale and aggregation effects. In addition to these problems, flow data are a neglected part of the GIS revolution, as most current GIS packages struggle to handle interaction data. The two-dimensional nature of the simplest of flow tables presents difficulties to GIS data structures. The inherent mathematical complexity of the related gravity models and network based decision support systems are more complex than in other areas of GIS application. The other key feature of flow information is that most of it cannot be remotely sensed. Flows are derived data that do not exist until other information is processed to make the interactions visible and even then they are extraordinarily hard to map. The data volumes are also potentially massive and the data structure is usually far more complex. Those who analyse and use flow data tend to treat them as a separate component, with GIS only being used as a mapping tool. As a result, most flow data sources have been ignored or remain grossly under-analysed in GIS; yet flow data are of increasing commercial importance via data warehousing systems.

This chapter addresses some of the GIS-relevant issues in the management and manipulation of spatial flow data. This is an important task because the information explosion occasioned by the computerisation of management systems has created many new flow data resources. Indeed in a human, as distinct from a physical context, most of the secondary data sources relating to people and organisations, their behaviours and activities in space are in essence flow data, of which some examples are listed:

- journeys to work, shopping, recreation, school, etc
- traffic and passenger flows by land, sea, air
- migration flows
- money flows
- commodity flows, raw materials, goods
- telephone traffic
- information flows
- gas, water, electricity flows

Most of this information exists, but nearly all of these datasets are, at best, only partially analysed. There are many reasons for this neglect of flow data richness:

1. lack of available GIS (and other) tools that can easily handle it
2. data storage and processing costs are N^2
3. complexity because of the nature of flow data and because they require an understanding of the relevant social, political, and economic processes,
4. flow data often exist in a highly disaggregate form necessitating considerable pre-processing to manipulate them into a usable form,
5. the resulting flow tables depend on the geography that is used,
6. a lack of models that can handle the dynamics of flows over time,
7. few major advances in the last 20 years in the analysis of flow data

Yet flow data are extremely important because they reveal details of how spatial processes operate and how people behave in a spatial setting. The future efficiency of many public and commercial organisations will almost certainly depend on their ability to make better use of their flow data resources.

This chapter considers a longstanding but currently largely neglected challenge of how to design output areas to provide good spatial representations of flow data sets. To do this, it revisits the spatial representation and spatial interaction literature from the 1970s and re-casts some of the methods into a modern GIS environment. Flow data present special challenges as interactions are only detected when a flow crosses a zoning system boundary. Although flows can also be defined within zones, it is the between-zone interactions that are hard to analyse and model in a GIS environment. Barras *et al.* (1971, p.140) wrote: “Zones are 'detectors' of trips. As the zone size decreases, the strength of interaction increases, where the 'strength of the interaction' is defined as the fraction of total trips which cross zone boundaries”. Conversely, the proportion of cross-boundary flows will tend to fall as zone size increases. As a result, the spatial patterns of flows are strongly dependent on the zoning system used to represent them. The quality of the resulting aggregate data in terms of the representation of underlying processes is therefore also heavily dependent on the geography of the zoning system that is used.

Once this geography decision was fixed and beyond the control of the end-users. This is no longer the case and new opportunities are emerging here, as elsewhere for users to be able to design their own geographies. There are also three extra complications with flow data. Firstly, most flows are directional and when spatially aggregated are non-symmetrical in that the flows from origin O_i to destination D_j will seldom, if ever, be identical to the flows from destination D_j to origin O_i. Secondly, the proportion of intra-zonal flows is highly dependent on the choice of zones. Intra-zonal flows are hard to model, indeed they can be regarded as very being poorly represented by the zoning system. Unfortunately, at a coarse spatial scale or if the zoning system is poorly designed, then the majority of flows can often be intra-zonal. Thirdly, the reliability of the zone-to-zone cost or distance measures widely used in flow modelling also depends on the size, shape, and configuration of the zoning system as this influences the degree to which it is well represented by a network or centroid points. As a result, zone design for the efficient handling of flow data is absolutely essential if good use is to be made of this data resource.

This chapter starts by a review of issues related to the nature of flow data, such as problems with spatial representation and their use in regionalization. The availability of flow data mainly through the UK Censuses of population is also reviewed. Despite the fact that only 10% of flow data are processed by the census, the wealth of information that can be obtained is of enormous potential. However, the difficulties in processing such data are also considerable, as experienced in early attempts of aggregating flows back in the 1970s. Consequently, a zone design system for interaction data (ZDeSi) is described and the first set of objective functions are explained. In the final part of this chapter journey to work flows are used to demonstrate the performance of ZDeSi. The performance of different objective functions is evaluated and the resulting zoning systems are compared. It is concluded that although it is hard to identify a universally applicable function, the results are fascinating in uncovering patterns obscured by the complexity of flow data.

9.2 UNDERSTANDING FLOW DATA

9.2.1 Spatial representation aspects

The two dimensional nature of flow data presents an immediate difficulty. Let T_{ij} represent the flow from origin zone O_i to a destination zone D_j. If the area of interest consists of N spatial units, then the resulting flow table contains N rows and N columns. If N is 1,000 zones, the resulting matrix table can be stored in the memory of a conventional PC. However, if N is 10,000, which is roughly the number of census wards in the UK and the finest resolution of 1991 census flow data, then a high performance workstation or supercomputer is necessary for storing and analysing the resulting flows (Turton and Openshaw, 1997). If N increases to 1.6 million, which is the number of postcodes in the UK, then a very high performance supercomputer is needed before any analysis can be performed. Finally, if N becomes 37 million, which is the number of households in the UK, then analysis becomes infeasible at present, although such datasets exist.

The latter problem would represent flows relating to individual households in the UK, which can now be given unique X,Y co-ordinates. To make matters worse, many flow data sources have a strong temporal dimension (e.g. hourly telephone traffic). Additionally, most flows can be disaggregated by the characteristics of the origin location O_i (e.g. age, gender, household, type, income, mode, etc) and by the characteristics of the destination location D_j (e.g. purpose, time of day, etc). So the N by N storage problem becomes rapidly worse as there is a N^2 pattern of flows for each of the disaggregations; for example, if there are 4 flow types, 10 age-sex groups and 5 income bands then $N^2\times200$ flows would need to be stored. If the flows are recorded daily then $N^2\times200\times365$ flows would need to be stored on a calendar year. A final complication is that as flow data are almost certainly reported for a given zoning system (e.g. electoral wards), the structure of the flows, the relationships, and the patterns strongly depend on this system or any subsequent aggregations.

In a very real sense flow data only exist when a zoning system is imposed on micro events; Figure 9.1 illustrates this process. The micro data represent discrete events that have certain characteristics; there is a start location, an end location, and a directional component. These micro events can be represented by straight lines connecting the origin location (O_i) with the destination location (D_j) or via a network path. Figure 9.1(a) demonstrates this concept for an imaginary area of interest, where 50 flows are represented by directed lines. In reality there may be millions of such observed micro events, for example when telephone or cash flows are considered. Very few computer models currently exist which can model these discrete events at this level of detail. The traditional solution is to aggregate the discrete events to a zoning system. In Figure 9.1(b) an arbitrary zoning system with three zones (A,B,C) has been imposed on these discrete events. The data can be aggregated to produce a flow table in Figure 9.1(d). Note that the diagonal elements are counts of discrete flows that crossed no zonal boundary and are therefore intra-zonal flows. The off diagonal elements are composed of discrete events that crossed one or more boundaries. The zoning system is providing an aggregate spatial representation of the underlying discrete flows. Clearly, some zoning systems with more or less zones in them may provide a better representation of the underlying data than other. Figure 9.1(c) shows an alternative zoning system consisting of three zones (X,Y,Z) that result in a different flow table. The flow values in Figure 9.1(e) are different due to aggregation effects. However, this is not the ordinary MAUP experienced with static data; here the zoning system is acting as a flow detector that makes the discrete events visible by recording the data to a coarser level of geographical detail. Without a zoning system there are no flow data, as all data would represent intra-zonal flows. Clearly some zoning systems will be better flow pattern detectors then others.

In the flow data context then, the zone design problem can have four different objectives:

1. as a data reduction and descriptive device to reduce discrete event or flow data volumes to a convenient level for spatial management and data reporting purposes
2. to optimise the spatial representation performance of a zoning system so that the partitioning minimises distortion and bias,
3. as an analysis and visualisation tool designed to make the flows visible in an understandable map form as a useful description of spatial organisation, and flow structure
4. to ease modelling problems or enhance model performance by tuning the spatial aggregation to generate more model friendly data

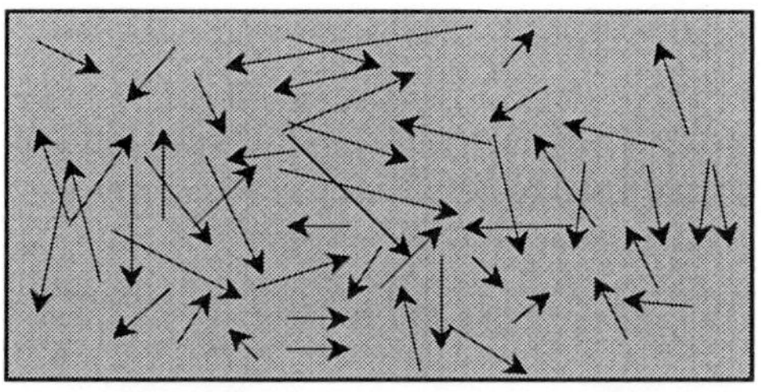

(a) Observed discrete flow

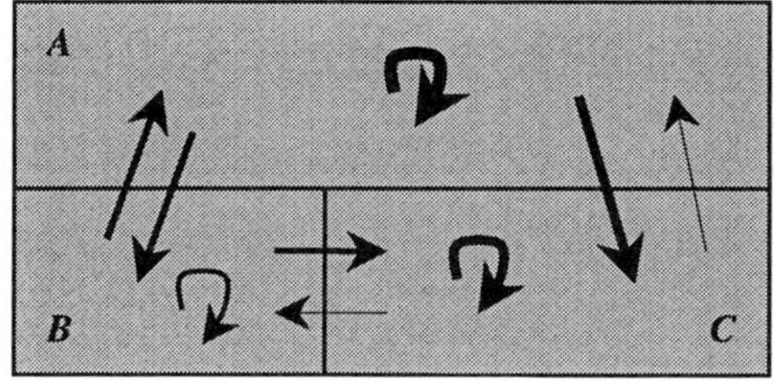

(b)

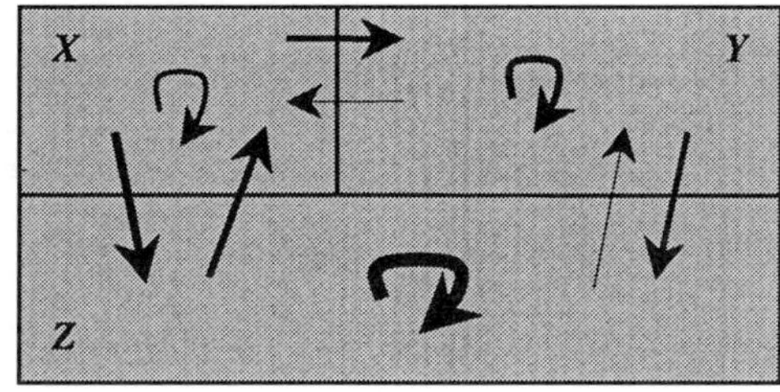

(c)

Flow table for zoning system ABC				
D_j / O_i	A	B	C	ΣD_j
A	14	4	7	*25*
B	5	4	3	*12*
C	1	1	11	*13*
ΣO_i	*20*	*9*	*21*	**50**

(d)

Flow table for zoning system XYZ				
D_j / O_I	X	Y	Z	ΣD_j
X	4	3	6	*13*
Y	2	5	5	*12*
Z	5	1	19	*25*
ΣO_i	*11*	*9*	*30*	*50*

(e)

Figure 9.1 The formation of flow data: (a) observed flows, (b)(c) different zoning systems (d)(e) Origin (O_i) - Destination(D_j) flow tables for the zoning systems ***ABC*** and ***XYZ***

9.2.2 Flow data regionalization and census flow data

Whilst GIS flow data functionality is underdeveloped, this is not true within other areas of geographical research. There is an extensive and long established literature describing the regionalization of flow data dating from the 1920s and 1930s. Various terms have been used to describe this research including: nodal regions, central place theory, labour market areas, catchment areas, trading areas, sales territories, travel to work areas, and functional regions. Most work has sought to identify structure in flow data designed to reveal useful knowledge about the underlying data (Coombes *et al.*, 1979). A wide range of techniques have been used to regionalise flow data including: graphic theoretic, markov chains, intramax, factor analysis, and various multi-stage grouping algorithms. Regionalization is another word for the classification of zones or regions on the basis of information about their spatial linkages provided by the flow table. All these methods are forms of zone design although often their primary purpose could be either analysis of data or the design of zones for gravity modelling purposes (Masser *et al.*, 1975). Masser and Brown (1978, Preface, p.v) explain the problem very clearly when they comment on "[...] the need for special aggregation and clustering procedures to be developed which enable data to be grouped in an efficient way for analytical purposes with a minimum loss of detail. In the case of interaction data, economy of representation is particularly important [...]".

The CURDS functional region definitions and the Official Government travel-to-work area (TTWA) definitions represent good examples of designing zoning systems for representing flow data, although typically other non-flow variables are also involved (Coombes *et al.*, 1978; 1979; 1985). The purpose was to create a geography that would be most useful for studying Britain using geographical entities that are comparable because they capture the essence of the daily functioning of the spatial economy. If one wishes to study unemployment levels then such regionalizations provide geographical objects that best represent the phenomena of which the data is a measurement. On the other hand, although administrative geographies such as Local Authorities (LAs) would be a useful geography for reporting variations in council tax, they would provide a very poor geography for reporting or comparing unemployment rates. LAs bear little or no resemblance to the geography of the spatial economy which, in general, does not follow administrative boundaries. TTWAs are specifically designed to represent local labour market areas within which the majority of residents work. As a result of this labour market self-containment, TTWAs provide a more useful geographical framework for computing workplace based statistical indicators. In essence, TTWAs are a good example of deliberate creation of special purpose geographies and hence of zone design for flow data albeit at a macro-geographic level of representation.

The 1966, 1971, 1981, and 1991 Census in Britain have all reported small area journey to work flows and, since 1971, migration flows. The journey to work data was based on a 10% sample to reduce the costs of re-coding destination addresses to the level of census geography. The migration flows were created by asking for addresses one year previous to the census, and as a result did not record intermediate moves. In both cases wards (in England and Wales) and postcode sectors (in Scotland) are the smallest level of output geography for flow data, although they bear little or no relationship to the processes that generate these flows, being merely a convenient administrative geography. The 1991 journey to work data form a 10,933 by 10,933 by 247 table of flows for the whole of GB. The special migration statistics (SMS) were fully coded in 1991 and provide a 10,933 by 10,933 by 12 flow table (Rees and Duke-Williams, 1995). Both data sets are provided at ward level so that users can re-aggregate them to their own geography. However, there are no zone design tools for flow data currently available that can assist the users in this process.

A number of software packages provided for downloading and managing the flow datasets (such as MATPAC91 and SMSTAB) do not provide mechanisms for designing appropriate regionalizations. Such software can be used for reformatting, extracting and aggregating the original ward level data only if the user is able to define what is needed. Despite the potential for research with flow data, the census related flow datasets are still difficult to access and manipulate (Flowerdew and Green, 1993). The difficulty in lies partly in their nature and partly in the absence of appropriate tools. An additional problem is the inadequacy of existing GIS packages to provide even the basic functions of packages like SMSTAB. The design of user defined flow geographies is far from realised under these conditions.

9.3 ZONE DESIGN WITH FLOW DATA

9.3.1 Early attempts

Previous research has mainly focused on the use of hierarchical or agglomerative clustering algorithms to progressively group flow data for *N* small zones to *N*-1, *N*-2 ,..., 2 regions so that some function is optimised. Masser and Brown (1975) describe this process using their *intramax* procedure, a hierarchical aggregation algorithm that seeks to maximise the proportion of within region interaction. The methods suggested by Masser *et al.* (1975) are designed to create regionalizations supposedly optimised for spatial interaction models. The main problem that Masser and Brown (1975) faced was how to best incorporate multiple criteria such as homogeneity and region size in a flow regionalization context. The task proved to be difficult to handle in the computing environment of the time.

On the other hand, Openshaw (1977) uses a function explicitly designed for gravity models. The problem then becomes one of identifying a zoning system which minimises those intra-zonal flows that gravity models have the greatest difficulty in handling. Openshaw (1977) described a series of experiments in applying his simple Automated Zoning Procedure (AZP) algorithm to flow data. The flow data used in Openshaw (1977) comprised only 73 zones, which were aggregated into 42 and 22 regions with the purpose of performing what was then termed *spatial calibration*. The zoning systems were re-engineered to explicitly optimise the goodness of fit of a simple spatial interaction model that was fitted to the re-aggregated data. The results showed that the flow data could be re-zoned to best fit a spatial interaction model with fixed parameters. In addition, it was possible to jointly estimate the zoning system and model parameters in order to optimise a combined objective function. The main difficulty with using the goodness of fit of a spatial interaction model as the zone design criterion, is that the inter and intra-zonal flows (and therefore costs) change during the aggregation process. The purpose of Openshaw's (1977) attempt, however, was to demonstrate modifiable areal unit effects rather than provide a general purpose tool for designing zoning systems for flow data.

9.3.2 A Zone Design System for interactions (ZDeSi)

This chapter revives Openshaw's (1977) original approach and upgrades the AZP algorithms to use the latest ZDeS methods (Openshaw and Rao, 1995; Openshaw and Alvanides, 1999). In addition, sparse flow matrix subroutines are used for handling the aggregation and disaggregation operations applied to flow data. These latter methods had previously been developed for the TTWA exercises and are extremely efficient, as they can handle very large data sets in the computer memory (Coombes *et al.*, 1985). The hardest aspect is the need to maintain the flow data and non flow data aggregations separate, as they involve different regionalization methods. The traditional AZP algorithm implemented by Openshaw (1977) and described in Opensaw and Rao (1995) can be summarised in the following steps:

1. Assume a set of N areas and a connectivity matrix Δ of dimension N×N
2. Generate a random zoning system of M regions, subject to the contiguity constraint
3. Compute the value of an objective function for the initial zoning system

4. Randomly select a region and make a list of its edge areas
5. Randomly select an edge area from this region's list
6. Check if the selected area can be moved to another region by preserving the contiguity
7. Recalculate the value of the objective function for the new zoning system
8. Accept the move if there is an improvement in the objective function
9. Update the list of edge areas and return to step 5 until all areas are processed
10. Repeat steps 4-9 until there is no improvement in the objective function

The above algorithm progresses by ensuring that the output zoning system consists of contiguous zones and that all areal units are allocated to at least one zone. The most time consuming part of the algorithm is *step 7*, where the objective function is recalculated. The computational cost depends on the complexity of the objective function, which may range from simple population equality to a sophisticated multiple regression model. In traditional zone design with non-flow data, the function does not need to be recalculated *per se*, as it can simply be updated instead. Unfortunately, such a shortcut cannot be applied in the case of zone design for interactions.

In order for zone design to be applied with flow data, every change in the zoning system needs the re-run of a network cost program, or some other way of measuring the effects of zoning system on inter and intra-zonal costs. This process is demonstrated in Figure 9.2(a), where a zoning system of four areas is imposed on the 50 flows discussed in Figure 9.1(a). The flows can be presented in tabular form with the diagonal elements representing the intra-flows. When small areas are aggregated into larger zones, as in Figure 9.2(b), the inter and intra-flows need to be recalculated. This happens because of the two-dimensional nature of flow data and the fact that intra-flows become intra-flows after aggegation. It is clear from Figure 9.2(b) that when area ***2*** is assigned to zone ***II***, it is not enough to sum the 5 intra-flows found on the diagonal of the initial flow table. Instead, a thorough search of the table is necessary to identify all the flows that involve area ***2*** either as an origin (O_i) or as a destination (D_j). In this example there is no interaction between non-adjacent areas, so there are no flows between areas ***2*** and ***3***. However, the flows between area ***2*** and all the zones in regions ***I*** and ***II*** need to be calculated every time area ***2*** is evaluated in the optimisation process. This process is not trivial as it involves the recalculation of the entire flow matrix, which consists of N^2 elements, unless a more appropriate algorithm is devised.

One way of overcoming this problem is by pre-calculating the sums of all origins (O_i) and destinations (D_j) for the original flow matrix in Figure 9.2(a) and storing these separately. The evaluation of the objective function then involves identifying the relevant sums of O_i and D_j for the edge area that is being considered for a swap. For example, consider the initial zoning system ***I-II*** in Figure 9.2(b), where area ***2*** is evaluated for a swap from region ***II*** to region ***I***. The sums of origins and destinations for area ***2*** are 12 and 9, including the 5 intra-flows. These values have been pre-calculated in the flow table of Figure 9.2(a) and are simply subtracted from and added to the relevant sums of the zoning system ***I-II***. The resulting zoning system ***I*-II**** in Figure 9.2(c), optimises the percentage of intra-flows using the summarised information, without any need for re-calculating the entire flow matrix. Note that the diagonal cells representing the intra-flows still need to be calculated.

A second practical issue concerns the choice of the zone design function. A number of objective functions have been suggested, although in practice they have only been embedded in multi-criteria hierarchical agglomerative procedures. Probably no previous work has viewed the problem as one of zone design in which aggregation effects are perceived as more important than scale. Instead all the historical applications of zone design to flow data were concerned with grouping spatial units in a progressive manner. This is likely to result in sub-optimal, sometimes poor results, because zones that have been joined together cannot be split at a later stage. A number of objective functions have been implemented in the first version of ZDeSi, drawn from the relevant literature in the 1970s. These functions do not necessarily result into similar regionalizations, as they are designed to provide very different zoning systems. However, the results are expected to be comparable and give some insight to the underlying patterns of the observed individual flows.

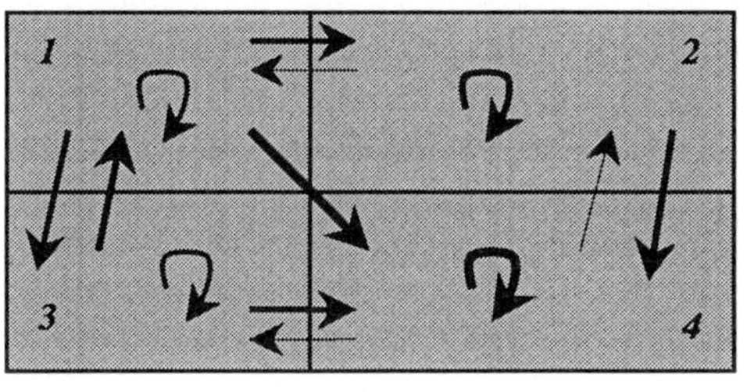

Flow table for initial zoning system 1234					
D_j / O_i	**1**	**2**	**3**	4	ΣD_j
1	4	3	4	2	*13*
2	2	5	0	5	***12***
3	5	0	4	3	*12*
4	0	1	1	11	***13***
ΣO_i	*11*	***9***	9	***21***	***50***

(a)

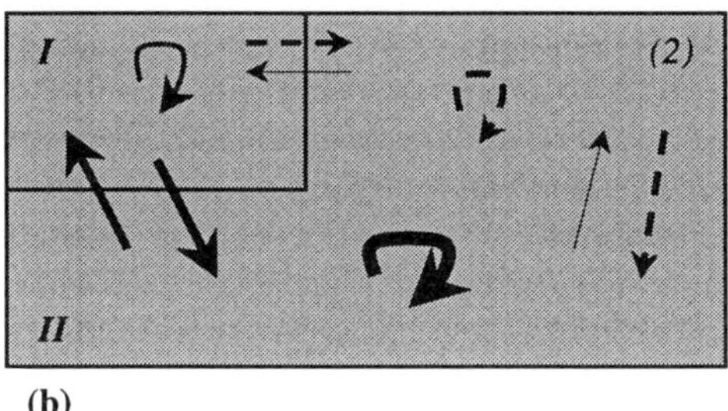

Flow table for zoning system I-II (area 2 assigned to zone *II*)				
D_j / O_I	**I**	**II**	ΣD_j	**2**
I	4	9	***13+***	*3*
II	7	30	***37-***	*1*
ΣO_i	***11+***	***39-***	**34**	*(9)*
2	*2*	*5*	*(12)*	*5*

(b)

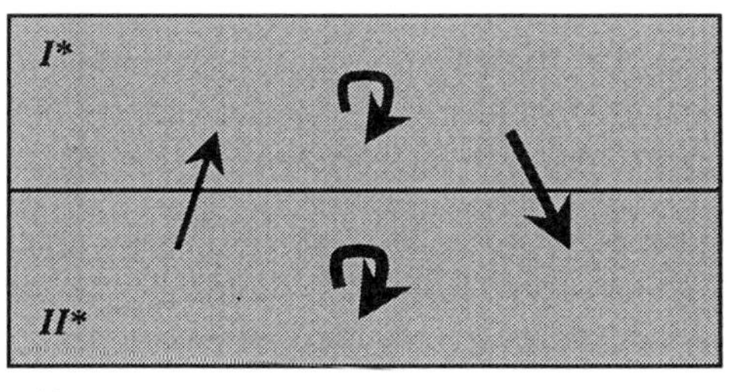

Flow table for zoning system I*-II* (area 2 assigned to zone *I*)			
D_j / O_i	**I***	**II***	ΣD_j
I*	14	11	***25***
II*	6	19	***25***
ΣO_i	***20***	***30***	**33**

(c)

Figure 9.2 The difficulty of evaluating flows in a zone design context: (a) initial zoning system and flow table, (b) sub-optimal aggregation *I-II*, (c) optimal zoning system I*-II* and final flow table

The first objective function considered here is the intramax suggested by Masser and Brown (1975). The intramax function can be expressed as:

$$\max F(Z) = \sum_i \sum_j (a_{ij} - a^*_{ij}) \quad \text{for } i \neq j \tag{9.1}$$

Where: $a_{ij} = T^*_{ij} / S$ and

$$a^*_{ij} = \sum_k a_{kj} \sum_k a_{ik} \tag{9.2}$$

Note also that: S is the sum of all the observed flows in the original flow matrix T_{ij}, while T^*_{ij} is the aggregated $M \times M$ flow matrix as defined by the M regions.

The function seeks to group together the N areas into a zoning system with M regions by maximizing the proportion of the total interactions within these regions. However, the process is slightly more complicated as the element a^*_{ij} in Equation (9.2) takes into account variations in the size of the row and column totals. Hirst (1977) suggests a modification necessary to handle variations in the row and column totals on the residual values. The modification simply involves dividing by a^*_{ij} to produce the version developed in ZDeSi:

$$\max F(Z) = \frac{\sum_i \sum_j (a_{ij} - a^*_{ij})}{a^*_{ij}} \quad \text{for } i \neq j \tag{9.3}$$

On the other hand, Broadbent (1970) noted that a basic objective should be to ensure that the amount of interaction that takes place within a region is small, compared with the amount of interaction between the regions. In other words a good zoning system would have most interaction between the regions rather than within them. Brown and Masser (1978, p.52) wrote: "These rule-of-thumb criteria assume that a satisfactory level of spatial representation is achieved when at least 85 or 90 percent of the total number of trips cross a zonal boundary". Based on this principle they formulated their Intramin function (Masser and Brown, 1975). The intramin function simply involves:

$$\min F(Z) = \frac{\sum_i T^*_{ii}}{S} \tag{9.4}$$

Where T^*_{ii} are the intra-region flows for the zoning system defined by the M regions and S are the total flows. This function, in common with intramax (Equation 9.4), was mainly used to create an initial zoning system for gravity modelling purposes where the requirement was to minimise intra-region flows. It is noted here that intra-flows are least well represented by spatial interaction models, as the models are more concerned with interaction between rather than within places.

However, from a data description point of view, the opposite function would be far more useful, to seek a regionalization that maximises the proportion of intra-regional flows. This function would, in theory, define regions of maximum self-containment, such as city regions, labour market areas, and retail catchment areas. Indeed it is this function that is most often used, particularly when the purpose is to reduce N areas to a far smaller number of M regions for a wide range of policy and applied purposes. This maximisation function has also been implemented in ZDeSi as:

$$\max F(Z) = \frac{\sum_i T^*_{ii}}{S} \tag{9.5}$$

Another set of ZDeSi functions could involve deliberately seeking a zoning system which maximizes the performance of a model of some kind that is fitted to the data. Openshaw (1978) provides some simple examples of this approach. It could well be developed further in a slightly different direction, namely to try and simplify the gravity modelling task by ensuring that the zoning system creates flow data that better match the theoretical requirements of the model being applied. This is a research direction worth exploring further given the increasing use of flow data and spatial interaction models by researchers and commercial bodies. It is clear that there is no single objective function that is likely to be universally applicable or even widely acceptable. Instead, a function needs to be chosen which is appropriate both to the nature of the observed data and the specifics of the problem investigated. The beauty of a system like ZDeSi is that it can provide a range of functions to chose from, while the resulting regionalizations can be stored in a GIS or produced as a series of maps that can then be animated. In accordance with other exploratory spatial data analysis methods, the user is given the freedom to try different options and uncover the hidden patterns. This is demonstrated in the next section where a case study of journey to work flows for England and Wales is considered.

9.4 CASE STUDY: OPTIMISING JOURNEY TO WORK FLOWS

The journey to work flow data from the 1991 census at the District level are used to demonstrate some of the effects that zone design can have on flow data. There are 402 Local Authority Districts (LADs) in England and Wales (excluding the Isles of Scilly) for which the 1991 census provides flow information. The population living in this area is about 50 million and the 10% sample data results in about 2 million flows, aggregated in a 402×402 flow matrix. One of the problems encountered is the large number of zero elements in the flow matrix. Out of the 161,604 elements, only 36,648 represent actual flows between LADs, while the rest contain zero values. However, even such a sparse matrix produces a wealth of information that is very difficult to process. In addition, it is almost impossible to comprehensively illustrate these flows on the conventional District boundary map. Figure 9.3 shows the conventional LAD boundaries together with major metropolitan Districts in England. These boundaries are the starting point for consequent aggregations, while the Districts are used as landmarks.

The experiments involve minimising and maximising the percentage of intra-regional flows, which should, in theory, produce descriptive results of the observed flows. They mainly serve as a means of summarising the vast amount of flows by creating zones that reflect the underlying patterns. The 402 LADs were grouped into five up to 65 zones with an increment of five. The three models described earlier were used as objective functions, namely minimising intra-regional flows (Equation 9.4), the original intramax model (Equation 9.3) and the newly developed model that maximises intra-regional flows (Equation 9.5). The performance of the objective functions is shown in Figure 9.4, where the percentage of intra-region flows is plotted against the number of output regions for each model. It is clear from the graphs that as the number of regions increases the intra-flows decrease, regardless of the function being optimised. If all LADs are aggregated to form a single super-region consisting of England and Wales (excluding external flows to and from Scotland, for example), then the intra-region value becomes 100%. On the other hand, if every LAD forms a zone on its own, the intra-region value reaches its minimum at 59% (about 1.2 million flows). The minimum figure of 59% is useful for comparative purposes when evaluating the results of the regionalizations in Figure 9.4.

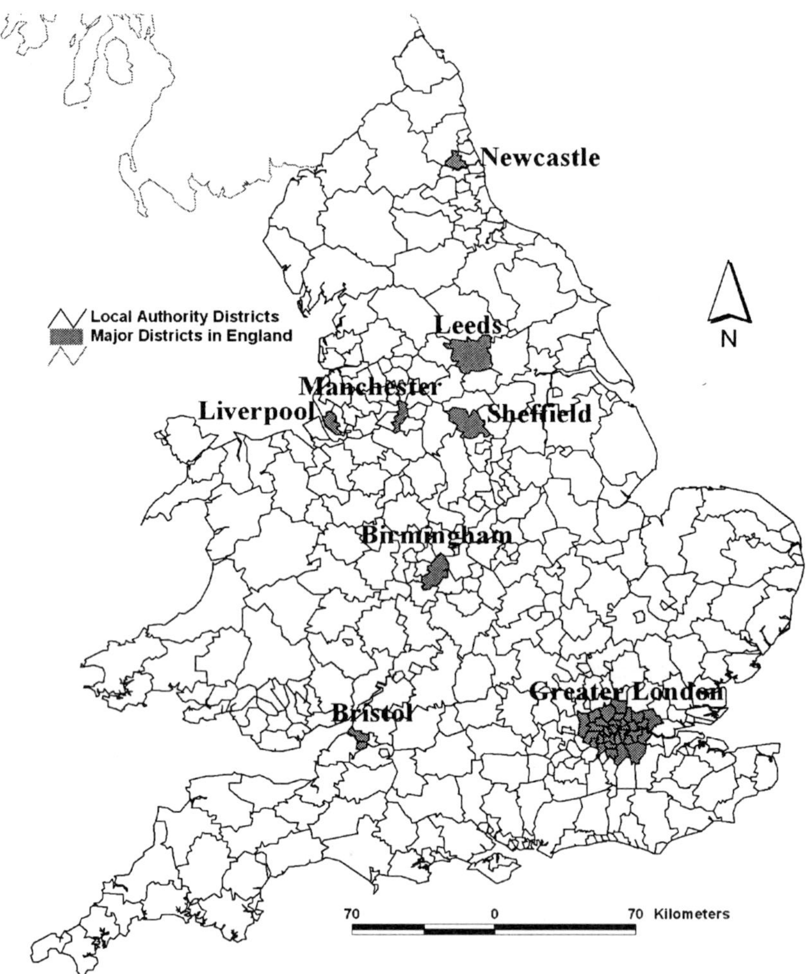

Figure 9.3 The 402 Local Authority Districts in England and Wales with the major Districts in England highlighted

The function that minimises the intra-region flows is represented by the bottom trend line in Figure 9.4 gradually decreasing from 77.6% to 67.0% as the number of regions decreases. The function seems to be performing reasonably well compared to the absolute minimum (59%). However, the resulting regionalization is very hard to interpret because of the irregular nature of the zones. A zoning system that minimises intra-regional flows should be ideal for a gravity model because the inter-regional flows would be maximum. Unfortunately, the zoning systems shown in Figure 9.5, would be very difficult to implement within a modelling framework. From a computational point of view, the function performs adequately as the major LADs fall in different regions, even when the intra-regional flows are as high as 76% for the 10 zones in Figure 9.5(a). For example, Leeds is known to attract commuters from Sheffield and in this case they are correctly grouped within different regions in order to produce minimum intra-flows. However, the results are not entirely consistent across different zoning systems. Note, for example, that although Leeds is initially grouped in the same region with Newcastle in Figure 9.5(a), it then forms the core of a separate region in Figure 9.5(b), and finally is grouped together with Sheffield in Figure 9.5(c). It can be argued that when major LADs are grouped together the interaction between them is very low. However, this is not the case between Leeds and Sheffield, which means that there might be inaccessible areas between these two cities, with very low interaction, forcing them two be grouped in the same region because of the contiguity constraint.

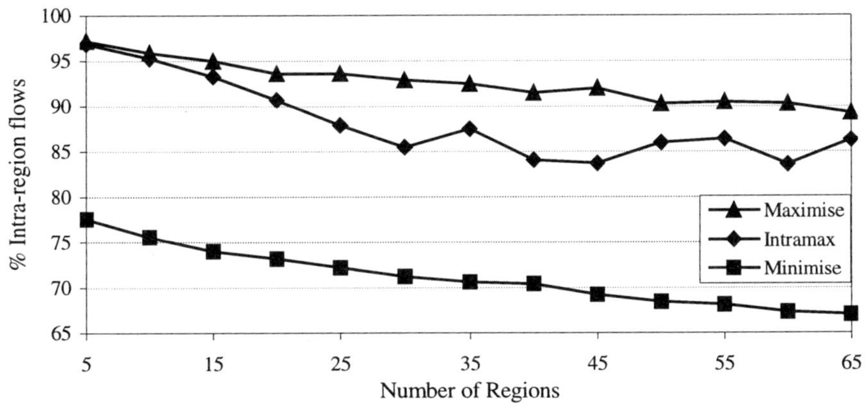

Figure 9.4 Regionalization of Districts using percentage intra-regional flow functions

The results of the intramax function suggested by Masser *et al.* (1975) and fully implemented by Masser and Brown (1975) into a hierarchical aggregation procedure are represented by the bottom trend line in Figure 9.4. The function is expected to maximise the proportion of intra-region flows, taking into account variations in the size of the row and column totals. It is evident from Figure 9.4 that the function does not perform well in an operational context, as it score very low for zoning systems 30, 40, 45 and 60. The 10-region system in Figure 9.6(a) which scores a high 95.3% groups together Leeds, Manchester and Liverpool contradicting the results obtained by the previous function in Figure 9.5(a). In addition, although zoning systems 30 and 50 produce comparable results of 85.5% and 86.0% respectively, the maps in Figure 9.6 hardly reflect these scores. Figure 9.6(b) shows a super-region that groups together most major cities in England from Leeds down to Outer London, while the LADs around Newcastle and Liverpool are fragmented into smaller regions. In Figure 9.6(c), Sheffield is outside this super-region, but Greater London is fragmented into LADs that form regions on their own. It becomes evident from these results that the intramax function gets trapped in local optima, producing low scores and fragmented regions.

In contrast, the newly developed function that simply maximises the intra-region flows is performing much better compared to the traditional and over-complicated intramax function. The result represented by the top trend line in Figure 9.4

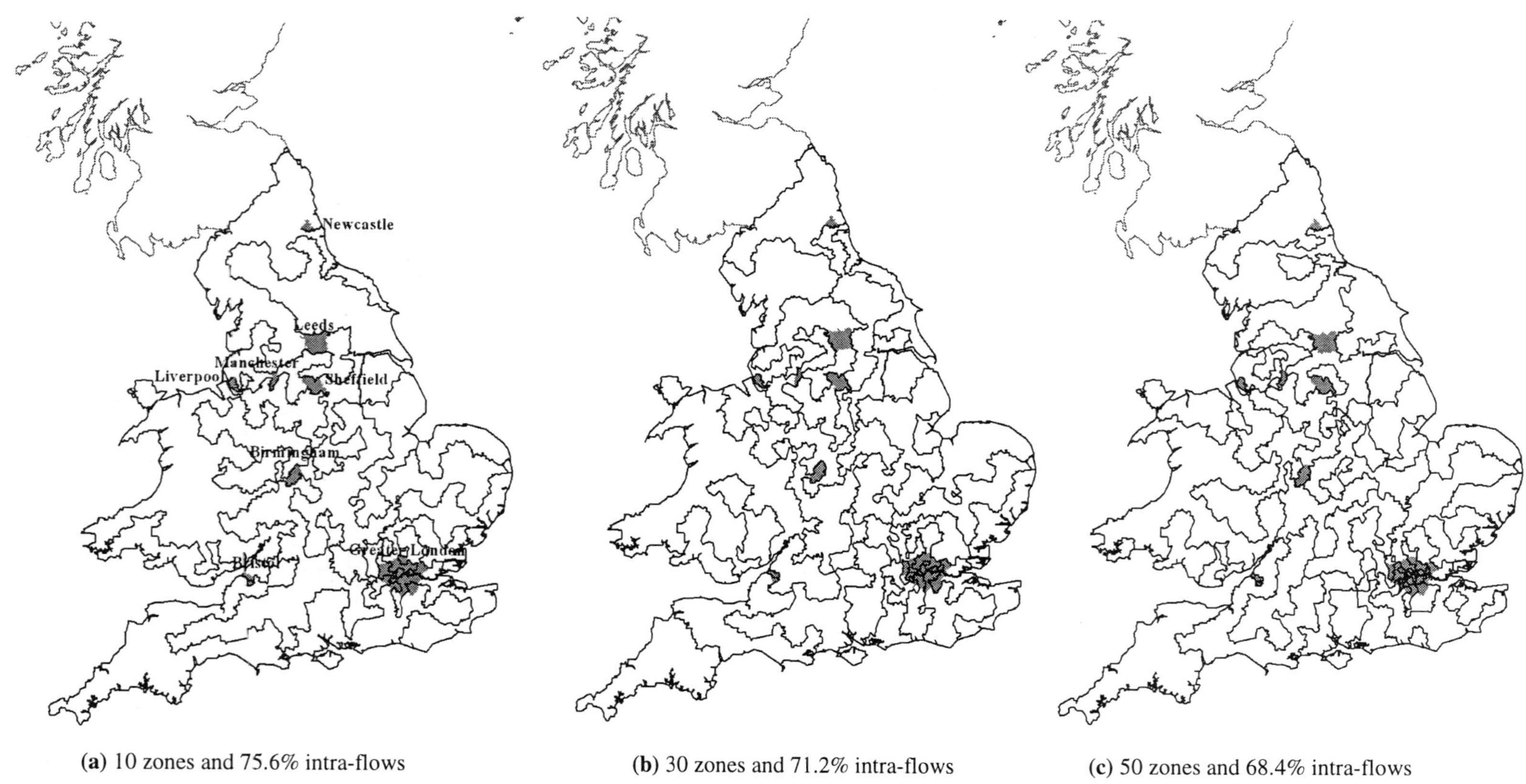

(**a**) 10 zones and 75.6% intra-flows

(**b**) 30 zones and 71.2% intra-flows

(**c**) 50 zones and 68.4% intra-flows

Figure 9.5 Resulting zoning systems from a function that minimises intra-region flows

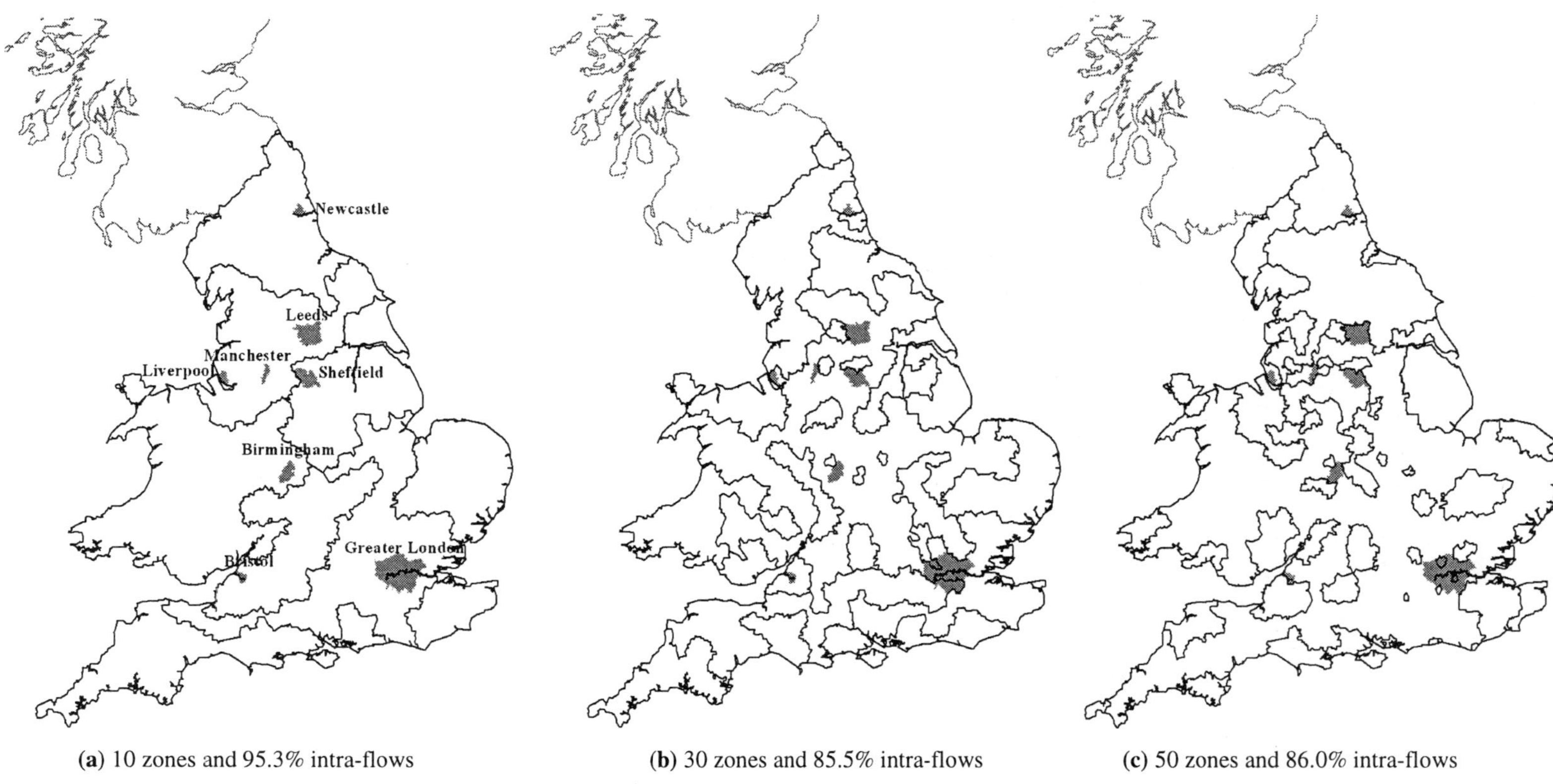

(**a**) 10 zones and 95.3% intra-flows

(**b**) 30 zones and 85.5% intra-flows

(**c**) 50 zones and 86.0% intra-flows

Figure 9.6 Resulting zoning systems from a the intramax function

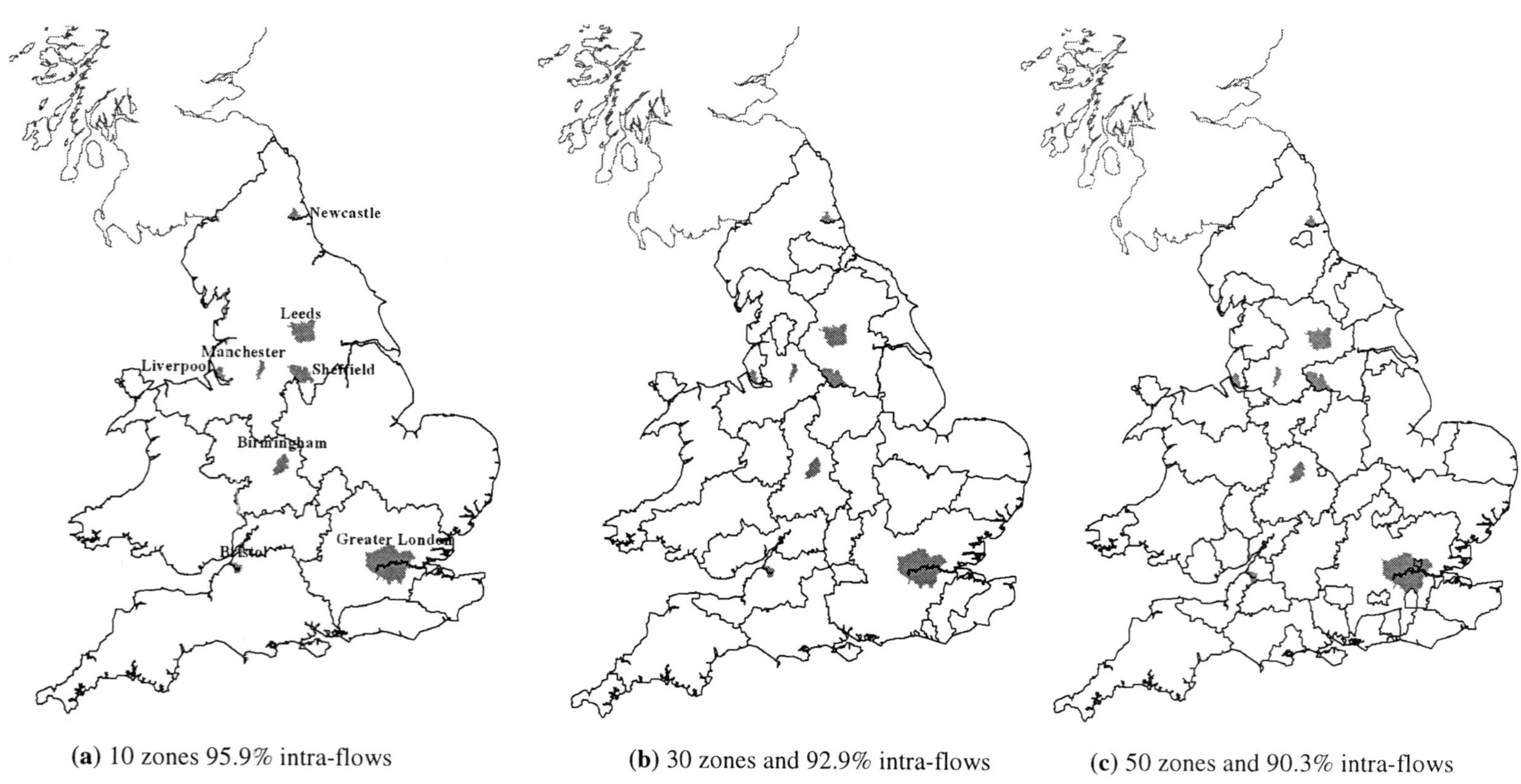

(**a**) 10 zones 95.9% intra-flows

(**b**) 30 zones and 92.9% intra-flows

(**c**) 50 zones and 90.3% intra-flows

Figure 9.7 Resulting zoning systems from a function that maximises intra-region flows

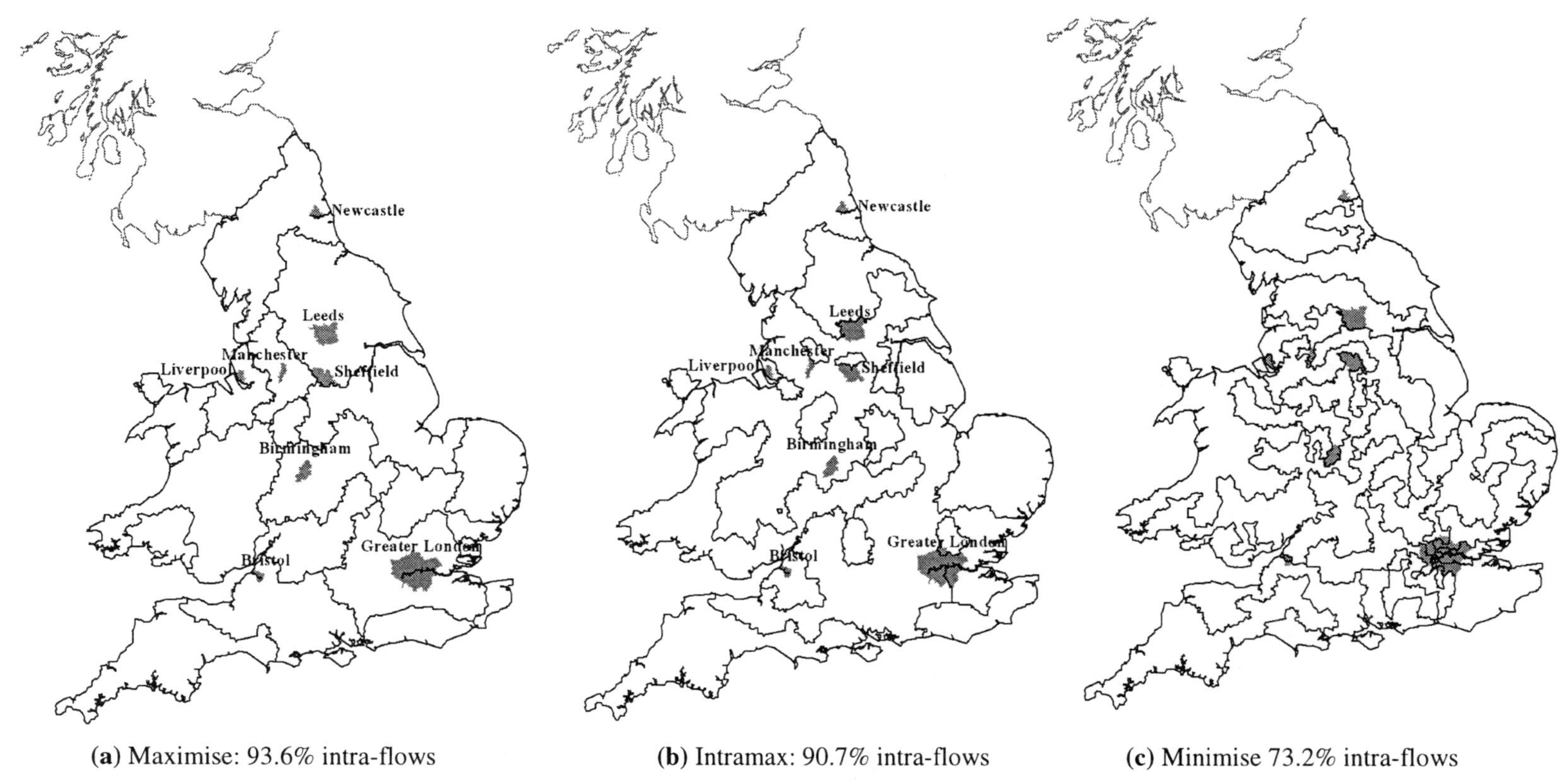

(a) Maximise: 93.6% intra-flows

(b) Intramax: 90.7% intra-flows

(c) Minimise 73.2% intra-flows

Figure 9.8 Comparison of the three functions for 20 zones

consistently outperforms the scores from the intramax function. At the same time, there is an expected gradual decrease from 97.2% to 89.3% as the number of regions decreases. The maps in Figure 9.7 demonstrate the ability of the function to identify areas of self-containment in terms of journey to work flows. Figure 9.7(a) clearly shows a north-south and east-west divide of England, while Wales seems to be forming a region almost on its own. The major northern cities fall in the same super-region, while Birmingham and Greater London form their own catchment areas. More interestingly, the maximisation function roughly produces the trends identified by the opposite function in Figure 9.5. As a result, Leeds and Sheffield remain in the same region, while Manchester, Newcastle and Bristol develop their own catchment areas in Figure 9.7(b). Eventually, Leeds comes together with Manchester, while Sheffield forms its own region in Figure 9.7(c), following the patterns identified in Figure 9.5(c). Finally, the first fragmented areas start appearing around Greater London in Figure 9.7(c) highlighting the problem identified with intramax at a much earlier stage in Figure 9.6(b). As the maximisation performance deteriorates from 90.3% for 60 zones to 89.3% for 65 zones in Figure 9.4, it is suspected that the fragmentation is a result of excessive number of regions requested.

A final comparison between the results obtained by these three objective functions is necessary in order to draw overall conclusions. The 20-region threshold is arbitrarily chosen for comparative purposes, as all three functions score relatively highly and the respective zoning systems are of low complexity. The resulting maps for the three functions are presented in Figure 9.8. The new function maximising intra-flows performs best with 93.6% of the total flows captured within the zones in Figure 9.8(a). The regional flow patterns are clearly visible, as the zones develop around major Districts. The only exception being Leeds and Sheffield which are placed in the same region, meaning that there is strong interaction at this scale. Intramax, produces comparable results with a high score of 90.7% intra-flows. The zones shown in Figure 9.8(b) are quite different to the results of the maximisation function. There exists a super-region capturing the flows between all the major cities, while smaller regions are formed with strong intra-flows. Although the intramax score is comparable to the new function, the resulting zoning system is of limited use. Unless a shape constraint is in place for limiting the formation of super-regions, intramax produces results that cannot be sustained. The same applies to the minimisation function, which produces the extremely complex zoning system shown in Figure 9.8(c). However, the principle behind the function is equally confusing and this is simply reflected in the resulting regions. Given that the function is attempting to minimise the intra-flows, the resulting zoning system can only be understood as the inverse of the map in Figure 9.8(a). Leeds, Sheffield and Birmingham, for example, are all placed in different regions, with additional zones disrupting the flows, while Greater London is fragmented by smaller regions radiating from its core. The boundaries of the zones in Figure 9.8(c) can be seen as imaginary dams preventing existing flows between major origins and destinations. They are useful in demonstrating that an optimal system for spatial interaction modelling, with minimum intra-flows can be constructed. However, as most gravity models make use of a zone centroid the complexity of the minimisation zones prevents them from being used in modelling. The only possible way of utilising these results in modelling is by applying a shape constraint, which would almost certainly worsen the performance of the function.

9.5 CONCLUSIONS

GIS removes many longstanding historical constraints on the form, management, and use of geographical information. It also broadens the scope of what can be considered to be geographical information. Unfortunately, flow data are a neglected GIS data type, as most systems find them hard to manipulate, lack facilities for mapping them, and contain few tools for analysis or modelling. The prospect of having user defined geographies for reporting flow data is even more important here than it is for most other spatial data types. Two basic needs can be identified. Firstly, zone design for basic spatial data management, where the purpose is the creation and reduction of flow data from observed individual flows. Secondly, re-engineering of already aggregated flow data, in order to serve a broad range of descriptive, analytical, statistical, modelling and visualisation purposes. This chapter has focused on the latter task by employing available flow data, such as journeys to work. It also emphasises the need to revive what is essentially a 1970s zone design technology, update the heuristics, and cast the tool-kit in a modern GIS context. It is interesting that more than 20 years later it is possible to re-invent these old tools and make them work faster and better than ever before in a very short time period; a matter of a few days to repeat research that originally took several years.

The chapter has also discussed the problems of extending existing zone design tools to handle flow data. A new version of the Zone Design System (Openshaw and Alvanides, 1999) called ZDeSi has been briefly described. The software can at present handle medium to large problems up to 10,000 zones, but the methodology can be extended to handle data of virtually any size. It was clearly shown that the methodology provides the means for producing useful descriptive representations of complex flow data. The resulting regionalizations prove that even at this level of geography, flow data are highly sensitive to the choice of the zoning system, perhaps even more so than is characteristic of other data types. In addition, the highly sparse nature of flow data means that computer run times increase as a sub-linear function of N and M rather than N^2. In other words, the problem is, in theory at least, only one magnitude harder compared to traditional zone design applications involving non-flow data as described by Alvanides and Openshaw (1999). However, better handling and optimisation algorithms are necessary if flow data zone design is to be widely adopted. The results presented here can be significantly improved (ranging from 1% to 5%) by using simulated annealing optimisation, although run times change from a few minutes to several hours. Still, the potential applications for this new tool in creating census output areas optimised for representing flow data are considerable. The range of potential applications also extends far beyond census data and could include various commercial data sets and telecommunications traffic. The next step is to improve ZDeSi by adding constraint handling mechanisms that already exist for the non-flow data. The creation of a web interface to the technology so that the research community can benefit in the light of the 2001 UK Census is also considered.

9.6 REFERENCES

Alvanides, S., and Openshaw, S., 1999, Zone design for planning and policy analysis. In *Geographical Information and Planning*, edited by Stillwell J., Geertman S. and Openshaw S. (Berlin: Springer-Verlag) pp. 299-315

Barras, R., Broadbent, T. A., Cordey-Haynes, M., Massey, D. B., Robinson, K. and Willis, J., 1971, An operational urban development model for Cheshire. *Environment and Planning A,* **3**, pp. 115-234

Broadbent, T. A., 1970, Notes on the design of operational models. *Environment and Planning A*, **2**, pp. 469-476

Brown, P. J. B. and Masser, I., 1978, An empirical investigation of the use of Broadbent's rule in spatial system design. In *Spatial Representation and Spatial Interaction*, edited by Masser I. and Brown P.W.J. (Leiden: Martinus Nijhoff) pp. 51-69.

Coombes, M.G., Dixon, J.S., Goddard, J.B., Openshaw, S. and Taylor, P.J., 1978, Towards a more rational consideration of census areal units: daily urban systems in Britain. *Environment and Planning A,* **10**, pp. 1179-1186

Coombes, M. G., Dixon, J. S., Goddard, J. B., Openshaw, S. and Taylor, P. J., 1979, Daily urban systems in Britain: from theory to practice. *Environment and Planning A,* **11**, pp. 565-574

Coombes, M. G., Green, A. E. and Openshaw, S., 1985, New areas for old: a comparison of the 1978 and 1984 Travel-to-Work areas. *Area,* **17** pp. 213-219

Flowerdew, R. and Green, A., 1993, Migration, transport and workplace statistics from the 1991 Census. In *The 1991 Census User's Guide*, edited by Dale A. and Marsh C. (London: HMSO) pp. 269-294

Hirst, M. A., 1977, Hierachical aggregation procedures for interaction data: a comment, *Environment and Planning A*, **9**, 99-103

Martin, D., 1998, Optimising census geography: the separation of collection and output geographies. *International Journal of Geographical Information Science,* **12**, pp. 673-685

Masser, I. and Brown, P. J. B., 1975, Hierachical aggregation procedures for interaction data. *Environment and Planning A,* **7**, 509-524

Masser, I. and Brown, P. J. B. (Editors), 1978, *Spatial Representation and Spatial Interaction* (Leiden: Martinus Nijhoff)

Masser, I., Batey, P. W. J. and Brown, P. J. B., 1975, The design of zoning systems for interaction models. In *Regional Science-New Concepts and Old Problems*, edited by Cripps, E.L., (London: Pion), pp. 168-187

Openshaw, S., 1977, Optimal zoning systems for spatial interaction models. *Environment and Planning A*, **9**, pp. 169-184.

Openshaw, S., 1996, Developing GIS-relevant zone-based spatial analysis methods. In *Spatial Analysis: Modelling in a GIS Environment*, edited by Longley, P. and Batty, M. (Cambridge: GeoInformation International), pp. 55-73

Openshaw, S. and Alvanides, S.,1999, Applying geocomputation to the analysis of spatial distributions. In *Geographical Information Systems: Principles, Techniques, Applications and Management*, edited by Longley P.A., Goodchild M.F., Maguire D.J. and Rhind D.W. (Chichester: Wiley) Vol.1, pp. 267-282

Openshaw, S. and Rao, L., 1995, Algorithms for re-engineering 1991 census geography, *Environment and Planning A*, **27**, pp. 425-446

Rees, P. and Duke-Williams, O., 1995, The story of the British special migration statistics. *Scottish Geographical Magazine,* **111,** pp. 13-26.

Turton, I. and Openshaw, S., 1997, Parallel spatial interaction models. *Geographical and Environmental Modelling,* **1**, pp. 179-197

10

Propagating updates between geographic databases with different scales

Thierry Badard and Cécile Lemarié

10.1 INTRODUCTION

Nowadays, geographic information systems are considered to be truly analysis and decision-making tools. For that reason, a growing number of organisations invest in such systems and add specific information necessary to the tasks for which they have the responsibility. However, the implementation of such systems is difficult and relatively expensive. That is why these institutions purchase reference data sets from geographic information producers. From these they are able to develop their own information systems. Thus, these organisations are considered, by the producer, as users of reference geographic data sets.

To carry out their assignments, these institutions (e.g. governmental agencies, departments of public works, environmental agencies, public utilities such as electricity, gas and water, private companies with a national remit, etc.) clearly need updates from the producer, in order to have the most faithful and realistic image of geographic reality. But these updates must allow for the preservation of consistency of all the knowledge added in the systems by users. Producers (e.g. national mapping agencies) may have similar preoccupations if they want to propagate the updates from their reference data sets to their derived products (e.g. cartographic products or other databases with different scales) in order to reduce the production costs. However, at the present time, such a propagation of the effects of spatio-temporal evolutions from a reference database to a user's or a derived geographic database, may induce significant risks of information loss or leave the database in an inconsistent state. A taxonomy of the problems hindering the integration of the updates for geographic databases has already been presented and detailed in Badard (1998).

This paper focuses on a generic updating tool for geographic databases under development at the COGIT laboratory of IGN (the French National Mapping Agency), which takes all these problems into account. This mechanism aims at automating the propagation of updates between databases with possible different levels of abstraction, representation and scale.

In section 10.2, some interesting experiences dealing with the updating of geographic databases are presented in order to point up the different elements necessary to the definition of a generic updating tool. The structure of such a generic mechanism is proposed and detailed in section 10.3. Section 10.4 focuses on the propagation step of this process and details how the effects of the updates on the different representations are controlled. In order to illustrate this updating process, examples stemming from two experiments are presented and discussed in section 10.5. The first one concerns the propagation of updates from BDTOPO® (a 1:25,000 topographic database with 1 m resolution produced by the IGN) to its derived cartographic database. The second one is about the updating of Route500®, a geographic database stemming from a generalisation of the road network of BDCARTO® (a 1:100,000 database with 10-meter resolution produced by the IGN).

10.2 SOME EXPERIENCES WITH THE UPDATING OF GEOGRAPHIC DATABASES

The purpose of this section is not to draw up a complete review of the different solutions which have been implemented to tackle the crucial issue of updating in multi-scale databases. Most of the approaches presented in the literature fulfil specific needs, hypotheses and constraints which often are not fully compliant with the requirements of a geographic data producer responsible for the maintenance of its different databases. Nevertheless, two interesting methods are detailed here because they allow the identification of the different elements necessary to the definition of a generic mechanism dedicated to the updating of multi-scale databases.

Thus, the mechanism for the propagation of updates in multi-scale databases defined in Kilpelaïnen (1995) provides an interesting example of what a generic method could be. This mechanism relies on the assumption that each generalisation process involved in the derivation of the different representations is formally defined. This is what she named an *incremental generalisation environment.* In such an environment, updating is performed as detailed below:

> 'In an incremental generalisation environment the generalisation process is done completely for the whole geodata base only once. After the following update transactions to the geodata base, the old generalised output is also updated in an incremental way, which means that the generalisation process is performed only for the modules influenced by the updates' (Kilpelaïnen, 1995).

This updating mechanism suits the specific case of multi-scale databases where the generalisation processes between all representations are explicitly defined and controlled. Nevertheless, the preservation of consistency in the different representations relies on a relevant identification of the *modules* which have been updated. This term is not precisely defined because it depends on the generalisation processes involved in the propagation. A *module* includes not only the modified objects in the base level but also all contextual information necessary to the production of an up-to-date representation compliant with the whole generalisation process previously performed. Indeed, if the set of selected entities is not complete, the propagation process may result in inconsistencies: effects of the update on the contextual information may not be completely performed. So, the identification of *modules* is a crucial point of the propagation process but it is far from being easy to implement.

Another interesting approach is defined in Uitermark *et al.* (1998). It deals with the construction of a prototype for propagating building updates between two topographic maps with different scales. This prototype is based on an update propagation strategy that can be decomposed in six steps:

1. The large-scale database is rolled back in time till the moment of actuality of the mid-scale data set.
2. Correspondences between the two data sets are determined via a map overlay process.
3. Correspondences are checked for inconsistencies due to the different specifications of the two data sets.
4. Aggregates are reconstructed and updates are filtered with respect to the mid-scale database specifications.
5. The relevant updated entities are generalised.
6. The generalised objects are integrated in the mid-scale database by means of a rubber-sheeting adjustment algorithm.

Although this process is limited to the updating of buildings, it emphasises the relevance of defining a strategy to perform a consistent and complete propagation of updates between databases with different scales. However, as in Kilpelaïnen (1995), spatial integrity is not fully controlled (e.g. intersections with another type of geographical entities may occur and are not detected by the process), which may leave the database in an inconsistent state.

Moreover, in both processes the updates are clearly identified, which is far from being a general case. Indeed, few stable and reliable systems of identifiers or time management are implemented in geographic databases. Therefore, for a mechanism for the updating of multi-scale databases to be considered as truly generic, it must include a process for the automatic retrieval of updates between two versions of the same geographic database.

Considering the different elements highlighted in this section, a generic mechanism dedicated to the propagation of updates in multi-scale databases can be defined. The structure of such a mechanism is detailed in the next section.

10.3 STRUCTURE OF THE UPDATING MECHANISM

As mentioned in the previous section, this generic mechanism allows the updating of geographic databases from the retrieval of changes in the reference data sets to the integration of the evolutions and to the propagation of their effects in derived or users' databases. The general structure of this mechanism is illustrated in Figure 10.1.
As illustrated in Figure 10.1, this mechanism may be divided in three main steps. The first step consists in the automatic retrieval of schema and data evolutions between two versions of the reference database (Badard, 1999). This step may be divided in two main sub-processes.

A geographic data matching process is first performed. Geographic data matching is akin to the notion of conflation defined in Saalfeld (1988) and Laurini and Thompson (1992). It consists in the computation of correspondence relationships between sets of geographical entities that represent the same phenomenon in two representations of the real world (Lupien and Moreland, 1987; Lemarié and Raynal, 1996; Lemarié and Bucaille, 1998). The two representations may have very different scales, levels of abstraction and/or representation. Such a mechanism has been developed at the COGIT laboratory of IGN (the French national mapping agency) for several years. It relies on the implementation of numerous algorithms involving all geographical information levels (i.e. semantic, geometric and topologic) and allows the processing of all types of geographical feature: points, lines and areas.

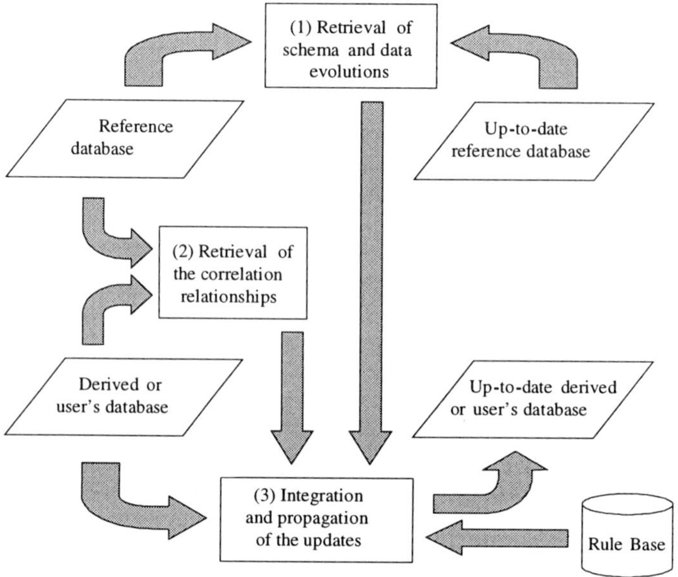

Figure 10.1 Structure of the updating mechanism for geographic databases

Finally, the correspondence relationships processed between the two versions of the reference database are analysed in order to retrieve updates. This analysis deals with the cardinalities of the relationships and in accordance with the typology of the spatio-temporal evolutions defined in Badard (1999), objects which have been created, deleted, merged, split, aggregated or semantically and/or geometrically modified are detected and extracted.

The second step deals with the preservation of the spatial integrity constraints. It aims at retrieving often implicit but relevant relationships between spatial entities in order to perform a complete and consistent updating of geographic databases. This minimal set of necessary relationships that we have named *correlation relationships* are established not only between the databases but also within the data sets themselves. They can be classified into four main categories as proposed in Badard (1998):

- Composition relationships: an entity is composed of several sub-entities (e.g. a road is comprised of road sections).
- Correspondence relationships: these are established between sets of entities that represent the same phenomenon in two representations of the real world.
- Geometric relationships: these are meant to translate spatial relationships between geographical entities and are especially concerned with the notions of geometry sharing, intersection, inclusion and connection as defined for instance in Egenhofer and Herring (1990) and Lemon and Masters (1997).
- Dependency relationships: these are relationships as established between objects designed in 'classical' databases. They are used to compute attribute values. For geographic databases, the geometry may be considered as an attribute.

This step involves mechanisms partially based on the same generic data matching toolbox used in the first step, which generates a set of geometric and topologic relationships. Such relations are analysed and classified by means of the semantics of the geographical entities. In order to automate this extraction process, specific metadata describing the nature of entities, the scales, and the thresholds of detection have been defined (Bonnani, 1998).

The third step deals with the updating of users' or derived databases. The correlation relationships previously extracted are used in order to propagate the effects of the updates detected in the reference data sets. To control the propagation process and allow a complete and consistent updating of geographic databases, a rule base is implemented. Such a rule base takes the available specifications and metadata (scales, resolution, etc.) into account and triggers the appropriate algorithms for the updating of the geographical entities stored in the databases. The rule base is fully upgradeable and then allows the management of specific conflicts. Unsolved conflicts are listed by the process and can be solved either interactively or by enhancing the rule base.

This final step is thus the core of the updating mechanism. Very different sources of information (types of update, correlation relationships, metadata, rules of propagation, etc.) and algorithms are involved in it. In order to better understand how these different sources of information interact and how algorithms are triggered, the propagation step of the updating mechanism is detailed in the next section.

10.4 THE UPDATING MECHANISM

As mentioned before, this step follows the retrieval of the updates and of the different relationships necessary to a complete and consistent updating of the data sets. Due to the differences between the databases in terms of content and structure, this step can be very difficult to perform. As the different modifications performed by users in their databases can be considered as a derivation process of the reference database, in this section, the term *derived database* will refer either to a derived or a user's database. Most of the difficulties that the propagation process has to take into account, result from:

- Differences between schemas of the databases: two classes in the reference database may have for instance been merged in a single class in the derived or user's database by means of an attribute.
- Differences between data stored in the databases: users may have already integrated some updates in their databases or have added specific information, which are not included in the reference data set.
- Differences between the levels of representation: databases stemming from a derivation process (e.g. generalisation) do not include the same data as in the reference data set.

Despite these difficulties, an updating mechanism which performs a complete and consistent updating of any kind of derived or users' databases has been defined and its implementation is in progress. The process can be decomposed in five main steps:

1. Update filtering.
2. Integration of the updates.
3. Propagation of the effects of updates.
4. Management of updating conflicts.
5. Checking of database consistency.

These different steps are presented and detailed in the following sections.

10.4.1 Update filtering

This step concerns the selection of the relevant updates which have to be integrated in the derived database. This step is very important because each modification performed in the derived database may induce conflicts with the data modified or added by the user. It is then necessary to narrow the integration of the updates in the derived database only to the objects which need to be updated. To select which modifications have to be integrated in the derived database, the following information is used:

- the available metadata of the derived database to determine if the modification is relevant in comparison with the resolution of the database. For instance it is not necessary to integrate a geometric modification of 10 centimetres in a derived database where the geometric resolution is estimated to 10 meters.
- the derivation process (when it is formally defined) in order to check if a modification in the reference database has an impact in the derived database. For instance it is not worth integrating a splitting due to the modification of an attribute value if this attribute is not in the derived database.
- the correspondence relationships which have been extracted between the reference database and the derived database to determine if the updated object in the reference database is represented in the derived data set.

In order to automate this filtering step, all selection criteria are expressed by means of rules and stored in a rule base. Such a rule base is fully upgradable and allows not only automation of the selection process but also to control the integration and the propagation of the updates. These different steps are detailed in the following sections.

10.4.2 Update integration

Once the updates have been filtered, each relevant update has to be integrated in the derived database. Each update is processed separately. First, the objects which have to be updated in the derived database are detected by means of the correspondence relationships extracted during the correlation relationships retrieval step. These objects from the derived data set are in correspondence with objects from the reference data set involved in the update. Update integration in the derived database depends on:

- The type of evolution (deletion, creation, splitting, merging, aggregation, semantic i.e. attribute and/or class—modification, or geometric modification).
- The cardinality of the correspondence relationships between the reference and the derived data sets (i.e. 1-to-1, 1-to-n, n-to-1 or n-to-m).
- The derivation process between the reference and the derived data sets.

These three types of information determine which algorithms have to be triggered in order to integrate the updates in the derived database. The first and the second are independent of the derived and the reference data sets whereas the derivation process is completely dependent on this pair of databases. However, two situations may occur:

1. The derivation process may be formally defined: in this case, it may be entirely expressed by means of integration rules and a fully automated integration of the updates can be performed.
2. The derivation process may be partially determined because manual interactions are involved: it is then difficult to express the derivation process by means of integration rules in order to automate the whole step of update integration. As a result, a manual interaction may be required in the integration step in order to process the unexpressed part of the derivation process.

Therefore, the rule base previously defined is enhanced by including some generic rules independent of the reference and the derived databases and some specific rules which express the derivation process between both data sets. For the expression of the integration rules, two strategies are possible. The derived objects are updated by applying the same type of update as was detected in the reference database. Otherwise, the derived objects are updated by performing anew the derivation process on the up-to-date objects which are stored in the reference database.

The first solution is implemented when the derivation process is not fully determined (i.e. manual interaction is involved) but partially expressed by means of integration rules. The second solution is preferred when the derivation process is formally defined by the integration rules. The latter solution allows the preservation of the consistency of the derived database. Nevertheless, these two strategies are often combined in order to apply the most appropriate method to each type of evolution.

Most of the time, the expression of the integration rules allows the automation of an important part of the integration process. Specific updates for which it has not been possible to define integration rules, are processed manually as described in section 10.4.4.

10.4.3 Update propagation

Once the updates have been integrated in the derived database, the objects which are correlated with the updated entities are processed. This step which aims at propagating effects of the updates relies on two processes:

1. The retrieval of objects which are correlated with the up-to-date objects within the derived database. This retrieval is performed by using the correlation relationships which have been computed in the second step of the whole process (see section 10.3).
2. The update propagation. Two strategies can also be defined depending on the knowledge of the derivation process. If the derivation process is formally defined, then it is possible to determine if the effects of the update have to be propagated to the correlated objects. Rules of propagation are then expressed to define how the propagation is performed. If the derivation process is partially determined, constraints which have to be checked between entities in order to maintain spatial integrity can be expressed. These constraints do not necessarily imply that the propagation rules are fully defined. Some correlated objects will not be automatically updated but the process allows for their detection. These unsolved updates are processed interactively by the user as described in section 10.4.4.

10.4.4 The management of updating conflicts

The previous steps have demonstrated that an important part of the process can be automated. Nevertheless, the process has often to deal with specific integration conflicts. Due to the specificity of some updates or the lack of relevant integration and/or propagation rules, the propagation and/or integration processes of updates may result in inconsistencies or in impossibilities. These conflicts are detected by the process during the integration and/or the propagation steps because spatial integrity constraints are violated or because some updates have not been processed. The user is then compelled to either solve these conflicts interactively or enhance the rule base and trigger anew the integration and propagation steps (as described in sections 10.4.2 and 10.4.3).

10.4.5 Checking database consistency

Once the updates have been propagated, several tests are performed to check the consistency of the whole derived database. These different tests consist in checking the topology of the database and compliance with the metadata available on the derived database which have been updated. Possible conflicts detected in this step are listed by the process and can be solved either interactively or by enhancing the integration and the propagation rule base by triggering anew the step described in section 10.4.4. The update which is the source of this conflict is thus rolled back.

In order to illustrate how the different steps of the updating mechanism interact, an overview of the process is presented in the following figure:

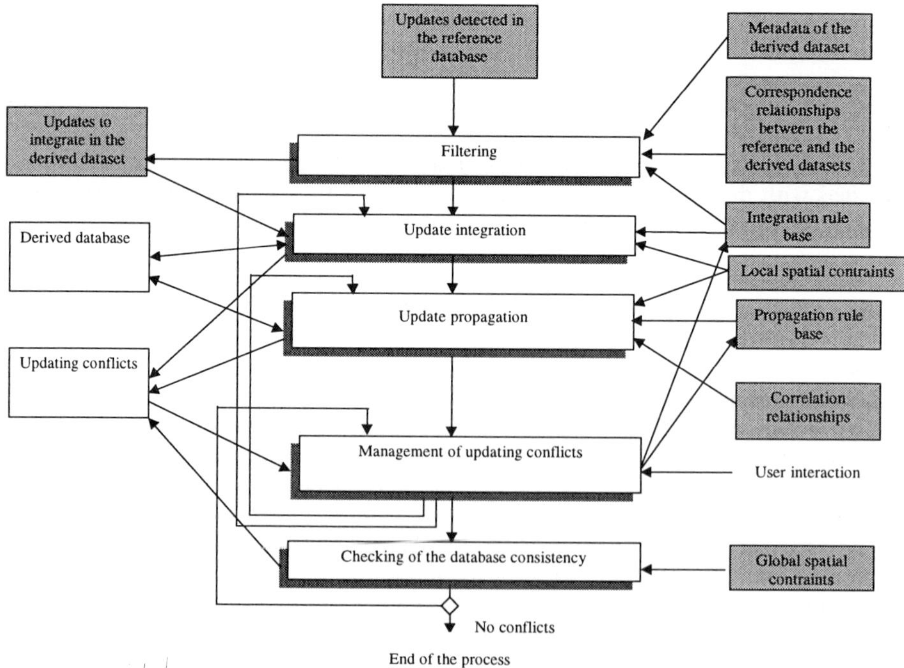

Figure 10.2 Overview of the process of update propagation

In order to prove the feasibility of such an updating process, examples of strategies and rules stemming from two experiments are presented in the next section.

10.5 EXPERIMENTS

The first experiment concerns the propagation of updates from BDTOPO® (a 1:25,000 topographic database with 1 m resolution produced by the IGN) to its derived cartographic database (TOP25®). The second one is about the updating of Route500®, a geographic database stemming from a generalisation of the road network of BDCARTO® (a 1:100,000 database with 10 m resolution also produced by the IGN).

These different tests which aim at validating the updating mechanism previously presented in this paper are in progress. The updates and the different correlation relationships have already been retrieved. Some steps of the propagation mechanism have been also defined and implemented. Some of these steps are thus reviewed in the following sections in order to illustrate the implementation of the rule base.

10.5.1 Update filtering

As detailed in section 10.4.1, this step allows the selection of the relevant updates which have to be integrated in the derived database. This selection relies on various types of information: the metadata describing the derived database, the derivation processes when they are formally defined and the correspondence relationships established between the reference and the derived data sets. In order to illustrate how the selection criteria are expressed, examples stemming from the experiment with BDCARTO® are presented below:

- Selection due to the metadata: As the geometric resolution of Route500® is estimated to 100 meters, all displacements under 25 m are not selected in BDCARTO®.
- Selection based on the derivation process: All elements of the secondary road network which have been updated in BDCARTO® are not selected because Route500® only includes a representation of the primary road network.

10.5.2 Update integration

In this section, we assume that the correlation relationships are established from the reference database to the derived one. For each type of evolution, algorithms involved in the integration of updates for the both experiments are defined. These algorithms are not completely detailed but principles of the method are provided. The detection of conflicts which may occur in this integration step is also presented below:

- Destruction: several cases have to be taken into account. Objects in the derived database involved in n-to-1 or n-to-m correspondence relationships are deleted only if the n objects of the reference database are also deleted. In all other cases (i.e. 1-to-1 and 1-to-n correspondence relationships), objects of the derived database are deleted.
- Creation: new objects of the reference database are created in the derived database by applying the derivation rules when they are defined or by requiring an interaction with the user.

- Semantic modification: objects of the derived database involved in 1-to-1 or 1-to-n correspondence relationships are automatically updated. Otherwise, if the reference objects are not semantically updated in a similar way, the user is prompted to perform the updating interactively.
- Splitting: this type of update is automatically integrated if objects are referenced in 1-to-1 or 1-to-n correspondence relationships or if all objects of the reference database involved in another type of correspondence have undergone this modification for the same reason. Causes of the evolution have then to be retrieved in order to determine if this splitting is due to an attribute modification or to an intersection with a new object. The splitting operation is then performed only if the attribute is also updated in the derived database or if the new object is also inserted. Nevertheless, this emphasises the role of derivation process in the definition of a strategy for the integration of updates. For instance, the derivation process between BDTOPO® and TOP25® is not formally determined, so the splitting point is located by using ratio on curvilinear abscissa for the reference object and then inserted in the derived objects. Finally, spatial integrity constraints have to be checked. Conversely, the derivation process between BDCARTO® and Route500® is completely defined. Objects involved in a relevant splitting operation are then derived by applying anew the derivation process. The consistency of the whole derived database is thus preserved.
- Merging: as before, this type of update is automatically integrated if objects are referenced in 1-to-1 or 1-to-n correspondence relationships or if all objects of the reference database involved in another type of correspondence have undergone this modification for the same reason. The integration is performed by merging the geometry of the derived objects when they are represented by contiguous lines or surfaces. The new attributes of the resulting object are then determined. Nevertheless, the integration is useless when the reference objects involved in the update are represented in the derived database by only one object.
- Aggregation: this type of update is processed with respect to the same requirements defined for the merging operations. However, the integration of these evolutions may be difficult to control so that the deletion of the old derived objects is first performed and the derivation rules are triggered on the reference objects involved in the aggregation to deduce the new version of the derived objects.
- Geometric modification: this type of update may be processed if objects involved in the different correspondence relationships fulfil the requirements defined before. Otherwise, the user is prompted to perform the updating interactively. With this type of update, the knowledge of the derivation process allows the determination of the most suitable algorithm. For instance, in BDCARTO®, derived objects involved in the geometric modification are deleted and the update is integrated by applying the derivation rules to the reference object. This derivation process consists in the succession of a filtering algorithm as presented in Douglas and Peucker (1973) and of a smoothing algorithm. In BDTOPO®, the update is integrated by using a 'morphing' algorithm, the distance between the derived and reference database is evaluated and the same difference between the updated reference object and the derived objects is performed.

For each type of evolution, the integration of update is possible when the updated objects are involved in 1-to-1 or 1-to-n correspondence relationships. With other types of correspondence (n-to-1 and n-to-m), the update can be automatically integrated only if the n reference objects are involved in the same modification. Otherwise the updating is performed by requiring an interaction with user. 1-to-0 correspondence relationships have to be checked by the user to ensure that the reference object is really not represented in the derived database.

10.5.3 Update propagation

In the experiment with BDTOPO®, some interesting correlation relationships have been detected. Indeed, as TOP25 is dedicated to cartography, some cartographic symbols which are derived from geographic objects have been added in it. For instance, bridges are symbolised on maps by short lines bordering the roads, and place names of rivers which are printed on maps follow the shape of rivers they name.

In this experiment, correlated objects have to be updated only when reference objects are involved in destruction or geometric modification. For instance, in the case of geometric modification, correlated objects have to be geometrically updated. However, propagating this type of update in TOP25® may be problematic because the correlated objects are produced by a derivation process and then checked interactively. So, the whole derivation process cannot be formally defined. Nevertheless, it is possible to define propagation strategies for specific geographical entities. For instance, the propagation of the updates performed on the bridges of the TOP25® consists in applying the same geometric displacement to bridges' symbols. In addition, modifications performed on named objects may induce the update of their place names. As a place name is depending on several geographical objects, it is then necessary to first define if the modifications induce effects on it. If so, the place name is computed anew by applying the different propagation rules.

10.6 CONCLUSION AND OUTLOOKS

A generic updating tool for geographic databases which aims at automating the propagation of updates between databases with possible different levels of abstraction, representation and scale has been presented. This automation is provided by means of the implementation of a rule base which allows the control of the update integration and propagation. This rule base is fully upgradable and allows the consistent and complete updating of any kind of derived geographic database.

The implementation of this mechanism is in progress: the steps dealing with the management of conflicts have not been fully completed yet. Nevertheless, first results of the experiments presented in this paper have demonstrated its feasibility and about 80% of the updates have been consistently propagated. Once the implementation of the updating mechanism will be completed, the rate of automatic updating is expected to reach 90%.

10.7 REFERENCES

Badard, T., 1998, Towards a generic updating tool for geographic databases. In *Proceedings of GIS/LIS'98, Annual Conference and Exposition*, (Bethesda, MD: ASPRS), pp. 352-363.

Badard, T., 1999, On the automatic retrieval of updates in geographic databases based on geographic data matching tools. In *Proceedings of the 19th International Cartographic Conference*, (Ottawa: International Cartographic Association), pp. 47-56.

Bonnani, L., 1998, *Etablissement de liens de correlation dans un but de mise à jour des bases de données géographiques.* Technical report IGN/DT/SR/980017, M.Sc. thesis in computer science (Paris: University of Paris-Dauphine)

Douglas, D. and Peucker, D., 1973, Algorithms for the reduction of the number of points required to represent a digitised line or its caricature. *The Canadian Cartographer*, **10**, pp. 112-123.

Egenhofer, M. J. and Herring, J. R., 1990, A mathematical framework for the definition of topological relationships. In *Proceedings of the 4th International Symposium on Spatial Data Handling*, (Columbus, OH: International Geographical Union), pp. 803-813.

Kilpelaïnen, T., 1995, Updating Multiple Representation Geodata Bases By Incremental Generalization. *Geo-Informations-Systeme*, **8**, pp. 13-18.

Laurini, R. and Thompson, D., 1992, *Fundamentals of spatial information systems*, (London: Academic Press).

Lemarié, C. and Raynal, L., 1996, Geographic data matching: first investigations for a generic tool. In *Proceedings of GIS/LIS'96, Annual Conference and Exposition*, (Bethesda, MD: ASPRS), pp. 405-420.

Lemarié, C. and Bucaille, O., 1998, Spécifications d'un module générique d'appariement de données géographiques. In *Actes du 11ème congrès Reconnaissance des Formes et Intelligence Artificielle* (Aubiere: Laboratoires des Sciences et Materiaux pur l'Electronique, et d'Automatique), pp. 397-406.

Lemon, D. and Masters, E., 1997, The nature and management of positional relationships within spatial databases. In *Proceedings of the 2nd Annual Conference of Geocomputation'97*, (Dunedin: University of Dunedin), pp. 15-23.

Lupien, A.E. and Moreland, W.H., 1987, A general approach to map conflation. In *Proceedings of the ACSM/ASPRS Annual Convention & Exposition, Technical Papers, AutoCarto 8*, (Bethesda, MD: ASPRS), pp. 630-639.

Saalfeld, A., 1988, Conflation – Automated map compilation. *International Journal of Geographical Information Systems*, **2**, pp. 217-228.

Uitermark, H., van Oosterom, P., Mars, N. and Molenaar, M., 1998, Propagating updates: Finding Corresponding objects in a multi-source environment. In *Proceedings of the 8th International Symposium on Spatial Data Handling (SDH'98)*, Vancouver, BC, Canada, edited by Poiker, T.K., and Chrisman, N., (Burnaby: International Geographical Union), pp. 580-591.

11

A simple and efficient algorithm for high-quality line labelling

Alexander Wolff, Lars Knipping, Marc van Kreveld, Tycho Strijk and Pankaj K. Agarwal

11.1 INTRODUCTION

The interest in algorithms that automatically place labels on maps, graphs or diagrams has increased with the advance in type-setting technology and the amount of information to be visualised. However, though manually labelling a map is estimated to take fifty percent of total map production time (Morrison, 1980), most geographic information systems (GIS) offer only very basic label-placement features. In practice, a GIS user is still forced to invest several hours in order to eliminate manually all label-label and label-feature intersections on a map.

In this chapter, we suggest an algorithm that labels one of the three classes of map objects, namely polygonal chains, such as rivers or streets. Our method is simple and efficient. At the same time, it produces results of high aesthetic quality. It is the first that fulfils both of the following two requirements: it allows curved labels and runs in $O(n^2)$ time, where n is the number of points in the polyline.

In order to formalise what good line labelling means, we studied Imhof's rules for positioning names on maps (Imhof, 1975). His well-established catalogue of label placement rules also provides a set of guidelines that refers to labelling linear objects. (For a general evaluation of quality for label-placement methods, see Van Dijk *et al.*, 1999.) Imhof's rules can be put into two categories, namely hard and soft constraints. Hard constraints represent minimum requirements for decent labelling:

(H1) a label must be placed at least at some distance ε from the polyline.
(H2) the curvature of the curve along which the label is placed is bounded from above by the curvature of a circle with radius r.
(H3) the label must neither intersect itself nor the polyline.

Soft constraints on the other hand help to express preferences between acceptable label positions. They formalise aesthetic criteria and help to improve the visual association between line and label. A label should:

(S1) be close to the polyline,
(S2) have few inflection points,
(S3) be placed as straight as possible, and
(S4) be placed as horizontally as possible.

We propose an algorithm that produces a candidate strip along the input polyline. This strip has the same height as the given label, consists of rectangular and annular segments, and fulfils the hard constraints. In order to optimise soft constraints, we use one or a combination of several evaluation functions.

The candidate strip can be regarded as a simplification of the input polyline. The algorithm for computing the strip is similar to the Douglas-Peucker line-simplification algorithm (Douglas and Peucker, 1973) in that it refines the initial solution recursively. However, in contrast to a simplified line, the strip is never allowed to intersect the given polyline. The strip-generating algorithm has a runtime of $O(n^2)$, where n is the number of points on the polyline. The algorithm requires linear storage.

Given a strip and the length of a label, we propose three evaluation functions for selecting good label candidates within the strip. These functions optimise the first three soft constraints. Their implementation is described in detail in Knipping (1998). We can compute in linear time a placement of the label within the strip so that the curvature or the number of inflections of the label is minimised. Since it is desirable to keep the label as close to the polyline as possible (while keeping a minimum distance) we also investigated the directed label-polyline Hausdorff distance. This distance is given by the distance of two points: (a) the point p on the label that is furthest away from the polyline and (b) the point p' on the polyline that is closest to p. Under certain conditions we can find a label position that minimises this distance in $O(n \log n)$ time (Knipping, 1998). Here we give a simple algorithm that finds a near-optimal label placement according to this criterion in $O(nk + k \log k)$ time, where k is the ratio of the length of the strip and the maximum allowed discrepancy to the exact minimum Hausdorff distance.

If a whole map is to be labelled, we can also generate a set of near-optimal label candidates for each polyline, and use them as input to general map-labelling algorithms as described by Edmondson *et al.* (1997), Kakoulis and Tollis (1998), and Wagner et al. (1999). Some of these algorithms accept a priority for each candidate; in our case we could use the result of the evaluation function.

In his list of guidelines for good line labelling, Imhof also recommends the labelling of a polyline at regular intervals, especially between junctions with other polylines of the same width and colour. River names e.g. tend to change below the mouths of large tributaries. This problem can be handled by extending our algorithms as follows. We compute our strip and generate a set of the e.g. ten best label candidates for each river segment that is limited by tributaries of equal type. Then we can view each river segment as a separate feature, and again use a general map-labelling algorithm to label as many segments as possible. Prioritising each label candidate with its distance to the closer end of the river segment would give candidates in the middle of a segment a higher priority and thus tend to increase label-label distances along the polyline.

This chapter is structured as follows. In the next section we briefly review previous work on line labelling. In Section 11.3 we explain how to compute a buffer around the input polyline that protects the strip from getting too close to the polyline and from sharp bends at convex vertices. In Section 11.4 we give the algorithm that computes the strip and in Section 11.5 we show how this strip can be used to find good label candidates for the polyline. Finally, in Section 11.6 we describe our experiments. Our implementation of the strip generator for x-monotonous polylines and the three evaluation functions can be tested on-line at *http://www.math-inf.uni-greifswald.de/map-labeling/lines*.

11.2 PREVIOUS WORK

For an extensive bibliography about map labelling in general, see Wolff and Strijk (1996). The problem of automated line labelling has been dealt with before. Doerschler and Freeman (1992), Barrault and Lecordix (1995), Alexander and Hantman (1995); Edmondson et al. (1997), and Kramer (1997) allow only rectangular labels; curved labels are not considered. Freeman (1988) lists a set of label-placement rules similar to those of Imhof (1975), and roughly outlines an algorithm. An analysis of Figure 8 in Freeman (1988) shows that river names are broken into shorter pieces that are then placed parallel to segments of the river. Each piece ends before it would run into the river or end too far from the current river segment.

Barrault (1997) takes curved labels into account. First, he splits an input polyline into sections depending on its length and junctions (forks) with other polylines. For details of this step, see Barrault and Lecordix (1995). Then the polyline is treated with an adaptation of an operator from morphological mathematics, closure, that is a mixture of an erosion and a dilation. This operator yields a baseline for label candidates where the polyline does not bend too abruptly. It is not clear how this is done algorithmically; no asymptotic runtime bounds are given. Finally, simulated annealing is used in order to find a good global label placement, i.e. a placement that maximises the number of features that receive a label and at the same time takes into account the cartographic quality of each label position.

Poon et al. (1998) analyse a more theoretical problem. They label an instance of axis-parallel line segments with rectangular labels of common height. While the length of each label equals that of the corresponding line segment, the label height is to be maximised.

While the restriction to rectangular labels is acceptable for technical maps or road maps (where roads must be labelled with road numbers), we feel that curved labels are a necessity for high-quality line labelling. The method we suggest is the first that fulfils both of the following two requirements: it allows curved labels and its runtime is in $O(n^2)$. The runtime thus only depends on the number of points of the polyline, and not on other parameters such as the resolution of the output device. Note that the time bound holds even if the approximate Hausdorff distance is used to select good label candidates within the strip as long as we choose the parameter k linear in n.

11.3 A BUFFER AROUND THE INPUT POLYLINE

In order to reduce the search space for good label candidates, we generate a strip along the input polyline that is (a) likely to contain good label positions and (b) easy to compute. Generating our strip consists of two major tasks. First, we compute a buffer around the polyline that our strip must not intersect. Second, we generate an initial strip and refine it recursively. Each refinement step brings the strip closer to the polyline, but also introduces additional inflections.

The input to our algorithm consists of a polyline $P = (p_1, \ldots, p_n)$ with points $p_i = (x_i, y_i)$, a minimum label-polyline distance ε, a maximum curvature $1/r$, and a label height

h. It makes sense to choose r relative large in comparison to ε but the algorithm does not depend on this. We assume that P is x-monotonous, i.e. $x_1 < \ldots < x_n$. Non-monotonous polylines can be split up into monotonous pieces of maximum length in linear time by a simple greedy algorithm. That algorithm goes sequentially through the edges of the polyline. Whenever adding the current edge to the current piece would make that piece non-monotonous, a new piece is started with the current edge.

For ease of presentation we direct P from p_1 to p_n and only label the upper (i.e. left) side of the polyline. We use r-disk (r-arc) as shorthand for a disk (arc) of radius r. We say that p_i is *at a right turn* of P if p_{i+1} lies to the right of the directed line through p_{i-1} and p_i, see p_3 or p_4 in Figure 11.1.

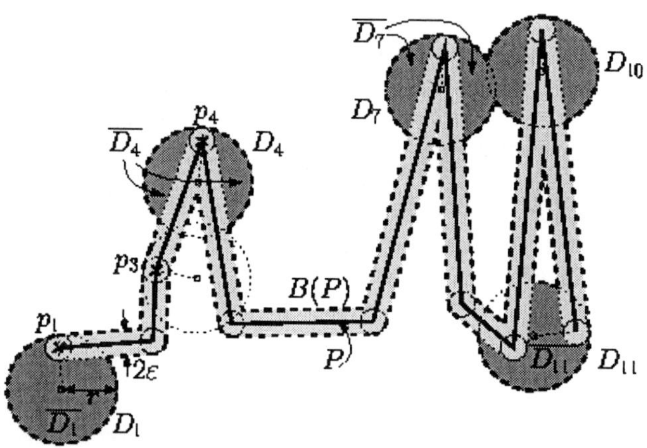

Figure 11.1 The boundary of the (ε, r)-buffer $B(P)$ (bold dashed line) of the input polyline P (bold solid line)

We define the *(ε, r)-buffer* $B(P)$ in two steps. First let the *ε-buffer* be the union of all ε-disks whose centre lies on P, see the light-shaded area in Figure 11.1. Second we add certain pieces of r-disks D_i placed at right turns p_i of P. Their task is to bound the curvature of our strip. The centre m_i of D_i is placed on the angular bisector b_i of the adjacent edges of P such that D_i touches and contains the ε-disk centred at p_i (see Figure 11.2). Let $\overline{D_i}$ be the part of D_i that is left of the ε-buffer and touches the ε-disk, see the dark-shaded areas in Figure 11.1. Then $B(P)$ is the union of the ε-buffer and the $\overline{D_i}$ for each right turn p_i.

To simplify the calculation of the strip, we also place r-disks D_1 and D_n at the endpoints p_1 and p_n of P, respectively. Let b_n be the normal to the edge $p_{n-1}\,p_n$ in p_n. Then the centre of D_n lies on b_n such that D_n touches and contains the ε-disk centred at p_n, see D_n in Figure 11.2. The placement of D_1 is analogous.

In order to compute the boundary of the (ε, r)-buffer we first compute that of the ε-buffer. This is simple since the x-monotonicity of P guarantees that the ε-buffer does not have any holes.

For computing the candidate strip it is important that we have access to the elements of the outer face of the (ε, r)-buffer in the order in which they occur. We compute the (ε, r)-buffer in two phases.

In the first phase, for each right turn p_i we follow the boundary of the ε-buffer from t_i to the right until we intersect the boundary of D_i for the first time. This intersection point is denoted by r_i, (see Figure 11.2). The arc from t_i to r_i, oriented clockwise, is the right arc R_i, one of the two parts of the boundary of D_i in which we are interested. The left arcs L_i that go counterclockwise from t_i to l_i can be computed analogously. A special case arises if t_i lies in the interior of the ε-buffer. Then R_i or L_i is empty, and we have to follow P from p_i in both directions until we arrive at a point or edge that corresponds to an arc or line segment on the upper part of the ε-buffer. From there, we can continue as usual.

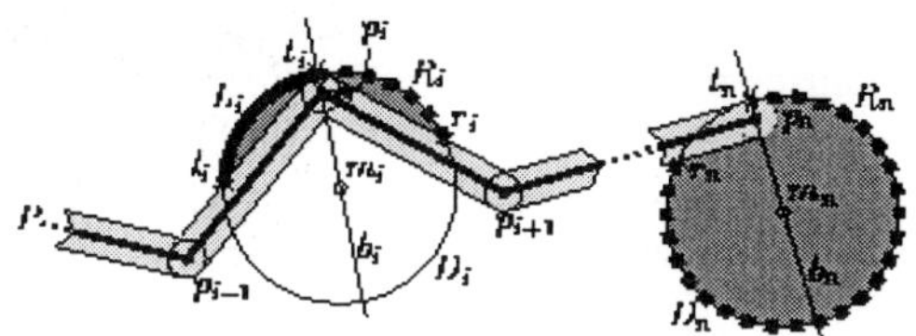

Figure 11.2 Placing r-disks D_i at right turns p_i of the input polyline P

Clearly, this procedure has a worst-case runtime of $O(n^2)$. The worst case occurs if there are a linear number of right turns p_i where we have to walk over a linear number of segments of the ε-buffer until we hit l_i or r_i, i.e. if r is large compared to the length of the edges of P. However, in practice one can expect to walk only over a constant number of segments of the ε-buffer; then the running time is $\Theta(n)$ (see Section 11.6). The worst-case running time can be improved using more sophisticated data structures, but we omit this improvement here as it makes the algorithm more complicated.

In the second phase, we incrementally extend the ε-buffer to the (ε,r)-buffer using the left and right arcs we just computed. We maintain B_{curr}, the outer face of the union of the ε-buffer and the areas $\overline{D_i}$ we have processed so far. Initially let B_{curr} be the boundary of the ε-buffer and let the interior of B_{curr} the interior of the ε-buffer. Let the r-arc A_i be the union of L_i and R_i. Note that A_i is the part of the boundary of $\overline{D_i}$ that is a potential part of the outer face of the (ε,r)-buffer. For each right turn p_i we check whether A_i lies completely in the interior of B_{curr}. If this is not the case we extend B_{curr} by using the appropriate parts of A_i.

The boundary of the ε-buffer consists of a linear number of line and arc segments to which we add $O(n)$ arcs of type A_i. One can prove that each of these arcs can contribute at most three pieces to the outer face of $B(P)$. Our implementation does not depend on this result, but it shows that the outer face of $B(P)$ has linear complexity. Due to the incremental construction this is also an upper bound for the size of B_{curr}.

Given these observations it is easy to devise an $O(n^2)$-algorithm that computes the boundary of the outer face of $B(P)$. We store B_{curr} in a doubly connected list. Since the length of this list is linear we can afford to scan the whole list when we search for intersections with the current arc A_i. If we consider carefully whether we enter or leave the interior of the area delimited by B_{curr}, we can update B_{curr} in linear time for each right turn. We omit details here.

In our implementation of the second phase we use a similar trick as in the first phase to avoid a quadratic runtime in many cases. We exploit the fact that an arc A_i usually spans only a constant number of elements of B_{curr}.

11.4 A CANDIDATE STRIP

Once we have the outer face of the (ε, r)-buffer, we compute the baseline of the label candidate strip and refine it recursively. We refer to the line and arc segments that delimit the buffer on the upper side between l_1 and r_n as baseline objects. We have access to these objects in the order in which they appear on the boundary of the buffer's outer face. We start with an arc A that touches the first and last object O_i and O_k, respectively. We bend A towards the buffer until it hits a third object O_j. There, we split A into two pieces, its children. We connect the children of A with a piece of O_j that initially has length zero. Then we recursively bend the children further towards the buffer, see Figure 11.3. While we bend, the portion of O_j that connects the children of A is growing. Note that there are two phases: in the first, the radius of the arcs increases while it decreases in the second. The recursion ends where O_i and O_k are adjacent on the buffer (since there is no O_j then) and in the second phase where the curvature of an arc would exceed $1/(r+h)$, h the label height.

For each level of the recursion, the sequence of arcs we obtain in this way forms a continuous curve L. If we direct L from left to right, it becomes obvious that the radius of all arcs that turn right (i.e. towards the buffer) is at least r and the radius of arcs that turn left is at least $r+h$. By using L as the baseline of our strip of height h we ensure that all arcs that form the upper boundary and the baseline of the strip have at least radius r. Thus the strip fulfils the curvature constraint H2. Since the baseline of the strip cannot intersect the ε-buffer it is clear that the strip also fulfils the distance constraint H1. The non-self-intersection constraint H3 can easily be kept by ending the recursion where the distance between O_i and O_k is less than $2h$.

If the number of inflections is to be kept small, the recursion can also be stopped whenever the directed distance of a strip segment to the polyline is below a given threshold. However this is difficult to check without the Voronoi diagram of the points and (open) edges of P.

It is possible to add two interesting refinement levels. In both, an arc of the baseline does not necessarily touch three objects on the boundary of the buffer's outer face. For a strip with more rectangular segments one could add a refinement level between levels 2 and 3 of the leftmost strip segment in Figure 11.3. Note that the radii of the annular strip segments there increase up to level 2 and then decrease again. Rectangular segments in an additional refinement level can thus be viewed as annular segments with infinite radius. On the other hand, to make the strip follow P as closely as possible, a final refinement level could be added where all annular strip segments are delimited by two arcs with radius r and $r+h$. The baseline of this strip is part of the curve on which a disk of radius $r+h$ is rolled around the buffer if the disk must always touch the buffer but not intersect its interior.

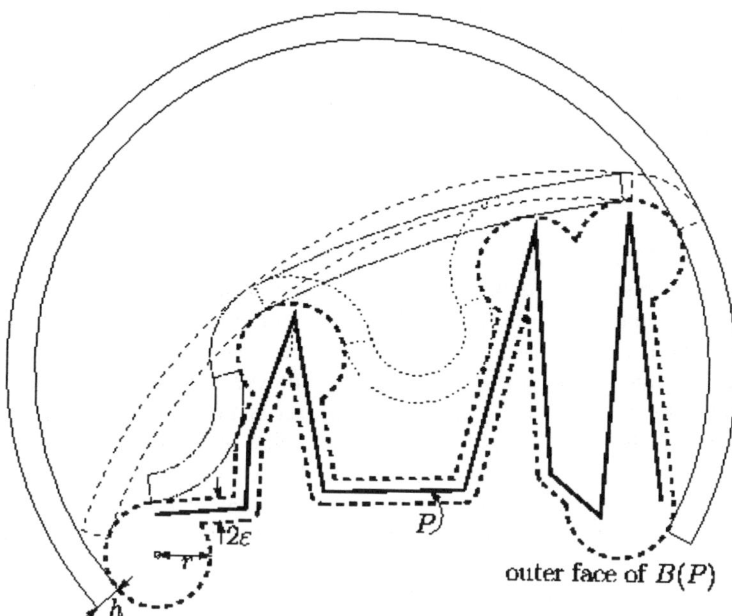

Figure 11.3 Refining the candidate strip: first level (solid), second level (dashed), third level (densely dotted), and fourth level (dotted)

In order to determine the third object on an arc, we test each object between the left- and rightmost object in constant time. Thus we need linear time for each level of the recursion. As with the Douglas-Peucker line-simplification algorithm, the number of recursion levels depends on the distribution of the input data and can vary from $\Omega(\log n)$ to $O(n)$. Given the outer face of the (ε, r)-buffer the strip can hence be computed in $O(n^2)$ time, while the average case can be expected to be in $O(n \log n)$.

11.5 FINDING GOOD LABEL POSITIONS

In order to satisfy the soft constraints, we evaluate label candidates within the strip according to curvature, number of inflections, or directed label-polyline Hausdorff distance. (We define the curvature of a label as the sum over curvature times length of each label segment. The curvature of a rectangular segment is 0; that of an annular segment with arcs of radius r_1 and $r_2 = r_1 + h$ is $1/r_1$.) For all three evaluation functions, the basic idea is the same. We discretise the space of label candidates such that the discrete space has linear size and contains minima. Then we search the discrete space for a minimum.

For curvature and number of inflections it is easy to see that there is a minimising label candidate that starts or ends with one of the rectangular or annular segments of the strip. In order to find a minimum, we push a label of the given length through the strip and stop whenever a new segment starts (or ends). To compute the measure of the current candidate, we only have to do a constant number of updates given the value at the previous position. This is how we can find a placement minimising curvature or number of inflections in linear time.

For Hausdorff distance, the discretisation is more difficult. We only take into account the baseline of the strip. In order to efficiently compute the distance between the baseline of a label candidate and the polyline P, we need to know the closest object (point or edge) of P for every point on the whole baseline. Intersecting the baseline with the Voronoi diagram of the objects of P would yield this information and lead to an $O(n \log n)$-algorithm under certain conditions (Knipping, 1998).

However, computing the Voronoi diagram for a set of points and line segments is not a trivial task in practice. Therefore we implemented a simpler algorithm that finds a near-optimal label placement as follows. Given an integer k, we split the baseline into k pieces of equal length. Let γ be the length of such a piece. We approximate the distance between each piece and P by the distance of the piece's midpoint from P. This can be done by brute force in $O(nk)$ time with $O(k)$ storage. Then we proceed as above: we push the label through the strip, stop at each midpoint and evaluate the current label position. Its Hausdorff distance to P is within γ from the maximum over the distances of all baseline pieces covered by the label. For fast access to this approximate maximum, we keep the appropriate distances in a priority queue. During the execution of the algorithm, we must insert the distance of each piece at most once into the queue. The same holds for deletions. Each such operation costs $O(\log k)$ time, hence we can compute an optimal placement among all those starting at a midpoint of a baseline piece in $O(nk + k \log k)$ time with $O(k)$ storage. The triangle inequality guarantees that this placement is at most γ further away from P than a placement minimising the exact directed Hausdorff distance. A detailed description of the implementation of the above evaluation functions can be found (in German) in Knipping (1998).

11.6 EXPERIMENTAL RESULTS

In order to analyse our line-labelling algorithm, we applied it to synthetic and to real-world data. The latter is taken from the CIA-map data at the URL *ftp://gatekeeper.dec.com/pub/graphics/data/cia-wdb/db.tar.Z*, see Figures 11.4 and 11.5. In both figures, labels were placed according to the approximated Hausdorff distance.

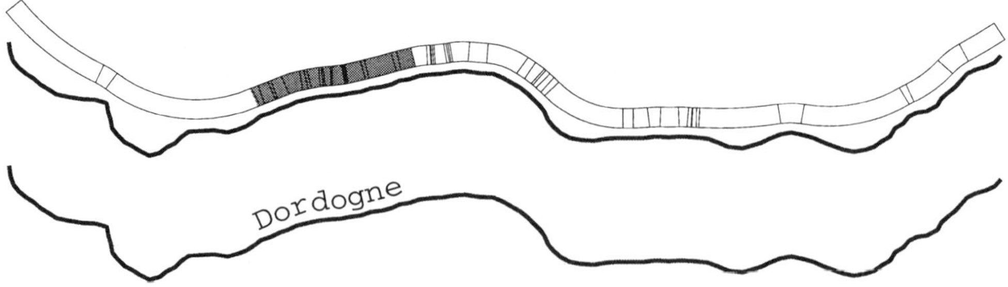

Figure 11.4 A piece of the Dordogne (109 points). Above with candidate strip and label placement (shaded grey), below with lettering

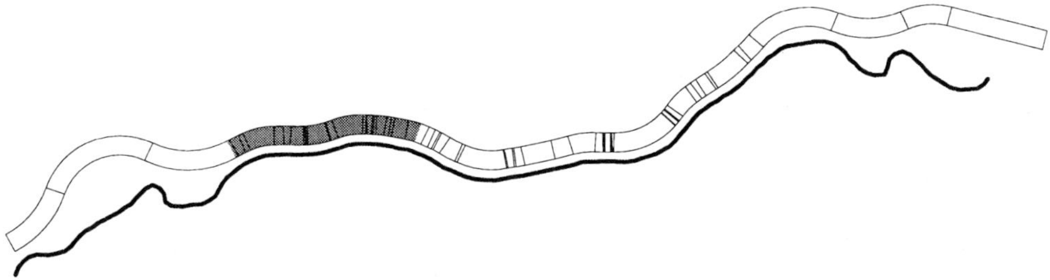

Figure 11.5 A piece of the Guadalquivir (130 points)

The synthetic data belongs to three different example classes. Due to lack of space we can only present our results on one class. For more detailed information including graphs depicting the frequency of crucial operations, see our Web page.

For the example class RandomWalk we use random numbers Δx_i and Δy_i that we draw from a normal distribution with mean 0 and standard deviation 1. In order to get an x-monotonous polyline we choose the x-coordinates as follows: $x_1 = 0$ and $x_i = x_{i-1} + |\Delta x_i|$. Then we scale all x_i by x_n such that $0 = x_1 < x_2 < \ldots < x_n = 1$. Similarly, we set the y-coordinates to $y_1 = 0$ and $y_i = y_{i-1} + \Delta y_i /100$.

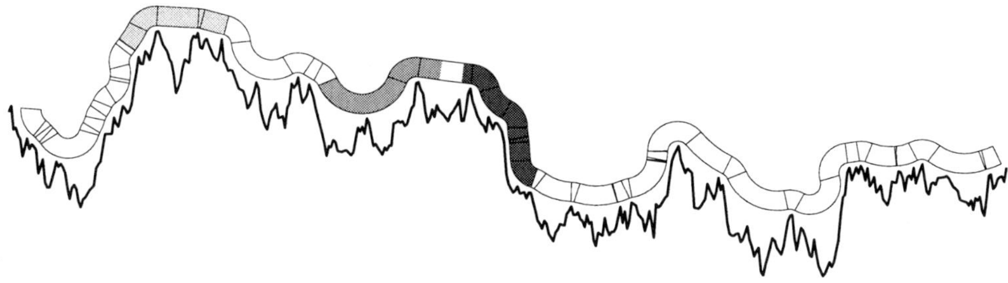

Figure 11.6 RandomWalk with 400 points

Figure 11.6 shows an instance of RandomWalk with the candidate strip of the last refinement level, not counting the additional levels mentioned in Section 11.4. The grey regions indicate an optimal label placement within the strip minimising curvature, number of inflections, and approximative Hausdorff distance (left to right, shaded light to dark). The parameters for the strip computation were minimum label-polyline distance $\varepsilon = 0.005$, curvature bound $r = 0.01$, and label height $h = 0.02$. More examples can be found in Knipping (1998) or generated on our Web page.

We generated 50 RandomWalk examples with 100, 200, ..., 1000 points to analyse the performance of our C++ implementation. We used the SunPRO-CC compiler with optimiser flags `-fast -O3` and measured runtimes on a Sun Ultra-Sparc 250. We prepared two graphs, see Figures 11.7 and 11.8. In both, the y-axis gives the average CPU time (in seconds) and the x-axis gives the number n of points of the polyline. The points on our graphs give the results averaged over all 50 examples; the extent of the vertical bars indicates the minimum and maximum runtime among these 50 examples.

Figure 11.7 shows the running times of the ε-buffer, (ε, r)-buffer and strip generation for RandomWalk. Note that the three curves are additive; i.e. the topmost curve corresponds to the total runtime. The two additional refinement levels mentioned in Section 11.4 were included.

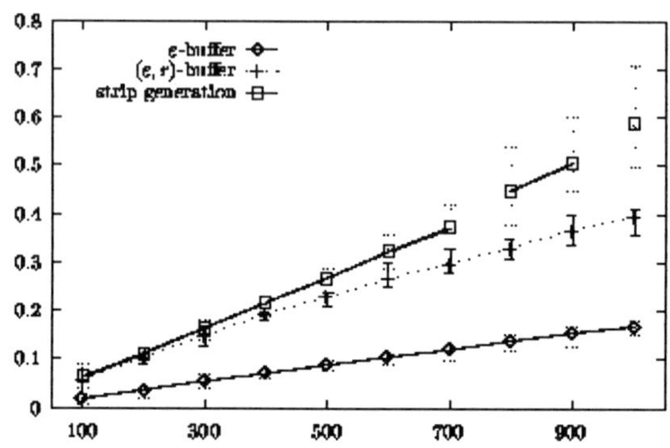

Figure 11.7 Strip generation time

In Figure 11.8 we give the runtimes for placing labels within the pre-computed strip according to curvature and approximated Hausdorff distance. Here the parameters were curvature $r = 8/n$, minimum distance $\varepsilon = 2/n$, label height $h = 10/n$, and label length $l = 50/n$. For minimising the Hausdorff distance we set the approximation parameter γ to $1/(2n)$. We omitted the curve for number of inflections since it is identical to that of curvature. Other than in the description in Section 11.5 we used lists instead of priority queues for the approximated Hausdorff distance, hence the quadratic runtime behaviour.

11.7 CONCLUSION

We have presented a new and conceptually simple method for high-quality line labelling. It is the first that fulfils both of the following two requirements: it allows curved labels and its worst-case runtime is in $O(n^2)$. We introduced a concept of gradual refinement that is similar to the idea of the Douglas-Peucker line-simplification algorithm. This concept allows to introduce additional application-dependent criteria and to stop the refinement when these criteria are met.

An experimental evaluation of our algorithm shows that it usually runs in subquadratic time and generally yields good results in practice. However, since we reduce the search space for good label candidates to a one-dimensional strip, it is clear that we cannot hope to find an optimal label placement in every case. As the following example indicates, a more flexible strategy in the buffer construction might help to overcome problems caused by the reduction of the search space.

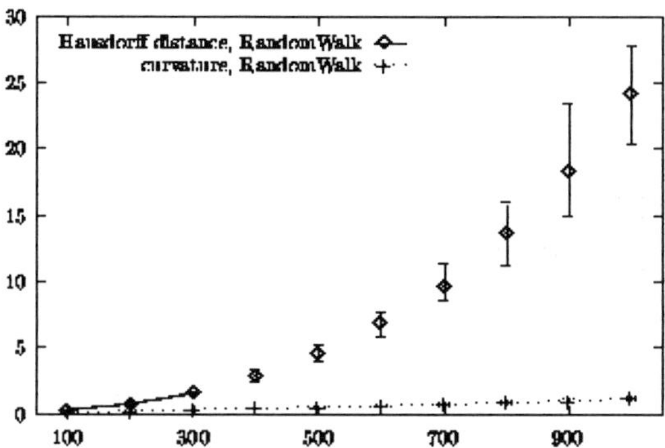

Figure 11.8 Running times for label placement

In Figure 11.9 we depicted all r-arcs at right turns of the input polyline P. The parameter r was chosen to be large compared to the average segment length of P. As a result, some of the arcs that contribute to the (ε,r)-buffer are quite distant from the input polyline P. They were caused by right turns incident to two very steep but short edges of P (see also the bends of the strip at both ends in Figure 11.5). It would be desirable to remove these arcs. However, we must ensure that the resulting strip does not violate the curvature constraint H2. This can be done as follows. After the first phase of the (ε,r)-buffer computation we compute the directed Hausdorff distance of each r-arc A_i to the ε-buffer between l_i and r_i. In order of descending distance we check for each A_i whether the corresponding ε-arc lies completely in the area $\overline{D_i}$ of another r-arc A_j. If this is the case, we remove A_i. Then we proceed to the second phase of the buffer computation as usual. Note that the resulting outer face of the buffer still consists exclusively of r-arcs and line segments. Thus the strip will still keep H2.

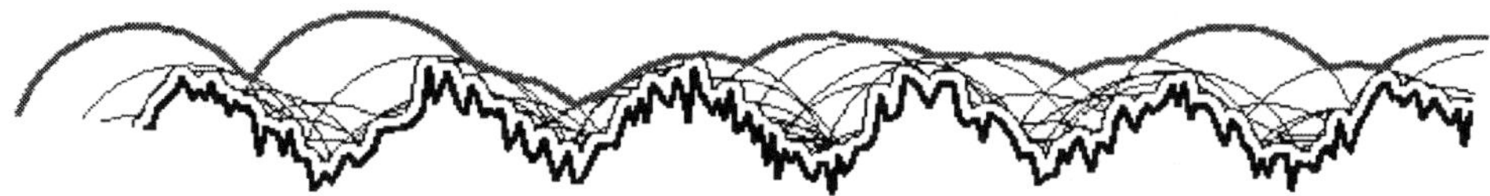

Figure 11.9 Disturbing effects of the definition of the (ε,r)-buffer. (The upper part of its outer face is marked by bold grey arcs; the input polyline below consists of bold black line segments.)

An alternative approach is as follows. We observed that our placement of the r-disks is good if the adjacent edges of the polyline are long enough. Then the directed Hausdorff distance between the arc A_i and the ε-buffer is minimised. However, in general the placement of the r-disks is too inflexible. It could certainly be improved if we tried to minimise the aforementioned distance during the placement. Then the placement of the r-disks would take into account not only the adjacent edges of the polyline but all of the polyline (or the ε-buffer) between l_i and r_i.

Finally we would like to acknowledge a simple and elegant idea of Mike Lonergan, University of Glamorgan, Pontypridd. He suggested putting the ε-buffer around the label (and thus simply thicken the strip by 2ε) instead of the polyline. Unfortunately, this does not solve the problem of placing the r-circles.

11.8 REFERENCES

Alexander, D. H. and Hantman, C. S., 1995, Automating linear text placement within dense feature networks. In *Proceedings Auto-Carto 12*, (Charlotte, NC: ACSM/ASPRS), pp. 311-320.

Barrault, M., 1997, An automated system for name placement which complies with cartographic quality criteria: The hydrographic network. In, *Lecture Notes in Computer. Science*, (Berlin: Springer-Verlag), **1329,** pp. 499-500.

Barrault, M. and Lecordix, F., 1995, An automated system for linear feature name placement which complies with cartographic quality criteria. In *Proceedings Auto-Carto 12*, (Charlotte, NC: ACSM/ASPRS), pp. 321-330.

Doerschler, J. S. and Freeman, H., 1992, A rule-based system for dense-map name placement. *Communications of the ACM*, **35**, pp. 68-79.

Douglas, D. H. and Peucker, T. K., 1973, Algorithms for the reduction of the number of points required to represent a digitized line or its caricature. *Canadian Cartographer*, **10**, 112-122.

Edmondson, S., Christensen, J., Marks, J., and Shieber, S., 1997, A general cartographic labeling algorithm. *Cartographica*, **33**, pp. 13-23.

Freeman, H., 1988, An expert system for the automatic placement of names on a geographic map. *Information Sciences*, **45**, pp. 367-378.

Imhof, E., 1975, Positioning names on maps. *The American Cartographer*, **2**, pp. 128-144.

Kakoulis, K. G. and Tollis, I. G., 1998, A unified approach to labeling graphical features. In *Proceedings of the 14th Annual ACM Symposium on Computational Geometry (SoCG'98)*, Minneapolis, MN, U.S.A., (New York: ACM press), pp. 347-356.

Knipping, L., 1998, Beschriftung von Linienzügen. Master's thesis. (Berlin: Fachbereich für Mathematik und Informatik, Freie Universität). Available at *http:// www.math-inf.uni-greifswald.de/map-labeling/papers/k-bl-98.ps.gz*

Kramer, J. C., 1997, Line feature label placement for ALPS5.0. Rutgers University, Piscataway, NJ, U.S.A.. Unpublished manuscript, available at *http://paul.rutgers.edu/~jckramer/academics/Report/*

Morrison, J. L., 1980, Computer technology and cartographic change. In *The Computer in Contemporary Cartography*. Edited by Taylor, D. (New York: Hopkins University Press).

Poon, C. K., Zhu, B., and Chin, F., 1998, A polynomial time solution for labeling a rectilinear map. *Information Processing Letters*, **65**, pp. 201-207.

van Dijk, S., van Kreveld, M., Strijk, T., and Wolff, A., 1999, Towards an evaluation of quality for label placement methods. In *Proceedings of the 19th International Cartographic Conference*, (Ottawa: International Cartographic Association) pp. 905-913.

Wagner, F. and Wolff, A., 1998, A combinatorial framework for map labeling. In *Proceedings of the Symposium on Graph Drawing '98*, edited by Whitesides, S. H. *Lecture Notes in Computer Science*, (New York: Springer-Verlag), **1547**, pp. 316-331.

Wagner, F., Wolff, A., Kapoor, V., and Strijk, T., 1999, Three rules suffice for good label placement. *Algorithmica, special issue on GIS*, editor Marc van Kreveld, to appear. Available at *http:// www.math-inf.uni-greifswald.de/map-labeling/papers/wwks-3rsgl-99.ps.gz*

Wolff, A. and Strijk, T., 1996, A map-labeling bibliography. *http:// www.math-inf.uni-greifswald.de/map-labeling/bibliography/.*

12

Modelling knowledge for automated generalisation of categorical maps - a constraint based approach

Alistair Edwardes and William Mackaness

12.1 INTRODUCTION

This research is concerned with the automatic derivation of multi scaled categorical maps from a single detailed database. The representation of categorical maps at increasing scales requires us to aggregate detailed categorical data, such as is found in capability maps or soil maps (Burrough, 1991) in order to produce a series of simplified, visually effective maps. As a prerequisite to implementation within an object oriented GIS, a model is required that enables us to represent these transitions in a systematic meaningful way. Such a model must take into account the special characteristics unique to categorical maps. A categorical map is a space exhaustive tessellation of space. The discrete boundaries we see on the map are rarely found in the landscape. The reality is that they represent inherently fuzzy phenomena (Burrough and Frank, 1996), and their form is very dependent on the classification used. Such characteristics present challenging environments for automated generalisation solutions, with strong spatial and semantic interdependencies existing between the polygons that define the various spatial extents. The research is pertinent due to the increasing demand for and use of Internet based products, generalised in both scale and theme. Given the lack of requisite cartographic skills, there is a need to capture the necessary cartographic knowledge in order to automate this process.

The paper begins with a discussion of categorical mapping and the challenges of automated generalisation in this field. The methodology utilised the Unified Modelling Language (UML) and the Object Constraint Language (OCL) to model the design constraints identified through empirical analysis. The results of formulating knowledge using these constructs is presented. Further work is discussed together with a conclusion highlighting the context and importance of this work.

12.2 PREVIOUS RESEARCH

Various research has examined methods for the automated generalising of categorical data, both manually (Hole and Campbell 1985) and in the raster (Jaakkola, 1998; Monmonier, 1983; McMaster and Monmonier, 1989; Schylberg, 1993) and vector domains (Peter and Weibel, 1999; Molenaar, 1998).

Hole and Campbell (1985) and Molenaar (1998) discuss the problem with an emphasis on the consistent application of an overall generalisation strategy, without which, they suggest, a "chaotic legend and a useless map" (pp. 132) will be produced. They describe three strategies: graphic generalisation, taxonomic generalisation and spatial generalisation. Graphic generalisation refers to the cartographic generalisation of the mapped units, reducing visual complexity without alteration to the original schema. Arguably, graphic generalisation involves little specific domain knowledge of the mapped data and consequentially is not sufficient alone to satisfactorily generalise a map. For example, alone it could not be employed to aggregate or amalgamate objects since this involves a more detailed knowledge of the phenomena. Taxonomic generalisation is a strategy that uses the data model and taxonomy to reduce the information complexity of the data by reclassification and aggregation of adjacent and nearby objects. The danger of this strategy is that it may emphasize the semantics of the information without reference to the real world spatial phenomena and thus the degree of generalisation will depend on the extent to which semantically similar units are found together. Generally, this strategy is also known as model generalisation. The strategy of spatial generalisation places the geographic phenomena as primary by generating new semantic categories based on recurring patterns in spatial arrangements of the geographic entities. The emphasis of this strategy is to enhance the information communicated about landscape and geographic process and hence it places no constraints on the need for physical or semantic similarity of the amalgamated entities. The danger of this approach is it may create a schema very different from the original and also very much localised to individual study areas. Molenaar's (1998) ideas complement these in suggesting a conceptual strategic stage to categorical generalisation followed by a stage of graphic generalisation. In discussing the conceptual stages he proposes four similar strategies and further suggests that any strategy must be composed of two sets of aggregation rules: one set defining what types of map elements can aggregate (e.g. elements of the same superclass or elements that are too small) and the other, here termed linkage rules, defining the topo-metric relationships under which objects may be aggregated (e.g. adjacent, touching and close proximity).

In terms of the graphic stage various methods have been implemented that deal with the generalisation of polygonal data, notably Muller and Wang (1992), Schylberg (1993) and Bader and Weibel (1997). What is generally acknowledged amongst these researchers is that intelligent automated generalisation methods require a range of information in order to produce workable solutions. Specifically, information relating to metric qualities relating to each object (size, orientation, compactness, frequency of occurrence), topological qualities (the neighbourhood of objects), and semantic qualities (soil classification, typical associations between soils, and between soils and their geomorphological characteristics). These ideas reflect the phenomenological approach to map generalisation (Ormsby and Mackaness 1999; Hangouet and Lamy, 1999).

What is also clear from the preceding discussion, is that there is likely to be a ceiling on the degree of generalisation that can be performed automatically with only the application of 'universal' cartographic principles for map design. The necessity for a strategic approach to the generalisation of the geographic information will ultimately distinguish unique generalisation requirements for different types of categorical mapped data, e.g soil, geology or land suitability. The approach in this chapter is to attempt to distil out requirements of graphic generalisation generally for categorical maps and explore the specific strategies that have been employed to generalise the geographic information in the particular example of maps describing land capability for agriculture. In this paper we thus explore the description of the process of generalising categorical maps through an empirical, case-based, examination of 'ungeneralised' source data and generalised target data. This approach attempts to determine what actions have been performed on the data and interpret why these actions have been performed. Through this analysis it is hoped to describe the nature of objects that may exist at the target scale; in other words by defining an ontology with which to describe and define the generalised information in both cartographic and geographic terms.

12.3 CATEGORICAL DATA

Frank *et al.* (1997) define this class of mapped information by saying 'a categorical coverage consists of two connected components: (1) a spatial component and (2) a thematic component. Using Sinton's terminology (Sinton, 1978), categorical coverages hold time fixed, control for theme, and measure the location, by searching for the largest areas with uniform properties. (This is in contrast to choropleth maps, in which location is controlled and theme is varied).' (Frank *et al.*, 1997, p. 217). The datasets used to test the methodology were 'Land Capability for Agriculture' paper maps, created by the Soil Survey of Scotland. The maps have been created from derived information gathered on a number of variables, such as soil, climate, wetness and erosion. 'The classification ranks land on the basis of its potential productivity and cropping flexibility determined by the extent to which its physical characteristics impose long-term restrictions on its agricultural use' (Bibby *et al.*, 1982 p.1). The Land Capability for Agriculture maps have been produced at 2 scales 1:50,000 or 1:63,360 (depending on the region) and 1:250,000. The classification is based on seven classes, some of which have been subdivided into finer levels, of at most three divisions. The divisions are additionally labelled with at most two of five possible unit symbols that are used to group areas by the specific limitations that determined their inclusion in one of the divisions. These units are only used at the larger scales. The classification system has been described extensively (Bibby and MacKney, 1969; Bibby *et al.*, 1982; Brown and Shipley, 1982). At the most general level, classes 1 to 4 describe land that is well suited to arable use. Classes 5 to 7 describe classes that are best suited for grazing. Within the arable classes, classes 1 and 2 are described as having low internal variability. This means they have been strictly defined and only land completely satisfying the descriptions are classified as such. Such distinctions are pertinent to the generalisation process as they serve to provide semantic limitations to the actions that can be performed.

12.4 METHODOLOGY

It is important to make explicit reference to the structure of the underlying data model, the classification system of this type of map and discuss empirical observations made from examining specific examples of generalisation. The methodology therefore proceeds as a three stage process. (1) Data Modelling in UML. This uses shared spatial primitives and the classification system of the Soil Survey of Scotland, in order to model the semantic information on the spatial components. (2) Case-based empirical analysis of specific examples of generalisation of categorical maps as a basis for defining design constraints. (3) Description of the 'solution state' of the generalised examples with reference to the constrained data model using OCL.

12.4.1 Constraint Based Problem Solving

A constraint based approach enables a problem to be defined in an intuitive manner. Constraint based systems are declarative as opposed to being imperative, describing what must be maintained and not how it should be maintained (Beard 1991). The constraints are therefore defined on the state of the objects and their inter-relationships and not on the actions that are applied in the process of generalisation. They define the domain within which a solution is acceptable. The notion of constraints has attracted significant attention in a number of disciplines, notably computer science and artificial intelligence. As a result of this interest, there has been the emergence of a 'constraint programming' paradigm (see Marriott and Stuckey, 1998). Within this paradigm two dominant branches of problems are found which can be described in terms of constraints - systems of *absolute* constraints, and *over-constrained* system; systems of absolute constraints (such as in the well known constraint satisfaction problems) are where a solution is sought that satisfies all constraints. Over-constrained problems are where no solution exists in which all the constraints can be met and hence some kind of preferential weighting or hierarchy must be used (Wilson and Borning, 1993). Map design has been variously described as a compromise between competing constraints. Thus this latter class of problem shares much in common with many problems associated with map generalisation. Of note however, is that in map generalisation the constraints are usually inter-related hence the solution of one constraint may often directly lead to another constraint becoming violated.

12.4.2 Constraint-based Generalisation

The usefulness of defining generalisation in terms of constraints has been extensively discussed in the domain of topographic maps (Beard, 1991; Ruas and Plazanet, 1997, Ruas, 1998; Weibel and Dutton, 1998; Harrie, 1999). Weibel (1997) also discusses categories of constraints with respect to polygonal map generalisation and within the context of line simplification.

Ruas defines constraints in map generalisation by saying 'A constraint describes explicitly either which information does not meet the generalisation specification or which information should be maintained in terms of the content during the generalisation process' (Ruas, 1998, p.228). This highlights the role of constraints in guiding the non-deterministic generalisation process. Weibel and Dutton (1998) use the definition of a constraint as 'a limitation that reduces the number of acceptable solutions to a problem' and regard constraints as 'design specifications to which solutions should adhere' (p.215). This definition highlights the role of constraints in determining the solution state of the generalised map. It is not the objective of this paper to extend or redefine these definitions. Whilst it is necessary to devise strategies in the application of methods to find solutions among a set of constraints, the focus of this paper is on the essential prerequisite to this stage - namely to precisely describe the constraints in a form that enables their representation within a computer based model.

12.5 INTRODUCTION TO UML AND OCL

To generate a formal ontological description, the data are first modelled using the notation of the Unified Modelling Language (UML). This is an object oriented analysis and design language, with which the entities in the system and their relationships may be modelled. UML uses a graphical syntax in which semantics are then embedded to facilitate a visual class based representation of the system. The UML notation provides a precise schema for the analysis and design of object oriented systems in terms of semantics and, a graphical abstract syntax (Booch *et al.*, 1997; Rational Rose Corporation, 1997a). UML is the product of several commonly used techniques and methods for OO analysis and design, primarily embodying the approaches of Booch (1994), Rumbaugh (1991) and Jacobson *et al.*, (1997). The language has looked at fusing these approaches and formalising the semantics as well as finding a common notation within which they can be expressed. UML has now been accepted as the standard notation for object oriented analysis and design by the Object Model Group (OMG). Specific examples of modelling with UML can be found in the extended version of this paper available at *http://www.geo.ed.ac.uk/~aje/GISRUK99.pdf*

The object constraint language (OCL) was defined as part of the UML specification (Warmer and Kleppe 1999; Rational Rose Corporation 1997b). It is not a programming language but a syntax and set of extendable semantics that can be used declaratively to state constraint expressions. As a declarative language it defines what must be true rather than procedures to describe what should be done. A constraint is defined as 'a restriction on one or more values of (part of) an object oriented model or system' (Warmer and Kleppe, 1999, p. 8). This generic definition is concerned more with aspects of maintaining the integrity of an object oriented system and so is not necessarily suitable for the needs of generalisation. An important part of the research was to evaluate the suitability of OCL with respect to automated map generalisation. The language has a number of advantages over other methods of describing constraints: it is precise and unambiguous, this is an advantage over the description of constraints in a natural language, where this cannot be assured. It is a typed language helping to prevent logical errors and it is directly related to the data model, this means the constraint are tied in with the semantics of the other stages of analysis and design. It is free of side-effect, this means that evaluating an OCL expression cannot change the state of the system since actions are not tied to the constraint evaluation. This can be relaxed if required, for example to define a system that triggers an action as a result of a constraint being violated, though this is an extension which is not supported by default within the specifications of OCL.

OCL is complementary to UML. The semantics of the language being the entities of the UML data model (e.g. classes, methods, properties and relationships). This provides a mechanism with which to constrain the state of the entities in the data model to concord more closely with their ontological description. The language may also be parsed using a Java parser (IBM, 1997) which reads in the UML and OCL descriptions and checks the OCL expressions are well-formed and correctly typed with respect to the data model and the specifications of OCL.

12.6 DATA MODELLING USING UML

The data model consisted of two main components: the spatial primitive classes and the semantic classes. The model of the spatial primitives was based on that of the OOGIS platform being used for implementation (Laser-Scan's LAMPS2). The semantic model used the class hierarchy as defined by the Soil Survey of Scotland. Figure 12.1 shows a section of the semantic class diagram. The section shown describes the linkage between the spatial and semantic components. In the section simple_area represents the areal spatial class and LCU (Land Capability Unit) represents the base classes for the semantic objects. The inheritance between these classes means that the semantic classes are supplied with spatial data structures and funtionality.

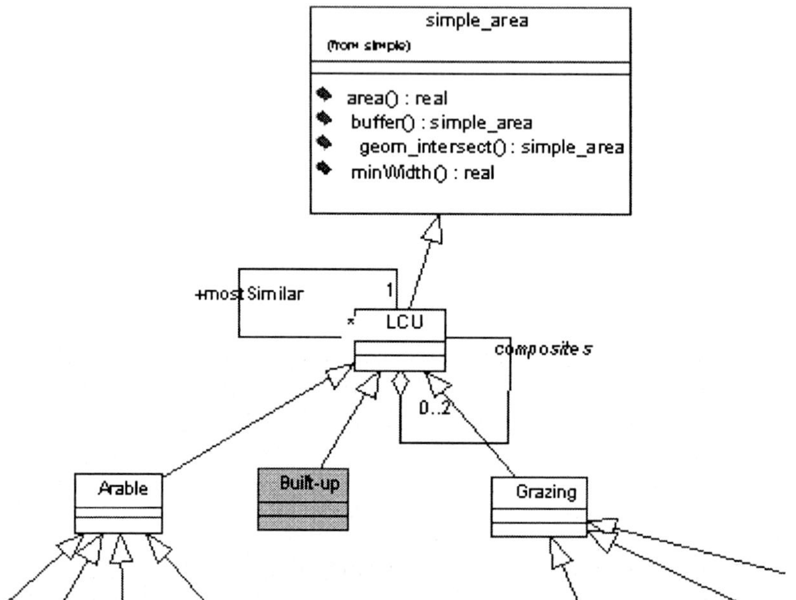

Figure 12.1 UML Class diagram for part of the classification system for categorical maps

In order to describe the state of a generalised map the objects were conceptualised as being aggregations of the objects of the 'ungeneralised' source map. This relationship is shown by the *composites* 'aggregation' relationship in Figure 12.1. The reason for adopting this approach is that the nature of categorical maps is space exhaustive. Therefore, any change in the boundary or removal of an object will necessarily result in a change to the adjacent objects through the addition or subtraction of area. This is unlike topographic maps, where this propagation of effects is limited by the existence of 'white space'. These changes in the areal extents of objects following generalisation have important semantic consequences that must be controlled for. This control is performed through the use of a global strategy (see section 12.2). In order to implement this strategy a binary composite relation was used, so objects could be considered to consist of two objects at any one time. The union of two objects into a composite was envisaged to generate a new object that assimilated the states of the simple objects. Since two composite objects each with a derived single state could also combine to form a further composite this binarization was not considered a restriction to the description of the generalised map. Essentially this results in the model as The Land Capability Unit (LCU) class being a recursive data structure, containing instances of itself. This approach also has the useful consequence of creating an indexing system. The precise mechanism by which adjacent objects schedule this process of becoming composites is part of ongoing research.

12.7 CASE-BASED ANALYSIS

The analysis of the generalisation of the Land Capability for Agriculture paper maps was based on the 1:50,000 map of Edinburgh (sheet 66, 1984) and the 1:250,000 map of South East Scotland (sheet 7, 1983). Empirical analysis was undertaken for a number of cases. The principal observations from the analysis of all the study areas are summarised in Table 12.1. A number of examples were examined. For reasons of brevity, a single example is illustrated. Figure 12.2, shows two maps, one at 1:50,000, with the equivalent area shown at 1:250,000, but enlarged to afford comparison shown second. The example has been annotated with observations of the differences between the two, and suggested reasoning that accounts for the differences. In the example in Figure 12.2 the classes are described as having comparatively 'high internal variability' in the definitions of the classes, this allows a greater degree of merging between classes. Other examples revealed that classes having 'low internal variability' make merging less appropriate and operators such as enlargement and enhancement become more important.

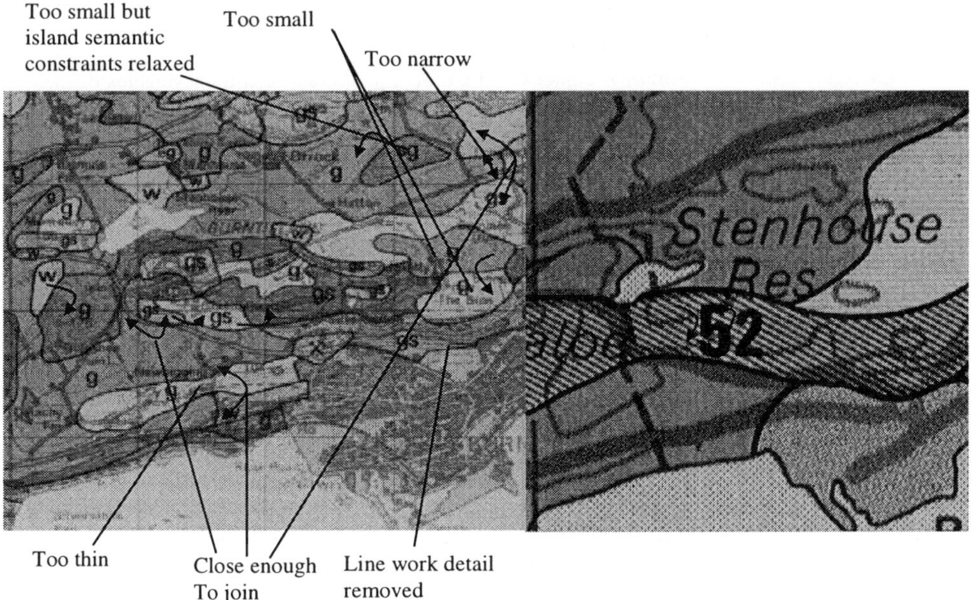

Figure 12.2: Example showing the generalisation of small objects with high variability within class definition. Crown copyright 1984 The Ordnance Survey and The Soil Survey of Scotland

Table 12.1 summarises key observations from the preceding analysis, of which Figure 12.2 and others are examples. This adopts a reverse engineering approach by describing the principle observations, describing the various contexts the observations were found within and interpreting these observations using theory from automated cartographic generalisation research. For reasons of space, the table is partial. The full table can be seen at *http://www.geog.ed.ac.uk/~aje/GISRUK99.pdf*

These observations concurred well with those of Monmonier (1983) who observed the operations of simplify, aggregate, gap bridge, erode smooth and dissolve. Having made observations, the next stage was to formalise the statements as constraints on the data model.

12.8 CONSTRAINT MODELLING

The constraint modelling was then performed using the knowledge acquired from the case based analysis and using the data model to describe the entities that were to be constrained. As outlined in the previous section, in the construction of constraints, an object was never considered in isolation but always as existing in context, and the generalisation was considered in terms of composite relations between objects, that define the ontology of the generalised objects that can exist at the target scale. The nature of the unions that could exist are defined by the constraints, for example semantic constraints may prevent certain classes of objects from merging. The list of constraints is by no means exhaustive and deals only with the metric, topological and semantic constraints as they affect the class of 'Land Capability Units'. The constraints deal with the class and the specialisation of it in terms of the constraints that are unique to this family of maps.

Table 12.1. Summary of observations and interpretation of empirical analysis

Observations	Observed Contexts	Interpretation
Land Capability Unit (LCU) object omitted after generalisation	**Within or adjacent to LCU of the same class but with a different unit attribute**	Semantic constraint violation defined by generalised data model, unit attribute suppressed, LCU objects aggregated
	Island / inner ring - LCU object isolated, distance between area and another area of a semantically similar class > ε	Minimum area constraint violation, Object insignificant with respect to overall pattern - Island merged with containing object
	Bounded by more than one other LCU object	Minimum area constraint
	-In semantic context, adjacent to LCU object of a semantically similar class	Objects merged, composite LCU type is that of the largest component object
	-In semantic context, within disjoint neighbourhood of LCU of a semantically similar class	Objects amalgamate. Composite LCU type is that of the largest component object
	- Semantically isolated, adjacent to no LCUs of a semantically similar class and outside the disjoint neighbourhood of any other semantically similar LCUs	Depending on gestalt constraints.dfdsf Object is split - area shared amongst adjacent objects. Or Object merges with most semantically similar class.
LCU object reduced	....	
LCU object enlarged	....	
LCU object enhancement	....	

12.8.1 Metric Constraint

The constraint defines the proximity for merging. It says that the composite LCU objects must be within a certain distance of each other for a parent LCU object to be valid. The constraint allows for two composite objects to be completely separated but constrains the extent of this separation. Thus two objects separated by another object may have become joined together but only if the distance separating them is less than a constant. The expression follows the association between the contextual object and its composites and tests that the geometric intersection between the buffers of the two objects geometries.

```
Land Capability Unit (LCU)
self.composites -> forAll(O1,O2 | O1.simple_area.buffer()
.geom_intersect(O2.simple_area.buffer()).area >
MIN_INTERSECTION_AREA)
```

The next example shows how the constraint dictates that one of the component objects must be greater than some constant that defines the minimum mappable unit. It reads - in the composites of an object self there exists an object O such that the area of O is less than the minimum area.

```
Land Capability Unit (LCU)
self.composites -> exists(O | O.area < MIN_AREA)
```

The expression states that the minimum width of an LCU area should be greater than a constant MIN_WIDTH defined by the map specifications. It uses the LCU query method 'minWidth', which is inherited from the simple_area base class to determine the minimum width value for the object.

```
Land Capability Unit (LCU)
self.minWidth() > MIN_WIDTH
```

12.8.2 Topological Constraint

This constraint dictates that for land capability unit (LCU) objects, the boundary should only separate distinct classes. This is because if a boundary has the same class of entity on either side, it holds no information. The expression starts by navigating through the UML model to find the collection of links that make up the boundary of the area LCU object. The condition tests that for each of the links there should be exactly one reference to an areal LCU and that it is of the same class as the contextual object, self. The operator 'select' performs this test by selecting out a subset of the collection which contains only those objects for which the class is the same as that of the contextual object. 'Size' then returns the size of this subset.

```
Land Capability Unit (LCU)
self.goth_area.goth-link -> forAll(link | link.goth_area ->
select(oclType = self.oclType) -> size = 1)
```

12.8.3 Semantic Constraint

This constraint is to ensure that the class of the LCU object is the same as the class of its largest composite. The constraint makes use of the logical connective 'implies' as defined in OCL.

```
Land Capability Unit (LCU)
self.composites -> if exists(O1 | self.composites
-> forAll(O2 | O1.area >= O2.area)) implies self.oclType
= O1.oclType
```

The OCL expression states that for at least one of the objects in the composite, the other must be the most semantically similar neighbour. The expression uses the composite's association to find the set of component objects. The condition that one or the other object is most similar is then tested.

```
Land Capability Unit (LCU)
self.composites -> forAll(O1,O2 |exists(O1 |
O1.mostSimilar -> count(O2)= 1)or exists(O2 |
O2.mostSimilar -> count(O1)= 1))
```

12.9 CONCLUSION

Within topographic map generalisation the use of constraints has been shown to be an extremely valuable tool in understanding the knowledge used in the process of cartographic generalisation. This chapter attempts to define a similar methodology to try to understand the knowledge that is used to perform categorical map generalisation. The chapter follows a methodology of empirical case-based evaluation of categorical map generalisation and attempts to capture the knowledge gleaned from this using a system of constraints to define the ontology of entities that can exist in a generalised categorical coverage. Using this methodology has proved useful since it allows for a description of the generalised information both in terms of its geographic properties and its cartographic representation. However, it is acknowledged that empirical analysis cannot be guaranteed in the first instance to furnish all the requisite knowledge for generalisation. Gaps in this understanding will need to be identified during the iterative stages of the development of a prototype system.

There is a need for descriptive, reliable and extensible tools that bridge the gap between the conceptual and the implementation levels of designing a system to solve any problem. This is undoubtedly one of the most significant reasons why object oriented systems have been so successful. Likewise, the popularity of UML and OCL is, in part, due to their unification and standardisation of a number of disparate, informal methods to design and analyse object-oriented systems. It was because of these reasons that the methodology as outlined in this paper was adopted. The tools offered the ability to analyse and design a system for generalisation that tightly integrated conceptual levels of knowledge acquisition with operational issues of knowledge representation. This has enabled the authors to seamlessly make the important step between the stages of modelling the geographic information and providing a complete ontological description of the entities that exist within the system in order to model the cartographic knowledge required to automate the generalisation process. The rigorously defined and unambiguous specifications of OCL and UML provide a distinct advantage to other methods which use several disparate systems of potentially ambiguous notation loosely coupled together to achieve what UML and OCL can achieve using a single integrated system. The provision of parsing tools for error checking and CASE tools, such as Rational Rose, that use the specification directly are also of enormous utility to the design process. The authors are conscious there are a number of improvements that could be made to better satisfy the characteristics of modelling in a GIS environment. One such improvement would for a spatial extension to UML (such as is proposed by Bedard (1999) and made publicly available at *http://sirs.scg.ulaval.ca/yvanbedard*), and to OCL, for example, by applying a spatial object modelling theory such as that outlined by Molenaar (1998). However, we are confident that the tools used are sufficiently descriptive and flexible in their semantics and syntax to satisfy the needs of geographic modelling. The aim of defining a methodology for the design of a constraint based, knowledge-based system for the cartographic generalisation of categorical maps has, in the authors' opinion, been achieved. Research is still in progress and it is likely that many of the limitations of the methodology cannot be known until a prototype has been constructed and the modelling process has undergone subsequent iterations.

12.10 ACKNOWLEDGEMENT

The authors are very grateful for the assistance in funding for this research from the European Union via the Esprit funded AGENT project, 24 939 LTR.

12.11 REFERENCES

Bader, M., and Weibel, R., 1997, Detecting and resolving size and proximity conflicts in the generalization of polygon maps. *Proceedings of the 18th ICA/ACI International Cartographic Conference*, (Stockholm: International Cartographic Association), pp. 1525-1532.

Beard, K., 1991, Constraints on rule formation In *Map generalization: Making rules for knowledge representation* edited by Buttenfield, B. P. and McMaster, R. B. (Harlow: Longman), pp.121-135.

Bedard, Y., 1999. Visual Modelling of Spatial Databases: Towards Spatial PVL and UML. *Geomatica*, **53**, pp. 169-186.

Bibby, J. S., Douglas, H. A., Thomasson, A. J. and Robertson, J.S., 1982, *Land Capability Classification for Agriculture*. (Aberdeen: Macaulay Institute for Soil Research).

Bibby, J. S. and MacKney, D., 1969, *Land Use Capability Classification, Technical Monograph No.1* (Aberdeen: Macaulay Institute for Soil Research).

Booch, G. 1994, *Object-Oriented Analysis and Design with Applications*, 2nd edition, (Redwood City, CA: Benjamin/Cummings).

Booch, G., Rumbaugh, J. and Jacobson, I., 1997, *Unified Modelling Language User Guide* (Reading, MA: Addison Wesley).

Brown C. J and Shipley B. M., 1982, *Soil and Land Capability for Agriculture: South-East Scotland.* (Aberdeen: Macaulay Institute for Soil Research).

Burrough, P. A., 1991, Soil information systems, In *Geographical Information Systems: principles and applications,* edited by Maguire, D. J., Goodchild, M. F. and Rhind, D. W. (Harlow: Longman), pp. 153-169.

Burrough, P. A., and Frank, A. U., 1996, *Geographic objects with indeterminate boundaries.* (London: Taylor and Francis).

Frank, A. U., Volta, G. S. and McGranaghan, M., 1997, Formalization of families of categorical coverages: *International Journal of Geographic Information Science,* **11**, pp. 215-231.

Hangouet, J-F. and Lamy, S., 1999, Automated Cartographic Generalisation: Approaches and Methods. *Proceedings of the 19th Int. Cartographic Conf.* Ottawa (Ottowa: International Cartographic Association), pp. 1063-1073.

Harrie, L. E., 1999, The Constraint Method for Solving Spatial Conflicts in Cartographic Generalization. *Cartography and Geographic Information Systems*, **26**, pp. 55-69.

Hole, F. D. and Campbell, J. B., 1985, *Soil Landscape Analysis.* (London: Routledge and Kegan Paul).

IBM, 1997, *OCL Parser version 0.3* Available at *http://www.software.ibm.com/ad/ocl*

Jaakkola, O., 1998, Multi scale Categorical Databases with Automatic Generalisation Transformations Based on Map Algebra. *Cartography and Geographic Information Systems*, **25**, pp. 195-207.

Jacobson. I, Booch, G. and Rumbaugh, J., 1997, *The Objectory Software Development Process* (Reading, MA: Addison Wesley).

Marriott, K. and Stuckey, P.J., 1998, *Programming with Constraints: An Introduction*, (Cambridge, MA: MIT Press).

McMaster, R., and Monmonier M., 1989, A Conceptual Framework for Quantitative and Qualitative Raster-Mode Generalization. *Proceedings of GIS/LIS '89*, Orlando, (Maryland: American Society for Photogrammetry and Remote Sensing), pp. 390-403.

Molenaar, M., 1998, *An Introduction to the Theory of Spatial Object Modelling for GIS*: (London: Taylor and Francis).

Monmonier, M. S., 1983, Raster-mode area generalisation for land use and land cover maps. *Cartographica,* **20**, pp. 65-91.

Muller, J. C., and Wang, Z., 1992, Area-patch generalization: a competitive approach. *Cartographic Journal*, **19**, pp. 137-144.

Ormsby, D., and Mackaness, W. A., 1999, The Development of Phenomenonological Generalisation Within an Object Oriented Paradigm. *Cartography and Geographical Information Systems*, **26**, pp. 70-80.

Peter, B., and Weibel, R., 1999, Integrating Vector and Raster-Based Techniques for the Generalisation of Categorical Data, In *Proceedings of the 19th International Cartographic Conference* (Ottawa: International Cartographic Association), pp. 1135-1145.

Rational Rose Corporation, 1997a, *Unified Modelling Language Specification, version 1.1 OMG documents* ad970802-ad970809 h*ttp://www.omg.com/uml/*

Rational Rose Corporation, 1997b, *Object Constraint Language Specification , version 1.1 OMG document* ad970808 *http://wwwomg.com/uml/*

Ruas, A. and Plazanet C., 1997, Stratergies for Automated Generalisation. In *Advances in GIS Research II* edited by Kraak, M. J. and Molenaar, M. (London: Taylor and Francis), pp. 319-336

Ruas, A., 1998. OO-Constraint Modelling to Automate Urban Generalisation Process. In *Proceedings of 8th Int. Sym. Spatial Data Handling,* edited by Poiker, T. K. and Chrisman, N., (Burnaby: International Geographical Union). pp. 225-235

Rumbaugh, J.,1991, *Object-Oriented Modeling and Design*, (Englewood Cliffs: Prentice-Hall).

Schylberg, L., 1993, *Computational Methods for Generalization of Cartographic Data in a Raster Environment*. Doctoral Thesis, Photogrammetric Reports No 60 (Stockholm: Royal Institute of Geodesy and Photogrammetry).

Sinton, D., 1978, The inherent structure of information as a constraint to analysis: systems of thematic data as a case study. In *Havard Papers on Geographic Information*

Systems, edited by G. Dutton. (Reading, MA: Addison-Wesley), pp. Sinton/1-Sinton/17.

Warmer, J. B and Kleppe, A. G., 1999, *The Object Constraint Language: Precise Modeling with UML.* (Reading, MA: Addison Wesley).

Weibel, R., 1997. A Typology of Constraints to Line Simplification. In *Advances in GIS Research II* edited by Kraak, M. J. and Molenaar, M. (London: Taylor and Francis) pp. 533-546

Weibel, R. and Dutton, G., 1998, Constraint-Based Automated Map Generalisation. In *Proceedings of 8th International Symposium on Spatial Data Handling,* edited by Poiker, T. K. and Chrisman, N., (Burnaby: International Geographical Union), pp. 214-224

Wilson, M. and Borning, A. 1993, Hierarchical Constraint Logic Programming, *The Journal of Logic Programming*, **16,** pp. 227-318.

13

Preserving density contrasts during cartographic generalisation

Nicolas Regnauld

13.1 INTRODUCTION

Cartographic generalisation is a two step process: an abstraction of the source data (to select or derive those required for the target product), followed by a representation process that produces a readable graphical image (a map). Research on the automation of cartographic generalisation has until recent times focused mainly on developing specific algorithms and measurements to detect, qualify and solve cartographic conflicts. The results have shown that solving basic conflicts is not sufficient to preserve the coherence of the map overall. This has led to the emergence of phenomenological studies, which focus on considering logical sets of objects rather than individual ones. Individual objects can be strongly modified (even eliminated) during generalisation, and if no phenomenological control is performed, the logical pattern of the objects can be lost, making the map inconsistent. The need of a phenomenological approach to generalisation has been discussed by various authors (Mark, 1991; Nyerges, 1991; Richardson and Müller 1991; Ormsby and Mackaness 1999). A complete discussion of geographical phenomena is presented by Hangouët (1998).

In this chapter, we focus on ways of controlling the density of buildings across a map. The density is an important source of contrast in a map. Usually, generalisation induces an increase in the density of buildings (due to enlargement), which makes it difficult to preserve all the nuances and variation in local densities. Our aim is to preserve the main differences, and preserve the logic of the pattern (for example, the closer to a town centre, the greater the density of buildings).

The first issue in preserving density variation is to detect that variation. The first section of this chapter discusses the need for high level spatial information to control automated generalisation processes. The second section discusses the different types of density measures that can be used. Then the third section focuses on tools and structures that can be used to represent or evaluate the different types of density. In the last section we discuss the different applications that this information on density can have in the field of automated cartography.

13.2 HIGH LEVEL INFORMATION CONVEYED BY TOPOGRAPHIC MAPS

Nowadays, the basic source of information describing the land cover is no longer the topographic map, but the topographic database. One of the issues is to maintain one generalist database (here we do not consider federated databases where data are stored on different databases) and from it, being able to derive different kinds of maps depending on specific needs. These needs can be addressed by considering principal themes, target scale, conditions of use, etc. A database has no scale, but a resolution. This represents the accuracy of the data, depending on the method of acquisition and possible digitisation processes. In theory this database could generate maps with the same resolution at any scale just by printing and scaling data, although this would assume that the resolution of the printer is sufficient. But even with this assumption, there is a limit of scale below which the map would become unreadable by a human reader because of their physical perception capabilities. These capabilities can be expressed by a set of thresholds specifying minimum size of objects, of detail, or minimum distance between objects for a human reader to be able to perceive and distinguish them. Thus decreasing scale generates competition for space between the objects. The generalisation is supposed to arbitrate among a set of constraints. There is a need to specify the influences which generalization decisions have on this process. Among them there are:

- Semantic priorities. The map specification can provide some information about the importance of the different themes. In addition, some order of importance between themes can be implicit. For example, roads are often considered as a very important theme, as they constitute the most usual infrastructure for people to access 'place'. Richardson and Müller (1991) and Richardson (1994) show a method to control the selection of objects depending on information of the relative importance of the different themes for the target map.
- Emergent structures. Between objects of the real world, many relationships exist. They can be as simple as the alignment of a building with a street, or far more complex such as the characteristics of the road network of a city, one that has a star shape for example. All this information can be perceived by looking at a map, but it is not explicitly stored in the database. Thus to enable the generalisation process to preserve objects, the first step is to identify them by analysing the database. For road network generalisation, Mackaness (1995) proposes a method of selecting the streets based on graph theory.
- Contrast. The contrasts are all kinds of similarity or difference that can be observed in a map. As for the emergent structures, this kind of information is not explicitly stored in the database and necessitates an analysis of the data to detect the ones we want to preserve. These contrasts can appear at any level. A bend in the middle of a straight road shows a contrast on that road. But this road can be globally non-sinuous, relative to the other roads in a mountain region. Detecting these contrasts requires a high level of characterisation of the initial data, for example using the techniques developed by Plazanet (1996).

In this chapter we study the use of density to help to control the generalisation process. Density can be studied from the points of view of both contrast and structure. Figure 13.1 and Figure 13.2 show two maps of the same area at different scales. In the smallest scale map, there are fewer buildings represented, but the differences of density are retained, and even exaggerated. In addition, on both maps, the distribution of the zones of different densities follows a characteristic structure: the most dense areas are in the city centre, they are surrounded by middle density areas, which themselves are surrounded by lower density areas, etc. There is a relation of order for the distribution of areas with regard to their density level. This means that in the analysis stage, when characterising the density of an area, it is important to takes into account where it is situated. It will force it to be homogeneous with its neighbours if its density value is close to theirs, and exaggerate the difference if it is sufficiently different.

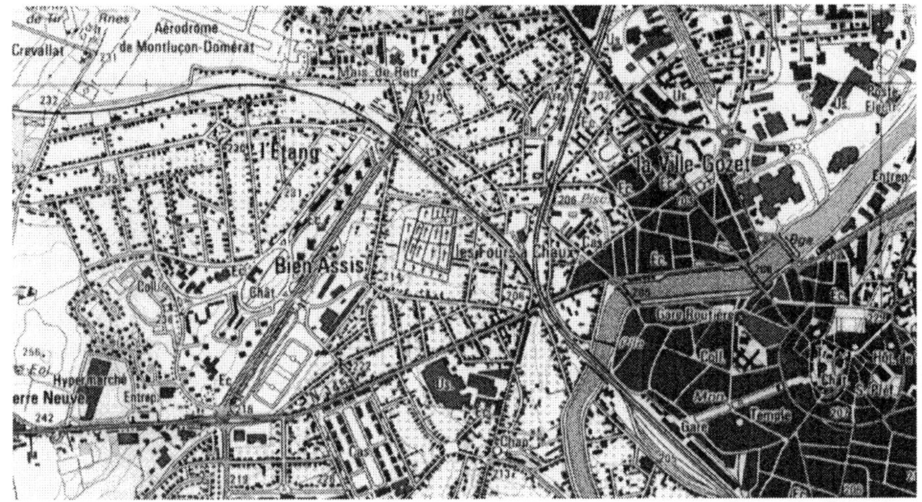

Figure 13.1 Extract of the map 2428 O from the IGN, Montluçon, 1:25,000 scale © Institut Géographique National Num. 9085

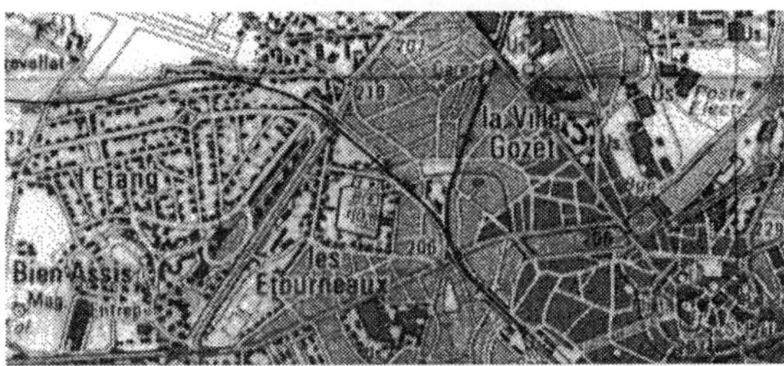

Figure 13.2 Extract of the map 2428 from the IGN, Montluçon, 1:50,000 scale © Institut Géographique National Num. 9085

13.3 DENSITY IN URBAN AREAS

Density is information that plays a key role in the perception and interpretation of a map. A specific region can most of the time be described by its density with regard to one or several themes. For example an urban area can be recognised by its high density of roads and buildings. An agricultural region can be recognised by its high density of fields, while a mountainous region can be identified by its high density of contour lines. In this chapter, we want to discuss the usefulness of density measures to control generalisation in urban area. This section focuses on describing some different ways of evaluating the density of an urban area. We explain for each one its advantages and its limits. We limit the discussion to buildings and roads, as they represent the vast majority of objects represented on maps in urban areas.

- Parcel density. A cluster of building stands on a parcel delimited by a set of connected road sections. The ratio between the cumulated area of the buildings and the area of the parcel gives density information. This is the most common way of computing density of buildings, but the result can be can be interpreted in different ways depending on the pattern of the buildings inside the parcel.
- Street frontage density. When travelling on a road, the idea of the density of the area crossed is given by the distance between the buildings along the road. A section of road between two crossroads can be qualified with regard to the density of buildings on its borders. It can be measured on each side, by taking the mean of the distances between the buildings. As it is a statistical measure, tools exist to qualify the distribution of the values and then segment the road section according to variations in the building density.

  This kind of density may seem odd, but in the context of building density, it is appropriate, as the most common type of organisation of buildings is alignment along a road (Regnauld, 1996). In the generalisation context, it can be used to control the density of buildings along a road, when the density of the parcel they belong to is low. For example, Figure 13.3 and Figure 13.4 show the same area at two different scales in the Ordnance Survey style. In the 1:50,000 scale map extract, some parcels have been completely coloured to represent the buildings, while other are coloured only on bands along the street. If we compare with the 1:25,000 scale map, we can deduce that those coloured bands are used along roads lined by a high density of buildings, but in a parcel of medium density, where free space needs to be shown.

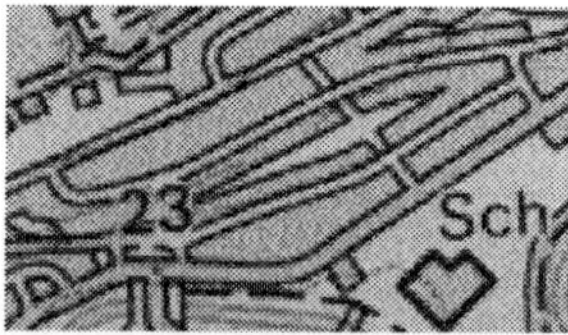

Figure 13.3 Elgin, 1:50,000, sheet 28 (Reproduced by kind permission of Ordnance Survey, © Crown Copyright MC/99-361)

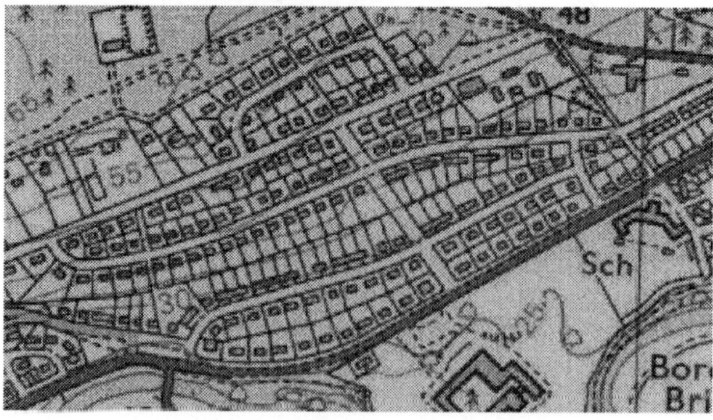

Figure 13.4 Elgin, 1:25,000, sheet NJ 16/26 , (Reproduced by kind permission of Ordnance Survey, © Crown Copyright MC/99-361)

- Building density. This kind of density gives information on the free space around the buildings. Given a region, we can compute the mean free space around each building. The free space for each building could be computed by a Voronoï diagram, which is well suited for this kind of analysis. In addition, studying the shape of the

Voronoï cell, the ratio of size between a building and its Voronoï cell and the position of the building inside gives a lot of information about the neighbourhood of a building. Ahuja and Tuceryan (1989) present the potential of the Voronoï diagram to detect different kind of dot pattern. Hangouët and Djadri (1997) give an algorithm to compute Voronoï on more complex objects than dots, and discuss the potential of the model for supporting map generalisation.

13.4 COMPUTING DENSITY

In this section we discuss how the three types of density that we have presented in the previous section (parcel density, street frontage density and building density) can be computed. The aim is not to precisely define a formula for their computation, but to present a model or structure that can be helpful for their representation. The exact computation formula can vary depending on the specific use intended for that density value.

13.4.1 Calculating the parcel density values

The parcels are created by partitioning the entire space using the linear features relevant for the application. For generalisation purposes, the most common features used (Ruas, 1995; Brazile and Edwardes, 1999) are roads, railways and rivers, omitting cul-de-sacs. To compute the value of the density, two options are possible: computing the original density value or simulating the density value for a specific map. The second solution takes the readability constraints appropriate for the target map. This should take the reduction of the usable space due to the enlargement of the road symbol (Brazil and Edwardes, 1999), and the enlargement of the size of the buildings, especially when they are small. The width of the linear objects is usually fixed for a given map, except for large rivers which may not need enlargement. The size of the building is more complex to estimate. The lowest size threshold depends on the target map, but to preserve some differences of sizes between the small buildings, some can be enlarged more than the threshold, leading to the enlargement of buildings initially larger than the threshold required to preserve the size hierarchy. A function to increase the building sizes with propagation is discussed in (Regnauld, 1999), but for our purpose, an estimation of the target size can be the maximum between the initial size of the building and the threshold for the target scale.

13.4.2 Calculating the street frontage density values

The frontage density of one side of a street can be computed by projecting each building onto the centreline of the street, and then computing the ratio between the 'built-up intervals' and the entire length of the line (see Figure 13.5). This model for representing the frontage of a street can allow further analysis, such as research on holes, or characteristics of the pattern of the buildings along the road. In addition, comparing the intervals obtained on the two sides of the road can help to detect similarities of frontage density or characteristics of facing buildings, such as symmetry.

The problem of deciding which buildings are part of the frontage of a street can be simply treated by means of a threshold distance. A threshold equal to the minimum readable width for a building of the source map seems to be adequate. Most buildings in a city will be closer to a street than this threshold, and it will avoid taking into account any building behind the first front line of the street.

Figure 13.5 Street frontage

13.4.3 Calculating the building density

The building density aims to qualify the neighbourhood of each building. These individual measures can serve to define homogeneous groups of buildings with respect to density. This is needed as a complement to the parcel density when its value is not significantly high. To qualify the space around a building, we propose two different methods:

1. The building and its Voronoï cell. By computing a Voronoï diagram on a set of buildings and roads, we get for each object a 'Voronoï garden' representing its own surrounding free space. For each building we can compute its individual density, defined by Hangouët (1998) as the ratio between the area of the building and the area of its Voronoï garden. Having computed this individual density for each building, a classification can be done. However, grouping buildings into homogeneous density groups needs to take additional information into account. The adjacency between building Voronoï gardens must be taken into account. In addition, buildings at the periphery of a group have a lower individual density, because there is usually some free space on one of their sides, demarcating the end of the group. Such buildings at the periphery of a group have a decentralised position inside their Voronoï garden. Figure 13.6 shows a Voronoï diagram on a set of buildings and a road. The group of four buildings on the left can be identified in two steps. First the three on the left have similar individual density and their Voronoï gardens are connected. Then the fourth building is connected to them, and its position in its Voronoï garden is decentralised in the direction of the group.

The main drawback of this method is that the individual density is sensitive to the size of the building. This is relevant in terms of pure density, but not for our purpose, which is to qualify the free space between the buildings in order to guide the generalisation process. An alternative indicator has been defined by Hangouët (1998) to substitute the size of the building by a combination of readability thresholds. But in this case, the indicator depends on the target scale (as do the thresholds).

2. The building and its nearest neighbour.To detect groups of homogeneous density, another approach (which is not so different), consists of linking buildings to their nearest neighbours. Regnauld (1996) present a method to link the buildings using a Minimum Spanning Tree (MST) structure, which is then segmented to define homogeneous groups with regard to the space between the buildings. Figure 13.7

shows the result of a Minimum Spanning Tree built on a set of buildings and then segmented. This result has been obtained by an automatic procedure developed in the *Stratège* platform, developed at the Cogit laboratory, at the IGN (French national mapping agency). The narrow edges show the final groups while the wide labelled ones show the edges of the original MST that have been eliminated during the segmentation. They are labelled by the order of elimination, as the process is incremental and first segments the most emergent 'holes' in the initial MST.

In each group, the density can be inferred by the mean distance between the consecutive buildings in the graph.

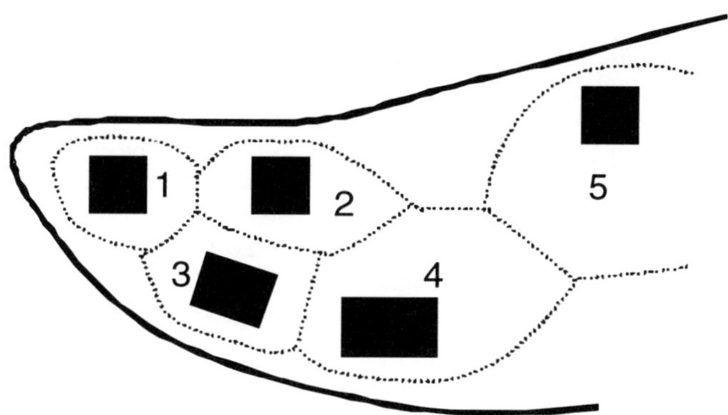

Figure 13.6 Voronoï on buildings and roads

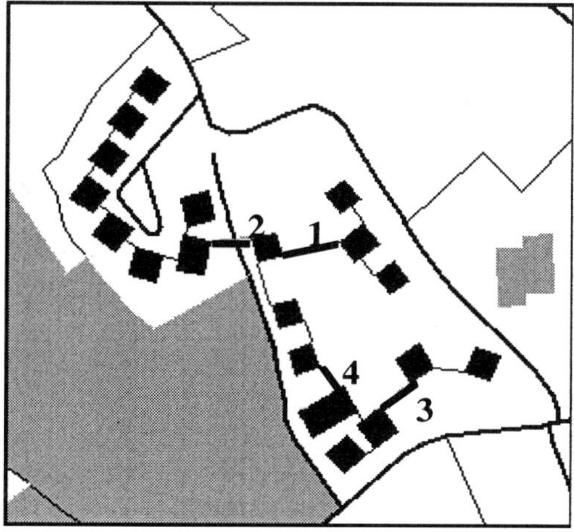

Figure 13.7 Segmented MST on buildings

13.5 USING DENSITY FOR GENERALISATION

In this section we describe the different ways of utilising density information. By discussing each of the types of density described in the two previous sections, we show the different levels of abstraction at which density information should be used to control the generalisation process.

13.5.1 Using parcel density

The parcel density can be used to maintain the differences of density in the different parts of an urban area. We cannot ensure the preservation of density differences by preserving the density in each parcel, as during reduction of scale, the density of objects drawn in the map increases considerably. We must focus on controlling the variations in density. A strict preservation of the density variations would mean that the difference of density observed on two original parcels would be the same on the target scale. However, simulating the density of a parcel at a target scale can show from slight to dramatic increases of density from the original one. The ratio between the original density and the simulated density is an indicator of the natural 'densification' of a parcel for a given change of scale. If we want to preserve the original density differences, then this indicator should be used for each parcel to control its level of object elimination.

However, controlling the densification phenomenon is not always relevant at the parcel level. The relevant level of detail varies with the scale. Generalising a map from 1:25,000 to 1:50,000 scale in urban area will induce dramatic changes. It would be irrelevant (and probably impossible) to preserve the local parcel density variations, but it is important to preserve the variation of density between the different districts of the city. For the detection of the district borders, the parcel density can be used. A three step process can be used to complete the district segmentation:

1. Classification of the parcels. Parcels should be classified with regard to their original density. For the problem of district determination, very few classes of density should be relevant (three or four classes).
2. Amalgamation. All adjacent parcels should be amalgamated.
3. Reclassification. A reclassification is then necessary to create coherent districts. Rules should be defined to reclassify isolated or small amalgams of parcels of a class into the class of one of their surrounding districts. Rules should ensure that the highest class is represented by only one district at the end of the process (the city centre).

Then for each class of density, a target density range can be defined to control the densification of each parcel in order to maintain the district coherence.

13.5.2 Using the street frontage density

The frontage density provides local information on the pattern of the buildings along a road. It can be used to supply the parcel density when the parcel is outside an urban area. In such a case, the parcel can be very large, with a very low density of buildings. However the buildings are most likely to be grouped along the roads, and the density of those groups can still be high. Figure 13.8 shows an area where the characteristic is the homogeneous density frontage for most of the streets. The frontage density for the streets at the border of the homogeneous area can help in generalising the last rows of buildings in large parcels of low density (left of the figure). The parcel density, as it is low, would not suggest the reduction of the number of buildings, which is however necessary since those buildings must stay in a small area along the street (this reduction can be noticed in the 1:50,000 map of Figure 13.8)

1:25,000, IGN 2743 ET

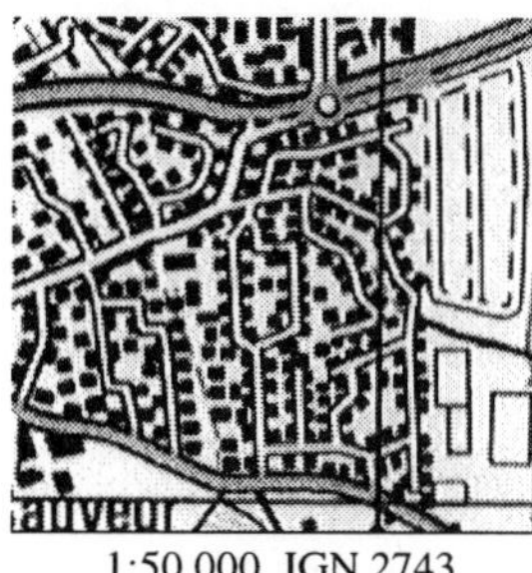
1:50,000, IGN 2743

Figure 13.8 Homogeneous front street density © Institut Géographique National Num. 9085

Furthermore, the street frontage density carries information suitable to control the similarities and differences of density in adjacent parcels. This kind of control has not yet been fully investigated as studies on roads and buildings together are still focusing on strategies to solve conflicts inside a parcel while preserving the geographical organisation of the objects (Ruas, 1999).

13.5.2 Using the building density

The building density is needed once again to supply the parcel density when it is not homogeneous, and when the street frontage density is not relevant either. This means that we need it to describe the groups of buildings inside a parcel and not along a road. Figure 13.9 shows an area where a lot of buildings are not organised right along the streets. We need a tool to detect the perceptible groups inside a parcel, to describe their internal and relative distribution. Group density is a component of this description. It should provide information on the inter-distances between buildings inside a group, and the distance between the groups. Such density information has already been used for analysing building pattern and performing some building typification (Regnauld, 1999). Some of those results are shown in Figure 13.10. The buildings have been enlarged in order to be readable at the target scale, and a selection has been done to preserve as far as possible inter and intra-group densities.

1:25,000, IGN 2743 ET

1:50,000, IGN 2743

Figure 13.9 Buildings in the middle of a parcel © Institut Géographique National Num. 9085

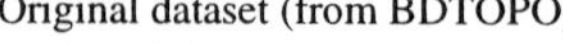

Original dataset (from BDTOPO)

Result after automatic typification

Figure 13.10 Building typification controlled by inter-distance between buildings © Institut Géographique National Num. 9085

To conclude on the usefulness of density control during the generalisation process, we can affirm that such control is needed at three levels. At the highest level, the parcel density provides the basic information needed to compute a segmentation of a large urban area into districts of homogeneous density. Then these districts can be classified according to their density level, which constitutes a constraint in order to maintain the global pattern of the city. At the lowest level, inside a parcel, some analysis should be done to detect the characteristics of the alignment of the buildings along the roads (which are the most common features structuring the building disposition), and the characteristics of those in the middle of the parcel. These characteristics include some local density information (either street frontage of building density) that must be taken into account during the generalisation. Their values influence the application of local operations of generalisation. At least, at a middle level, the frontage density can be used to check that the two sides of a road are generalised in coherent manner.

13.6 CONCLUSION

In this chapter we have shown the need for and the complexity of managing density information during the generalisation of urban areas. The need comes from the fact that nuances of density are very apparent to the map reader, so we must take care to preserve them during the generalisation process. The complexity of managing the density comes from the fact that there are several levels of density that must be managed at the same time. Density must be controlled at a high level to preserve the pattern of the city, and then at medium and low levels to preserve the pattern of the districts and parcels.

We have defined three types of density measurement that should take place in the control of the generalisation process: the parcel density, the street frontage density, and the building density. For each one we have discussed how to compute them and how they could be used. The big issue is now to combine them in a generalisation system, which needs some appropriate procedural knowledge. Acquiring this procedural knowledge is a complex issue as discussed by Weibel *et al.* (1995).

Our focus is now to integrate these propositions of density measures and control into a working system. The author is a member of the AGENT project (Lamy *et al.* 1999) which is providing a good framework just such a platform to integrate these controls. The system will manage the parcels as organisational entities in charge of controlling the generalisation of the geographical features that they contain. They behave depending on a set of constraints, and the different types of density that we have presented in this paper can be used to express constraints attached to such parcels.

13.7 ACKNOWLEDGEMENT

The author would like to thank William Mackaness for his help in correcting the manuscript. The author is very grateful for funding for this research from the European Union via the Esprit funded AGENT project, 24 939 LTR. Acknowledgement to the Ordnance Survey for access to their data through the CHEST agreement, and to IGN for access to their data under the Agent Project.

13.8 REFERENCES

Ahuja, N. and Tuceryan, M., 1989, Extraction of early perceptual structure in dot patterns: Integrating region, boundary, and component gestalt. *Computer Vision, Graphics and Image Processing,* **48**, pp. 304-356.

Brazile, F and Edwardes, A., 1999, Organizing Map Space for generalization through Object-primary Partitioning, *Workshop on Progress in Automated Map Generalization*, Ottawa *http://www.geo.unizh.ch/~fbrazile/tmp/partitions.pdf*

Hangouët, J. F. and Djadri, R., 1997, Voronoï diagrams on line segments : Measurements for contextual generalization purposes, *Proceedings of COSIT 97,* (Pittsburgh, PA: Springer-Verlag), pp. 207-222.

Hangouet, J. F., 1998, *Approches et méthodes pour l'automatisation de la généralisation cartographique ; application en bord de ville*, PhD thesis (Paris: Université de Marne La Vallée).

Lamy, S., Ruas, A., Demazeau, Y., Jackson, M., Mackaness, W.A. and Weibel, R. (1999). The application of Agents in Automated Map Generalisation. *In Proceedings of the 19th International Cartographic Conference)*, (Ottawa: International Cartographic Association), pp. 160-169.

Mackaness, W.A., 1995, Analysis of Urban Road Networks to Support Cartographic Generalization. *Cartography and Geographic Information Systems*, **22**, pp. 306-316.

Mark, D., 1991, Object modelling and phenomenon-based generalization. *Map Generalization*. (Harlow: Longman), pp. 103-118.

Nyerges, T.. 1991. Representing geographical meaning. *Map Generalization.*, (Harlow: Longman), pp. 59-85.

Ormsby, D. and Mackaness, W.A., 1999, The Development of Phenomenological Generalisation Within an Object Oriented Paradigm. *Cartography and Geographical Information Systems*, **26**, pp. 70-80.

Plazanet, C., 1996, *Analyse de la géométrie des objets linéaires pour l'enrichissement des bases de données géographiques*. PhD thesis (Paris: Université de Marne La Vallée).

Regnauld, N., 1996, Recognition of Building Cluster for Generalization. *Proceedings of the 7th International Symposium on Spatial Data Handling*, (Ohio: IGU Commission on GIS), pp. 185-198.

Regnauld, N., 1999, Contextual building typification in automated map generalisation, *Algorithmica, an International Journal in Computer Science. (*forthcoming)

Richardson, D and Müller, J.C., 1991, Rule selection for small-scale map generalization. *Map Generalization*, (Harlow: Longman), pp. 136-149.

Richardson, D., 1994, Generalization of spatial and thematic data using inheritance and classification and aggregation hierarchies. *Proceedings of Spatial Data Handling*, 2, pp. 957-972.

Ruas, A., 1995, Multiple paradigms for Automating Map Generalization : Geometry, Topology, Hierarchical Partitioning and Local Triangulation. *Proceedings of Auto Carto 12*, (Charlotte, NC: ACSM/ASPRS), pp. 69-78.

Ruas, A., 1999, *Modèle de généralisation de données géographiques à base de contraintes et d'autonomie*. PhD thesis (Paris: Université de Marne La Vallée).

Weibel, R. and Keller, S., Rachenbacher, T., 1995, Overcoming the knowledge acquisition bottleneck in map generalization : the role of interactive systems and computational intelligence. *Proceedings of COSIT' 95,* (Vienna: Springer-Verlag), pp. 139-156.

PART III

Spatial Information and Accuracy

14

Applying signal detection theory to spatial data

Brian Lees and Susan Hafner

14.1 INTRODUCTION

Extensive regions of Australia, Asia, Africa and South America, largely the regions which were not extensively glaciated during the Pleistocene, are currently suffering extensive land degradation and salinisation. In the Murray-Darling Basin in eastern Australia this problem is acute, and is reducing significantly the value of agricultural production. Managing land more effectively to minimise the impacts of salinisation is difficult without the means of mapping it cheaply and quickly. In this study, signal detection theory was applied to resolve complex spatial patterns in a satellite image of an area of intensive agriculture to try to identify areas subject to salinisation. In order to make the problem tractable and to minimise the cost of the solution, several assumptions were made. While these *tended* to be true, they were not strictly true, and a degree of error was thus introduced into the analysis. This necessitated the use of an optimising procedure which could cope with the resulting noisy data. A simple artificial neural network, using backpropagation, was used. Results were promising, despite the difficulty of the problem, suggesting that this form of spatial analysis may prove useful in other areas.

14.1.2 The Liverpool Plains

The Murray-Darling Basin is an important contributor to Australia's agricultural output. This paper describes the development of a low-cost method of providing farm-scale information on the problem of salinisation in one of the areas of current concern in the Basin, the Liverpool Plains. They form part of a highly productive agricultural area, increasingly affected by dryland salinity in the last 10 - 15 years. Much of the salinisation is not yet extreme, and is not visible as white crusts on the surface. Instead, it is evident in changes to the soil character and to vegetation vigour and health.

Cropping in the area is highly variable, temporally and spatially, as a result of opportunity, summer and winter cropping cycles, and strip and broadacre paddocks. The Liverpool Plains cover an area of 1.2 million ha.

The landscape of the Liverpool Plains has a complex history and in many places the geomorphology and stratigraphy violate many of the fundamental assumptions underpinning our most effective modelling tools. The Plains developed during a period of considerable change in the gross boundary conditions of this region. During the Cretaceous deep steep-sided north-south flowing river valleys were cut into the local Permian sandstones. The development of the major basins of Eastern Australia (Murray, Darling and Clarence) interrupted this pattern and left significant sections of abandoned river valley. The abandoned valleys here have been infilled, excavated and partially infilled over several erosional and depositional cycles.

During the Tertiary, basalt flowed westwards over the area mantling the prior landscape. As the basalts weathered, streams carried weathering products, both soluble and solid, away. During the late Tertiary and Quaternary these processes removed the basalt sheet from over much of the Liverpool Plains leaving the basalt Liverpool Range as the eroding remnant of this deposit (Figure 14.1).

The Liverpool Plains themselves constitute depositional material which has almost infilled the post Permian erosional landscape. The literature describes only two main subsurface units, a Pleistocene fluvial valley fill with its shoe-string sands, and the overlying 'basin' fill. Much of the fill comprises basalt weathering products, largely uniform clays. Whilst this may be accurate, it is too generalised for understanding the influence this basin fill may have on surface hydrology. It is important to add that the clay which infills the basin does not necessarily form an aquaclude. The combination of clays and solutes gives the upper basin fill, in many places, a hydraulic conductivity resembling that of coarse sands and gravels.

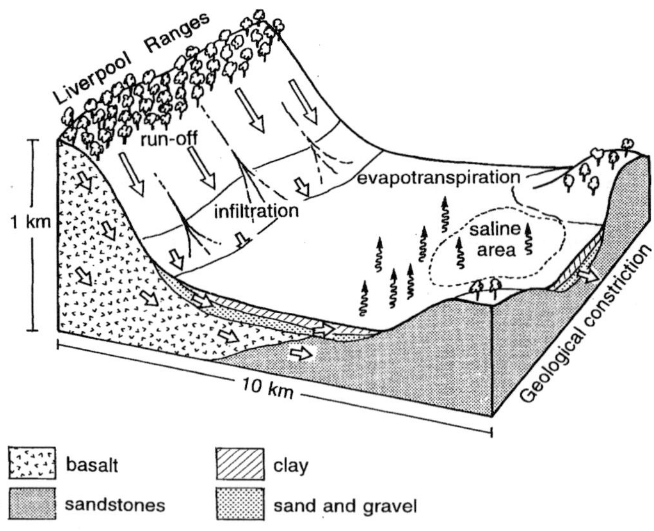

Figure 14.1 The Liverpool Plains Conceptual Model. Due to the constrictions on sub-surface flow imposed by the palaeo-landscape, 80% of the precipitation which falls on the basin is subsequently lost by evaporation and only 20% flows out into the Murray-Darling river system, thus concentrating salts produced by the weathering of the basalt in the depositional parts of the basin

There is considerable evidence that during humid times in the late Quaternary the flat areas of the Plains were extensive wetlands or shallow lakes. During arid times these wetlands and shallow lakes would have dried out and become playas. Looking west to the Willandra, where similar conditions have left a topography which has been little modified since the last of these episodes, one can see evidence of considerable and widespread reworking of the playa surfaces with extensive deflation of sediment into localised dune fields and deep gullying. It seems probable that similar reworking took place on the Liverpool Plains. With the return to wetter conditions in the Liverpool Plains, further erosion and transport of sediment from the basalt hills has led to the shallow burial of this reworked surface giving a highly variable sub-surface.

Hydrologically, the Liverpool Plains form a leaky basin. Most of the precipitation falling on the ranges and basin recharges the basin groundwater. Only about 20% flows northwards out of the basin with the balance lost by evaporation. Because of the subtle variability of the subsurface, evaporation, and the associated concentration of solutes, is highly variable across the plains.

14.2 MODELLING

On the basalt ranges the regolith to bedrock boundary tends to follow the trend of the surface topography. Within this unit the only inherited pattern is that associated with the outflow of the basalt. There appears to have been little tectonic activity in the area since the basalt was laid down, so fracturing and faulting are not significant. This provides the perfect setting for hydrological modelling of the SHE, TOPOG, or similar types of model which make the fundamental assumption that bedrock slope *tends* to parallel surface slope. These models will hold good down slope until they encounter the depositional part of the basin. Here, that fundamental assumption breaks down.

Within the depositional basin which forms the flat area of the Plains region, the continuity between any two occurrences of the boundary of a sub-surface unit cannot be assumed to be straightforward. Indeed, the natural sub-surface environment is characterised by extremely complex spatial relationships. The heterogeneities of an aquifer must be known in some detail in order to simulate or predict transport in ground water systems. However to determine these properties exactly every part of the region of interest would have to be drilled or excavated. Given this level of complexity, basin modellers have resorted to inferential modelling techniques.

The Liverpool Plains have been subjected to a wide range of geophysical investigations in attempts to map both the surface and sub-surface patterns with the aim of inferring the sub-surface processes from the surface pattern, if any. Analysing the surface patterns spatially, looking for 'connected-ness', lineaments and pattern density, allows the formation of hypotheses about the sub-surface in the same way that the Markov process models do - but knowledge of regional and district trends can be used to constrain the number of options generated.

Unfortunately, most of these techniques are too expensive to be used extensively over the wide areas of the Murray Darling Basin. The search for a practical, low-cost solution suggested that more conventional satellite remote sensing of the surface pattern is most likely to provide the basis for a suitable technique. But remotely sensed data do not by themselves provide an indication of areas at risk from salinity (Evans *et al.*, 1996) and can only detect, perhaps, the effects of salinisation on vegetation or soil.

14.2.1 Use of conventional classifiers for salinity mapping in the Liverpool Plains

McGowan *et al.* (1996) investigated the classification of single and multi-date Thematic Mapper data of salt affected land using procedures developed by CSIRO Division of Mathematics and Statistics (DMS) in Perth. The DMS methodology had been used with some success in Western Australia. McGowan *et al.* selected imagery at particular stages of the winter and summer cropping cycles from spring 1991 and summer 1991/2 when growing conditions were good (prior to the 1992-4 drought). Good growing conditions and knowledge of crops planted and their stages appear to be essential prerequisites for their methodology, because the basis of their salinity detection is in differentiating between healthy and salinity-affected vegetation.

DMS procedures were used for the calibration of multiple images to like values, canonical variate analysis, and enhanced maximum likelihood classification. The application of the methodology to the Liverpool Plains resulted in over-classification of saline areas in both single and multi-date imagery. However better predictions were possible by using a GIS to label and mask non-saline areas based on frequency of class labels. Over-classification was largely due to the inability to distinguish spectrally saline land from non-saline land with poor ground cover, which is at all times a significant proportion of the Plains. Areas under continued fallow could not be classified as saline or non-saline and the only means suggested to resolve this problem was to include additional imagery from another date in which this area was reduced. The success of the methodology is therefore dependent on the use of a set of rules to discard falsely classified saline areas or on the incorporation of imagery from more image dates with appropriate growing conditions and stage in cropping cycle. It would be fair to say that results from this project have been disappointing.

14.3 CONCEPTUAL FRAMEWORK

14.3.1 Hypothesis

Given the problems that the McGowan project identified, another approach seemed worth trying. One could look at other attributes of the signal in an attempt to resolve the complex surface patterns more cost-effectively. The exhaustive attempts discussed above all centre on analysis in spectral space.

Digital analysis of remote sensing data provides both spectral and spatial information. In many circumstances, the attribute information obtained from the analysis of spectral data is poor (see Figure 14.2). Such analyses usually require, as in the McGowan study, the addition of ground survey information to raise the quality of the attribute information to a useable level. On the other hand, spatial data from remote sensing can usually be readily transformed to high quality spatial information without the provision of very much supplementary data. If one looks at the type of analysis carried out in the McGowan study, it can be characterised as an analysis carried out in 'Spectral Space'. It seemed worth considering analysis of this data in 'Geographic Space' to investigate whether it might provide a better solution.

This view of data existing in 'domains' has proven useful in dealing with the complex analysis of datasets from disparate sources. It has been discussed before in the context of decision tree analysis of mixed datasets from GIS, remote sensing and ground point observation (Aspinall and Lees, 1994; Lees, 1994, 1996a). In these cases, the

Figure 14.2 Raw PAN imagery of the study area. The fields in the centre of the image are only 100 m wide. The image covers an area about 10 km across

analysis tools were envisaged as stepping from 'domain' to 'domain' to partition the dataset optimally. Later work with artificial neural nets envisaged this process taking place in parallel, rather than in series. The basic concept can even be extended to encompass Harvey's (1973) discussion of methodological problems at the interface of spatial and social analysis.

Putting it as simply as possible, data for, say, forest modelling, can be envisaged as existing in 'Geographic Space', 'Environmental Data Space', 'Spectral Space' and so on. Most remote sensing analyses for vegetation mapping are carried out in 'Spectral Space', with the results mapped into 'Geographic Space'. Most analyses in schema such as BIOCLIM or ISOCLIM are carried out in 'Environmental Data Space', with the results again mapped into 'Geographic Space'. Spatial analysis tools in GIS tend to operate in 'Geographic Space'. Within each of these spaces, the data has a particular set of spatial relationships. The topology of each is different although it is the same data set we are dealing with. As most of our analytical techniques involve some form of clustering, grouping, partitioning or segmentation, moving from space to space until the data can be partitioned optimally is a very useful thing to do.

There is often confusion about what can, reasonably, be used as an analytical 'data domain' or space. Most spatial analysis tools available in GIS are based on Waldo Tobler's 'First Law of Geography'. This states that 'everything is related to everything else, but near things are more related than distant things' (Tobler, 1970). The data domains are constructs within which this relationship is optimal. If Tobler's First Law is truer of our data set in 'Environmental Data Space' than it is in 'Geographic Space', then it suggests that the former is the most appropriate context for analysis.

In the case of the Liverpool Plains we set out to investigate whether we could resolve the separation of salinised land from non-salinised land by the analysis of spectral data in 'Geographic Space'. Looking at remotely sensed data over this area one is faced with a very complex spatial pattern. The signal, or spatial pattern, which relates to salinisation is only a minor contributor to this. The dominant patterns are related to the strong cultural pattern associated with agriculture. This breaks down into two main components, the broad division between agricultural land and non-agricultural land, and the detailed field pattern associated with intensive agriculture. There is also a great deal of within-field variability. This is associated with the differential responses of crops to salinisation, disease and insect attack. The latter two are highly correlated with the first. Stressed crops tend to be more vulnerable to disease and insect attack. We need to remove the strong cultural pattern, and control for crop-to-crop variability in sensitivity to salinisation, before we have a spatial pattern we can reasonably consider relates to the pattern of salinisation. This is not too different a procedure to that used in radio and radar circuitry for signal detection. In more conventional scientific terms, we need to control all these variables in order to isolate the variable of interest.

14.4 RESEARCH DESIGN

So, how can we implement this in a research design? The imposed pattern can be defined and removed if we accept some simple assumptions. The first assumption is that each field has been sown with one crop type in a uniform fashion. This is not strictly true, but *tends* to be true. The second assumption is that variation of the land cover reflectance within each field is due to the effect of soil geochemical variation on vegetation vigour, health and growth stage. This is certainly not true. Some of the variation is due to uneven sowing of the crop, storm damage, and disease or soil moisture availability. If one argues that increased salinisation of the soil *tends* to make the crops susceptible to disease, reduces their vigour and retards their growth stage (reasonable propositions) then one could argue that the second assumption also *tends* to be true. So, in simplifying the problem to make it tractable, uncertainty is being consciously introduced. The uncertainty is made up of the unknowable components of the variability, uneven sowing of the crop, storm damage or whatever.

The next source of uncertainty flows from our intention to develop a low-cost methodology that can be applied rapidly across a broad region. The methodology of the earlier remote sensing investigation relied on knowledge of the type of crops planted and their stages of growth. This is costly information to collect. If we can manage to get an acceptable mapping of areas affected by dryland salinity without going to this expense, then we have a satisfactory result even if we could have increased our accuracy or precision by doing so. So, after masking out the non-agricultural areas, a simple classification in spectral space will give a moderate number of major classes. Minor classes are folded back into the major classes and these are then treated as the major land cover types without the normal field checking and labelling procedure being carried out. The last stage in the logic of the procedure also adds a source of uncertainty. In order to remove the agricultural pattern the land cover reflectance needs to be standardised to one land cover type. To do this, relationships between the reflectance values of each land cover type and a selected reference land cover type, on soil with identical characteristics, need to be derived. As our aim is to produce a low-cost method that avoids intensive field sampling and laboratory analysis of soil characteristics, some assumptions again need to be made and some uncertainty introduced.

Given that the minimum practicable raster size we plan to use is 10 m, we can say with some confidence that the characteristics of soil in one cell *tend* to be the same as that in adjacent cells. Unfortunately, to build a land cover type to reference- land cover type relationship we cannot use adjacent cells. The fence-line between the fields of different crop or land cover types is always characterised by a line of mixels and we need to build the land cover type to reference- land cover type, or field to reference-field, relationship across this boundary. Our confidence in the characteristics of the soil remaining constant across a 40-50 m distance is very much lower. But, because we have two sources of variability here, the crop type and the soil characteristic, we need to control one to resolve the other. To build the land cover type to reference- land cover type relationship we need to accept as true the assumption that the characteristics of the soil remain constant across a 40-50 m distance even when we know that this only *tends* to be true. If this relationship were true we could construct a look up table to deal with the standardisation. Because it only *tends* to be true we need to use some optimising procedure that is tolerant of such noisy (or Fuzzy) data to obtain the best relationship between the reflectance values of the reference land cover type and each of the other land cover types in the area.

14.5 METHODS

Previous experience has shown that multi-layer feedforward neural networks using backpropagation, normally simply referred to as backpropagation, operate very effectively in this sort of situation (Fitzgerald and Lees, 1993, 1994, 1996; Lees, 1994, 1996). Whilst there are undoubtedly other ways of approaching this, backpropagation was chosen as the optimising procedure to build the relationship between the reference and non-reference land cover types.

Multi-layer feedforward neural networks using backpropagation are part of a suite of data-driven modelling techniques which are useful when the processes underlying a phenomenon are either unknown, only partially known, or would necessitate the generation of an impracticable level (scale, volume or cost) of input data. Within the suite of data-driven techniques they are useful for dealing with non-parametric data when there is insufficient data to use a more explicit technique such as decision trees. The sigmoid and hyperbolic tangent transfer functions used mean that they are rather better at dealing with fuzzy data than the crisp logic of decision trees. Both approaches give poorer results when dealing with parametric data than conventional parametric techniques and should not be seen as equivalents. The recursive nature of net training means that the introduction of spatial and temporal context is comparatively easy (Fitzgerald and Lees, 1994).

The multi-layer feedforward neural network using backpropagation is one of the commoner artificial neural net configurations in use in these sorts of application. Artificial Neural Nets are a field of quite unrelated algorithms. Many originated as projects to understand human information processing and were never intended as the analytical tools they are now sometimes seen as being. Backpropagation (Rumelhart and McClelland, 1986) is a supervised, rather than unsupervised, network. The input vectors are passed down through a multi-layered network. In the training phase, the output layer is compared to the known (or desired) output value or class associated with the input vector. If the output is in error the network weights are altered slightly to reduce the chance of this path being followed next time. If you were a Skinnerian dealing with rats, this could be described as punishing the network for its mistake. Samples are randomly drawn from the training data for as many iterations as are necessary. After a while, the network error rate will tend to stabilise and training can cease. This ability to use the training sample for as many iterations as are necessary is one of the most attractive features of neural nets.

Neural Nets are not classifiers, but tools to build classifiers. There is a common misconception that one can run, say, a maximum likelihood classification over a dataset and then compare the results with a neural network classification trained on part of the data. The two approaches are not directly comparable. The parametric classifier should give the same result each time it is run. This is not true of a neural network. Because we are dealing with a non-parametric approach one cannot make many assumptions about the data. Given a training sample which is representative of the data, and is of an adequate size, the system can learn the characteristics of the data. It is common to set the starting weights of the network randomly and select randomly from the training sample to avoid bias due to artefacts in the data. Many of the algorithms use some form of gradient descent procedure and thus can derive a 'first but not best' solution. Introducing some noise into the procedure can minimise this risk but it remains a significant problem. The procedure to avoid this is to run several training sessions, each from a randomised starting position. The resulting nets are then used to classify a representative, and independent, test sample. The net which performs optimally is then selected as the classifier. This is then checked using a third representative and independent data set, the validation sample, to give a measure of its performance. This requirement for three independent samples pushes the data demands of neural networks up significantly.

The Artificial Neural Network software used for this study was NeuralWorks v2.5. The data preparation described below relates specifically to this analysis tool.

14.5.1 Data Selection

As the crop strips in the Liverpool Plains are quite narrow (approximately 100 m in width), SPOT panchromatic data was necessary to provide adequate spatial resolution. Both SPOT multispectral and panchromatic data were obtained for the study area. The Pan data was captured on 10 November 1994.

Both datasets were registered to the Australian Map Grid, with an RMS of 0.58 for the Panchromatic data and an RMS of 0.34 for the multispectral data, using Nearest Neighbour resampling. The multispectral data was also sampled to a pixel size of 10 m in the same operation. No atmospheric correction or radiometric calibration was carried out.

14.5.2 Perform unsupervised classification

Unsupervised classification using the ER-Mapper ISOCLASS algorithm was performed on each of the Pan and XS sub-scenes. This is a 'migrating means' type of classifier. The classifications give reasonable differentiation of paddock boundaries. The single classes do not necessarily conform to crop types or stages of growth, but we tried to ensure that there were more spectral classes than crop types. Some of the smaller classes become redundant following the 'zonalmajority' step discussed below.

Six grids (1 raw Pan, 3 raw XS, and 2 classifications) were produced from the 2 classifications and 2 raw images (total of four bands).

14.5.3 Digitise paddock boundaries

Paddock boundaries were assumed to be in the centre of each mixel zone separating adjacent paddocks and were digitised from the image by bisecting the rows of mixels that separated purer signals. The regularity of vectorised paddock boundaries that emerged corresponded closely with crop boundaries visible in aerial photography over the area from September 1994 (Figure 14.3). A grid with a cell size of 10 m was created from the paddock boundary coverage.

14.5.4 Training data preparation

The training data needed to contain sufficient information so that the optimising procedure can both generalise over larger areas as well as consistently differentiate output classes from the input layers (which are likely to contain redundant or conflicting information). The training data were based on the relationships between a Pan derived reference class and other Pan classes. The multispectral data were used to prevent overspecialisation on the Pan data.

Within each paddock boundary, the majority Pan class was selected using the zonalmajority command (paddock boundary grid as zone grid and Pan ISOCLASS classification as value grid) (Figure 14.4).

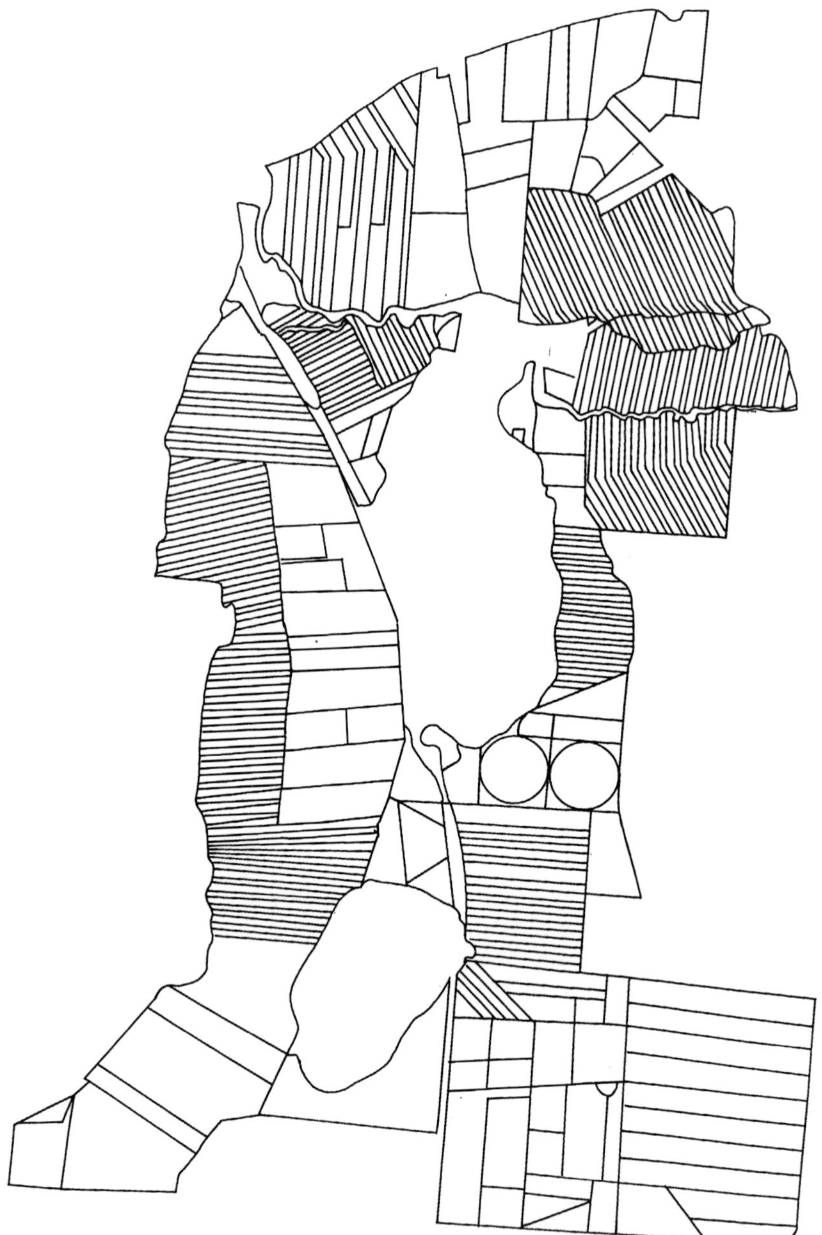

Figure 14.3 The field pattern derived from the imagery

Class 4 was then selected as the 'reference class' because:

1. It covered a large percentage of the total area
2. It was widespread within the study
3. It was adjacent to all other classes on at least one paddock boundary (this is a key requirement)
4. It covered a broad DN range (because neural networks are unable to extrapolate

outside the range of data on which they are trained)

5. It was present in various cropping patterns including those occurring in strip sequences as well as broader paddock sequences

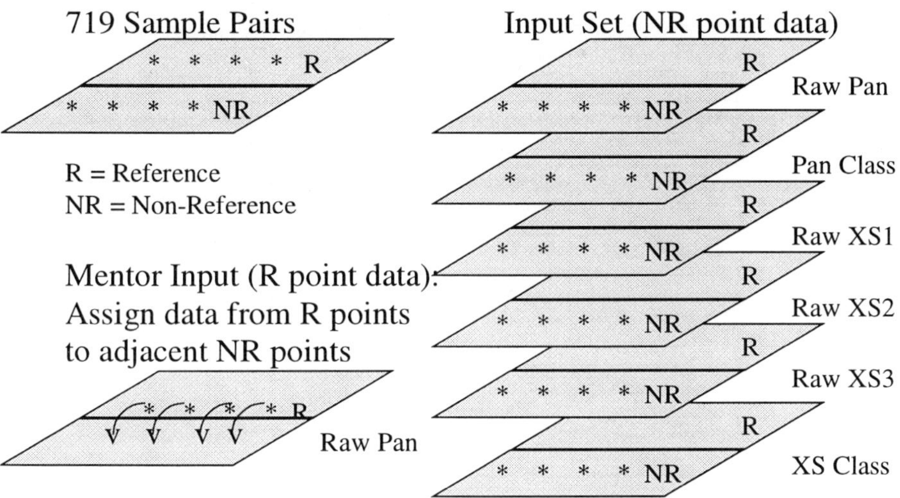

Figure 14.4 Training Data Preparation

14.5.5 Selecting sample points

There was no obvious way to automate sample point selection, due to the varied orientation and width of paddocks. Some 719 pairs of sample points were therefore selected manually against the backdrop of the zonalmajority grid and the paddock boundary coverage (Figure 14.4). Points were selected on either side of the paddock boundaries between the reference class and every other non-reference class. Care was taken to select these points as close to the paddock boundary as possible but avoiding mixels on any layer (i.e. at least 10 m from the paddock boundary).

The point coverage was converted to a raster (using Arc/Info Pointgrid and a 10 m cell size), with the value field (item) containing the point-id. A conditional statement determined whether the point-id was odd or even and two grids were created, one containing reference points (odd) and one containing non-reference points (even).

14.5.6 Making the cross-boundary assignation and frequency split

The raw Pan DN from each reference point was appended to the adjacent non-reference point. Each point-id on the reference point grid was incremented by one so that it matched the point-id of its pair in the non-reference point grid.

The DNs of the training data were then discretised into ten classes. If this had not been done we would have needed about 50 output nodes to cover the DN range of the reference grid and many would not have been represented in the training data set. Ideally, 99% of the data should be split into 10 classes of equal frequency, with the first and last half percentage of data being included in the first and last classes respectively. This is to prevent outliers influencing class boundaries. Given the large counts for the most frequent DN values, there was little flexibility in fine-tuning output class sizes.

Table 14.1 The distribution of the training sample data. The 'n' in each slice does affect training and therefore also the likelihood of prediction of output classes - the smallest classes (7 and 9) are probably under-predicted by the network

Mentor Input Class	Original DN range	Number of Samples
1	73-82	75
2	83-85	92
3	86-87	87
4	88	75
5	89	71
6	90	62
7	91	54
8	92-93	83
9	94-96	51
10	97-120	69

14.5.7 Input set

At the non-reference point locations, cell values from each of the following grids were retrieved:

1. Raw Pan (the DN range of the reference points which constituted the mentor input was 73-120, while that of the non-reference points was 66-146)
2. 10 class Pan classification
3. Raw XS1, XS2 and XS3
4. 15 class XS classification

14.5.8 Network architecture and parameters

The best architecture and parameters depend on the problem at hand and the data. The optimum architecture was set up based on previous experience (Fitzgerald and Lees, 1993, 1994, 1996; Lees, 1994, 1996) and the testing of other configurations. It consisted of:

1. one bias node for the mentor input
2. six input nodes (one for each of raw Pan, Pan class, XS1, XS2, XS3 and XS class)
3. ten hidden layer nodes and
4. Ten output layer nodes (one for each class in the mentor input which represents a

particular Pan digital number range).

Networks with fewer middle layer nodes (e.g. five) and fewer input nodes (removing Pan or XS classifications as these may have biased the network towards the agricultural pattern) were tried, however outputs seemed to either lose discrimination or there was no obvious improvement.

Initially, two groups of networks were developed. The first used the Sigmoid Transfer function and the Delta Learning Rule, while the second used the Hyperbolic Tangent Transfer function and the Normalised Cumulative Delta Rule. The second produced more promising outputs, and therefore two series of networks were based on it, which differed in their learning rates and momentum terms. Several runs were made with each configuration.

14.6 RESULTS

The aim of the analysis was to standardise the reflectance to a reference land cover type, which may or may not represent a particular crop. It was hoped that this would remove the grosser pattern, which is agricultural, leaving the more subtle patterns which are believed to correlate with salinisation. Patterns should be relatively independent of date, stage of cropping cycle and growing conditions - but not entirely vegetation independent.

The output data reflects training data design and neural network learning. The output is a normalised Pan image, not a classification, with the ten output classes representing the same DN ranges as those of the training data, that is, a ten class discretisation of the continuous raw Pan DN range.

The three networks that produced the best outputs were selected as classifiers and their outputs were field checked. These were selected because they retained a minimal pattern of agricultural features, displayed a maximum numbers of features that occur within, or cross, agricultural features, and which seemed geomorphically reasonable. One of these is shown in Figure 14.5.

Very detailed field checking of the soils and soil properties in the study area was carried out by McCarthy (1997) to determine relationships between the field sampled soil parameters and the patterns produced by the neural networks. She found three soil parameters to be important, soil structure, soil colour and soil cracking. Blocky soil structure and lighter soil colours both are associated with salinisation in the Liverpool Plains. The soil cracking is induced by crop demand for water and the presence of cracks greatly increases the amount of indirect or diffuse recharge to the shallow saline aquifer (McCarthy, 1997).

Because the field investigation examined soil characteristics, and the analyses were predicting relative reflectance values from crops which had long since been harvested, it is not possible to provide any quantitative accuracy assessment or validation of the technique. However McCarthy found that the relative *pattern* of soil structure was predicted very well. The variation in both soil colour and structure was very subtle but both an earlier reconnaissance check and McCarthy's work found a general trend between model output and soil colour. In the earlier reconnaissance investigation a correlation between apparent electrical conductivity and network values as high as -0.73 was identified in one transect. However this relationship could not be confirmed along other transects.

Figure 14.5 The output from the best of the networks. The darker areas correspond to higher salinity, the lighter to lower salinity. The residual field patterning in some places is the result of crop-induced deep cracking which allows direct recharge by fresh rainwater to the saline surface aquifer

Following the field investigations one can conclude that the model predicted the spatial patterning of salinity induced soil variability quite well, but no quantitative relationships were able to be derived. This is not unusual in this sort of investigation where the characteristics which create the patterning are physical and geochemical behaviours which are difficult to describe numerically. As the aim of the project was to extract the pattern of salinisation in as cost-effective a manner as possible, it was a success.

14.7 DISCUSSION AND CONCLUSION

Although the focus of this study has been the development of a low cost, rapid method of mapping salinity affected land in the Liverpool Plains, its true significance lies in a demonstration of the effectiveness of analysing spectral data in geographic space. As discussed in the introduction, remote sensing data provide two main types of information, attribute and spatial. The attribute information is often very poor and usually needs to be supplemented by additional, ground, data before it is useable. The spatial information, on the other hand, is usually of a very high quality but is rarely used in conventional analyses. By drawing on this spatial information content as much as possible we have managed to extract a useful result from the analyses without having to collect expensive ancilliary data.

The results of this investigation are comparable to that achieved by the McGowan *et al.* (1996) study, but were produced at a significantly lower cost. The amount of imagery required was much less, as was the associated processing load. The most significant cost-saving was the ancilliary ground data needed by the McGowan study to reinforce the weak attribute information derived from the spectral data. It seems clear that the 'data domain' schema used to provide the conceptual framework for this investigation (Aspinall and Lees, 1994; Lees, 1994, 1996a) can provide efficiencies in the analysis of natural resource datasets.

Looking specifically at the project itself, improvements in the result may be achieved if the network were trained with data that have an explicit relationship (direct or indirect) with salinity such as EM38 or depth to shallow water table data. However, this investigation set out to determine how well a very low-cost approach might perform and the results are surprisingly good despite the minimalist approach. It would certainly seem that the application of signal detection theory to help resolve complex spatial patterns has some wider promise in the land management of depositional environments where hydrological modelling based on topographic data is inapplicable.

14.8 ACKNOWLEDGEMENTS

This project describes work carried out as part of the Liverpool Plains Project by CSIRO Division of Land and Water. Susan Hafner was supported by project funds to carry out this work. Assistance from the following people is gratefully acknowledged : Dr. Joe Walker, Dr. Mirko Stauffacher, Peter Richardson, Shawn Laffan, Trevor Dowling, Peter Dyce, Ron Shamir and Dr. Bill Fitzgerald.

14.9 REFERENCES

Aspinall, R. and Lees, B.G., 1994, Sampling and analysis of spatial environmental data. In *Advances In GIS Research*, edited by Waugh, T. C. and Healey, R.G., (London: Taylor and Francis), pp. 1086-1099.

Evans, F.H, Caccetta, P. and Ferdowsian, R., 1996, Integrating remotely sensed data with other spatial datasets to predict areas at risk from salinity. In *Proceedings of the 8th Australasian Remote Sensing Conference*, Vol. 1, (Floreat, WA: Remote Sensing and Photogrammetry Association Australia Ltd), pp 18-25.

Fitzgerald, R.W. and Lees, B.G., 1993, Assessing the classification accuracy of multisource remote sensing data. *Remote Sensing of Environment*, **47**, pp. 1-25.

Fitzgerald R.W. and Lees, B.G., 1994, Spatial context and scale relationships in raster data for thematic mapping in natural systems. In *Advances In GIS Research*, edited by Waugh, T. C. and Healey, R.G., (London: Taylor and Francis), pp. 462-475.

Fitzgerald, R.W. and Lees, B.G., 1996, Temporal context in floristic classification. *Computers and Geosciences*, **22**, pp. 981-994.

Harvey, D., 1973, *Social Justice in the City*, (London: Edward Arnold).

Lees, B.G., 1994, Decision trees, artificial neural networks and genetic algorithms for classification of remotely sensed and ancillary data. In *Proceedings of the 7th Australasian Remote Sensing Conference*, Vol. 1, (Floreat, WA: Remote Sensing and Photogrammetry Association Australia Ltd), pp. 51-60.

Lees, B.G., 1996a, Sampling strategies for machine learning using GIS. In *GIS and Environmental Modelling: Progress and Research Issues*, edited by Goodchild, M.F., Steyart, L., Parks, B., Crane, M., Johnston, C., Maidment, D. and Glendinning, S., (Fort Collins, Co: GIS World Inc).

Lees, B.G., 1996b, Inductive modelling in the spatial domain. *Computers and Geosciences*, **22**, pp. 955-957.

Lees, B.G., 1996c, Improving the spatial extension of point data by changing the data model. In *Integrating GIS and Environmental Modelling*, edited by Goodchild, M. *et al.*, (Santa Barbara: National Centre for Geographic Information and Analysis), WWW, CD.

McCarthy, T., 1997, Spotting salinity with satellite imagery; the identification of soil properties associated with the incidence of salinity with SPOT imagery. Unpublished B.Sc. (Honours) Thesis, (Canberra, Australia: Department of Geography, Australian National University).

McGowan, I. and Mallyon, S., 1996, Detection of dryland salinity using single and multi-temporal Landsat Imagery. In *Proceedings of the 8th Australasian Remote Sensing Conference*, Vol. 1, (Floreat, WA: Remote Sensing and Photogrammetry Association Australia Ltd), pp 26-34.

15

Localized areal disaggregation for linking agricultural census data to remotely sensed land cover data

Alessandro Gimona, Alistair Geddes and David Elston

15.1 INTRODUCTION

An important problem in land resource analysis is concerned with finding ways of linking the available data sources, which take regard of their particular accuracy and resolution, and which present results for relevant spatial frameworks (Walker and Mallawaarachchi, 1998). This problem is particularly highlighted when the input data are drawn from census surveys, for which data are generally provided for irregular and arbitrary tracts, since often these do not match the geography of environmental resources. GIS-literate quantitative geographers in UK universities have paid considerable attention to studying this problem, and indeed their knowledge is now proving valuable in planning outputs for the 2001 population census (see, for example, Martin, 1997; Martin, 1998). It is thus somewhat ironic that the criticisms which have been levelled at the annual agricultural census by their predecessors and peers for more than a quarter century (Coppock, 1960, 1965; Clark, 1982; Robinson, 1988) remain valid today. Whereas the 'GIS revolution has changed the entire context in which users see and use [population] census data' (Openshaw, 1995, p.133), the agricultural census (which records crop/non-crop areas and livestock and labour numbers) has been 'made GIS-able' very much as an incidental to the ongoing production of annual 'text-and-table' summary reports (see, for example, MAFF, 1997; SOAEFD, 1998). This is achieved by the attachment of farmers' returns summed as parish-level totals to a parish boundaries coverage.

Recently, however, changes underway in the EU's financial support for agricultural production have stimulated initiatives to gather more detailed, digital, agricultural spatial data. For example, to assist with the checking and administering of area-based production subsidies, The Scottish Office Agricultural Environment and Fisheries Department (SOAEFD) has undertaken an extensive data conversion exercise over the last five years, which has required arable growers, eligible for payments, to submit field plans (based on large-scale OS mapping) for their holdings. Since SOAEFD also has responsibility for gathering data for the agricultural census (henceforth referred to as the JAHC - short for June Agricultural and Horticultural Census), there is an obvious internal overlap of effort, but as yet no indications of any major strategic plans to link or merge the census with this new database.

The work described here is set within this changing policy context. It stems from the notion that a more useful, sub-parish national picture of agricultural production could be obtained from the JAHC by linking it to a more detailed land cover map database. This was created earlier by the Macaulay Land Use Research Institute (MLURI) (the Land Cover of Scotland 1988 or LCS88 for short; MLURI, 1993) by interpretation of aerial photography. Using arable crops as an example, the question at the outset was "how to apportion the JAHC parish totals to those areas of the parish that have classified as arable areas within the LCS88 map database?"

In this paper, a disaggregation procedure is described, which was developed to tackle this question, by extending a quadratic programming method proposed for a land use study within the Tyne catchment in north east England (Allanson *et al., 1992*; Moxey *et al., 1995*). The method also draws on data from MLURI's Land Capability for Agriculture (LCA) survey, which gives a 7-class ranking of the land's potential suitability for crop-growing across Scotland (Bibby *et al., 1982*). In the following sections, attention is paid first to the main characteristics and comparison of each of the three survey data sets - an analysis which led to the conclusion that it was more relevant to concentrate on the joint spatial variation between the JAHC and the LCA using LCS88 as an ancillary data set. The procedure is then introduced in theoretical terms, and the stages of pre-processing carried out on the data are also described. Finally, two empirical examples are used to illustrate the results.

15.2 DESCRIPTION AND COMPARISON OF THE STUDY DATA SETS

This section describes briefly the data sets involved in the disaggregation and reports the results of a preliminary analysis. This provided the rationale for the creation of 'super classes' of land cover and for the use of LCS88 as an ancillary data set.

The origins of the JAHC presently administered by SOAEFD can by traced back to the first census carried out over Great Britain in 1866, before responsibility was devolved to the (then) Scottish Agricultural Board in 1912 half a century later (Clark, 1982). It is conducted by postal survey of the occupiers of all farms which SOAEFD recognises as 'main holdings'. Distribution of the forms is scheduled to coincide with the start of the census period at 1 June, and under the Agricultural Act 1947, the occupiers are legally obliged to complete the form accurately and to return it within two weeks. In recent years, the forms have included around 200 questions asking for the areas of crops, grass and rough grazing, the areas of set-aside ground and land under other uses, the numbers of livestock, and the number of full- and part-time staff.

Confidential treatment of the collected JAHC data has been of long-standing importance, to preserve the privacy of individual occupiers, and to help maintain a high response rate. The data are therefore only available in aggregated form, normally at the parish level.

Since the parishes vary considerably in size and shape, problems are posed when attempting to perform between-parish comparisons of the JAHC statistics. Amongst the 891 parishes, the range in area is from 1.3 ha (Cramond Island in the Forth estuary) to

112,019 ha (Kilmonivaig), with a size distribution which is positively skewed. Therefore, the degree of holding-to-parish aggregation, and hence the level of generalisation, varies depending on the size and shape of the holdings relative to the parishes within which they are incorporated. Furthermore, if the data are investigated using quantitative methods such as correlation or regression analysis, the sensitivity of the results to the definition of the zonal boundaries should be taken into account. The findings from several investigations which have explored this condition, known as the modifiable areal unit problem, point to the conclusion that it would be entirely possible that quite different, often unpredictable, relationships might be found between variables by altering the parish boundaries, or by aggregating them into different groups.

In contrast to the parish summaries which cover only agricultural land, the LCS88 is a complete geographical census of the entire Scottish land area interpreted from a mid-level (1:24000) air photography survey, the majority of which was flown in 1988. It categorises the bio-physical characteristics of the land surface according to a custom-designed land cover classification hierarchy which distinguishes principal, major and main cover features, sub-categories, and a spectrum of over 1,000 dual category mosaics. In comparison to other British / Scottish land cover surveys, the LCS88 is distinctive for its large spatial scale, high spatial resolution, and the extensive error-checking and validation procedures which have been applied to it. Further details about these are documented in the main report (MLURI, 1993) and have been summarised elsewhere in the GIS literature in relation to other studies (Aspinall and Pearson, 1995). The LCA represents the lowest of capability classifications derived from climate, soils and topography data. It offers three hierarchical levels of information, namely the class, division and unit, which are aimed at suiting national, regional, local or farm planning needs. Classes 1 to 4 are considered suited to arable crops, Class 5 only to improved grassland and rough grazing, Class 6 (covering 60% of the national land area) only to rough grazing, whilst land with extremely severe limitations which cannot be rectified is allocated to Class 7 (Bibby *et al., 1*982). Following requests from the then Agriculture Department, the classification was mapped and published on quarter-million scale OS base mapping in the early 1970's, and later on a series of 1:50 000 maps covering the main arable areas. Areas non suited for agriculture were given a class number greater than 7. We refer to Class 7+ in this paper to indicate all land with class label 7 or more. Subsequently, a database file of point samples was created from these maps by reading off the class and division ranks at the one-kilometre National Grid intersects. From the point sample file the national LCA one-kilometre point map of the top-level classes was created using GIS. For each parish, the proportion of points belonging to a LCA class was then used to estimate the proportion of the parish areas belonging to that class.

To provide temporal consistency with MLURI's land cover survey, only the summaries from June 1988 were used in the disaggregation stage.

In order to make LCS88 and JAHC comparable and to be able to investigate the use of the former as an ancillary data set, a common land cover classification was devised. This consisted of the aggregation of their land cover and crop types (respectively) into 'super-classes'. Table 15.1 provides their names and descriptions of the LCS88 features and JAHC variables from which their area totals were derived. Candidate LCS88 categories were selected in consultation with a MLURI colleague who had worked on the LCS88 photo-interpretation (Towers, 1998), and in order to simplify work with the classification system only the primary codes of the mosaic features (representing the more dominant of the cover types in the mosaic) was considered during the allocation. Note that the fourth type, 'Other land', is necessary within the disaggregation procedure to allow the relative apportionment of areas of land within each LCA class.

Overall, the LCS88 estimates proved to be strongly correlated with the JAHC areas for each summary group (returning Pearson *r* values of 0.875 for Arable, 0.885 for Improved Pasture and 0.911 for Rough Grazing). However, there are clear geographical influences in the nature of discrepancies between them. The area of arable land estimated from LCS88 is generally larger than the JAHC estimates for the smaller parishes situated on the fertile lowland fringe which runs from the Borders around the east coast up to Inverness. By contrast, in the larger parishes bordering the west side of this fringe - where physical conditions for crop-growing are poorer - the ratios are reversed.

Table 15.1 Derivation of the summary land types for which disaggregations have been performed

JAHC Variables	⇒	Summary Land Type	⇐	LCS88 Categories†
Total crops & fallow; Unspecified crops; Grassland for grazing under 5 years old		Arable		Arable land (4)
Grassland for mowing 5 years old or older; Grassland for grazing 5 years old or older		Improved Pasture		Improved pasture (8)
Rough grazing		Rough Grazing		Dry heather moor (8); Wet heather moor (8); Undifferentiated heather moor (12); Smooth grasslands (8); Undifferentiated smooth grasslands (4)
Total area of parish‡ *minus* Σ Area of Arable, Improved Pasture and Rough Grazing estimates		'Other'		Water features, developed land and land used for recreation, transport and industry (26); woodland and scrub (14); Bracken (4); Bog, marsh, maritime grass, wetlands, dunelands and undifferentiated montane vegetation (19)

[† The number of LCS88 single interpreted features covered by each description is shown in the brackets; ‡ Parish areas were derived from the digitised boundaries data set available for this study.]

Notwithstanding the problems associated with matching up the class and variable definitions, these differences point to (a) feature interpretation errors in the LCS88 resulting from the difficulties of distinguishing different cover types from the photographs and (b) form-filling errors associated the JAHC. We suspect that thc LCS88 Arable estimates in lowland parishes are mainly attributable to the mis-interpretation of fields of improved grassland because of their visual similarities, and indeed the possibilities of this having occurred were acknowledged during the creation of the data set (MLURI, 1993). Conversely, in the upland parishes where cropping is less widespread and livestock farming is more prevalent, it is more likely that arable land covers were mis-classified as grassland types. Such trends merit a more systematic investigation which could perhaps exploit earlier efforts to quantify the category confusion for the LCS88. However, such an analysis was beyond the scope of the current paper.

Almost the opposite pattern of differences was found for the Improved Pasture totals. Generally speaking, the LCS88 estimates of Improved Pasture are relatively smaller than the JAHC figures in the smaller, lowland arable parishes in the eastern part of the country, whereas the ratios are reversed in the larger upland parishes. Again, this would suggest that fields of improved grass have been mis-classified as arable land in the former area; elsewhere it seems that improved grass may have been under-interpreted, most likely due to its confusion with rough grazing and moorland. However, census-taking errors may also be quite important in this example, especially with regard to holdings in upland parishes where the distinction between what is considered as improved and unimproved grassland may be less clear. Additionally, their relative areas may be known with less precision in the absence of clearly demarcated field boundaries.

The relative sizes of the Rough Grazing totals were found to be highly variable from one parish to the next. This is not surprising given the spectrum of agriculturally poor land covered by this category, and the correspondence between the data may vary as much with the farmers' interpretation of rough grazing as anything else. On the 1988 census form, farmers are asked to estimate this as the sum of area covered by mountain, hill, moor and deer forest situated within the farming unit, whether enclosed or not. Compared to fields used for crops and improved grassland which tend to be clearly bounded and hence easier to measure (from maps or direct measurements), such areas are known to be difficult to ascertain. There is no LCS88 Rough Grazing category, so we attempted to reflect this diversity by grouping several moorland and semi-natural grassland cover types for the LCS88 estimate (see Table 15.1). This of course ignores the distinction between individually- and communally-owned grazings, whereas the census question specifically asks the farmers to exclude any common share of such grazing that they may have.

These comparisons show that the LCS88 area estimates are correlated with the JAHC summaries for the Arable and Improved Pasture, but much less so for Rough Grazing. Also systematic differences between the two data sets have emerged. For this reason, and because the areas of LCA classes were expected to be less variable in time, it appeared more useful to use LCA classes as target zones for our disaggregation.

15.3 DISAGGREGATION PROCEDURE

The principal objective was to generate disaggregated estimates of the areas under different farmland uses *within* the parishes. These estimates had to satisfy two requirements, namely (a) to preserve the JAHC parish summary totals whilst (b) to capture the sub-parish spatial variation in actual or potentially suitable land cover types identifiable from the LCS88. The method was developed on the new variables created reported in Table 15.1.

The approach used in this study was localised in the sense that disaggregated area estimates are generated separately for small groups of neighbouring parishes, identified using a cluster analysis divided in two steps The method devised used a constrained quadratic minimisation algorithm solved by iterative proportion fitting to adjust a matrix of estimated weights such that they match, for each parish group, the total areas of the summary types as recorded in the JAHC (giving the column totals), *and* the relative amounts of land within each of the LCA classes (giving the row totals). As such, it was similar to the method applied for the River Tyne catchment in north-east England (Allanson *et al., 1*992). However, whereas the focus of that study was on a relatively small pre-defined area, this study covers the whole of Scotland, and therefore has had to take account of much wider geographical variation in the environmental conditions that affect agricultural land use patterns. It was preferable therefore to avoid producing a single set of 'global' weights for the whole country, as these would be very generalised and would tend to 'smooth' out interesting local variations in farmland use. Cluster analysis was therefore used to identify existing regional differences. Working instead on the collective summaries for statistically similar parishes has merit for two reasons. Firstly because the disaggregation parameters for these parishes are likely to be similar. Secondly when the parishes are also geographically adjacent, as is often the case, this is expected to minimise the holding-to-parish land allocation errors inherent in generating the JAHC summaries, assuming that land 'lost' from one parish is 'gained' by a neighbour.

The sequence of processing is depicted in Figure 15.1. Cluster analysis was performed in two stages. Both clustering stages were carried out using SAGE, a suite of tools which has been developed recently for performing exploratory spatial data analysis on ARC/INFO polygon coverages (Haining *et al., 1*998). The first stage produced nine main regions. The second stage split each of these regions into several sub-regional groups which form the basis for the disaggregation procedure.

The nine regions were identified by classifying the parishes using the entire set of 1988 crop and grassland area variables recorded in each parish summary (30 variables in total). Each input variable was expressed as a proportion of the total agricultural area in the parish. Each resultant agricultural region was processed separately in the second-step of the cluster analysis. In this case, the objective was to find sub-region groups of eight to ten parishes having (a) a heterogeneous mix of distribution of land cover classes in LCA classes, so that the disaggregation procedure could achieve a good degree of spatial discrimination, and (b) which were as geographically compact (see below) as possible. In SAGE terminology, this is expressed as a 'regionalisation' problem (in that spatial proximity as well as agricultural similarity are aimed for). This was obtained by using *k*-means classification. The LCA profile for each parish were derived from a LCA-LCS88 cross-tabulation. These were included to evaluate the former criterion, with the objective function being to minimise the difference between the total value for all zones in the parish group and the mean total for all other groups.

Geographical compactness was evaluated by a simple measure of the proximity between the parish centroids (Wise *et al.*, 1997), the co-ordinates of which had been extracted from the digital boundaries coverage.

The crosstabulation between LCA and LCS88 provided an estimate of the expected proportions of each LCA class into each of the super-classes devised in this study. These proportions were subsequently used as starting values and constraints in the estimation of our model parameters (see penalty function below). The disaggregation problem consisted of finding sets of parameters that represent the proportion of each LCA class to be allocated to each super-class, within a sub-regional parish group. Following Allanson *et al.* (1992), the general form of the problem is illustrated in Table 15.2. The notations LCA_{ip} and $AIRO_{jp}$ represent, respectively, the areas of land falling within each LCA class ($i = 1,\ldots,M$ where $M = 7$) and the totals for *Arable, Improved Pasture, Rough Grazing* and *Other* from the JAHC summaries ($j = 1,\ldots,N$ where $N = 4$) for each parish p

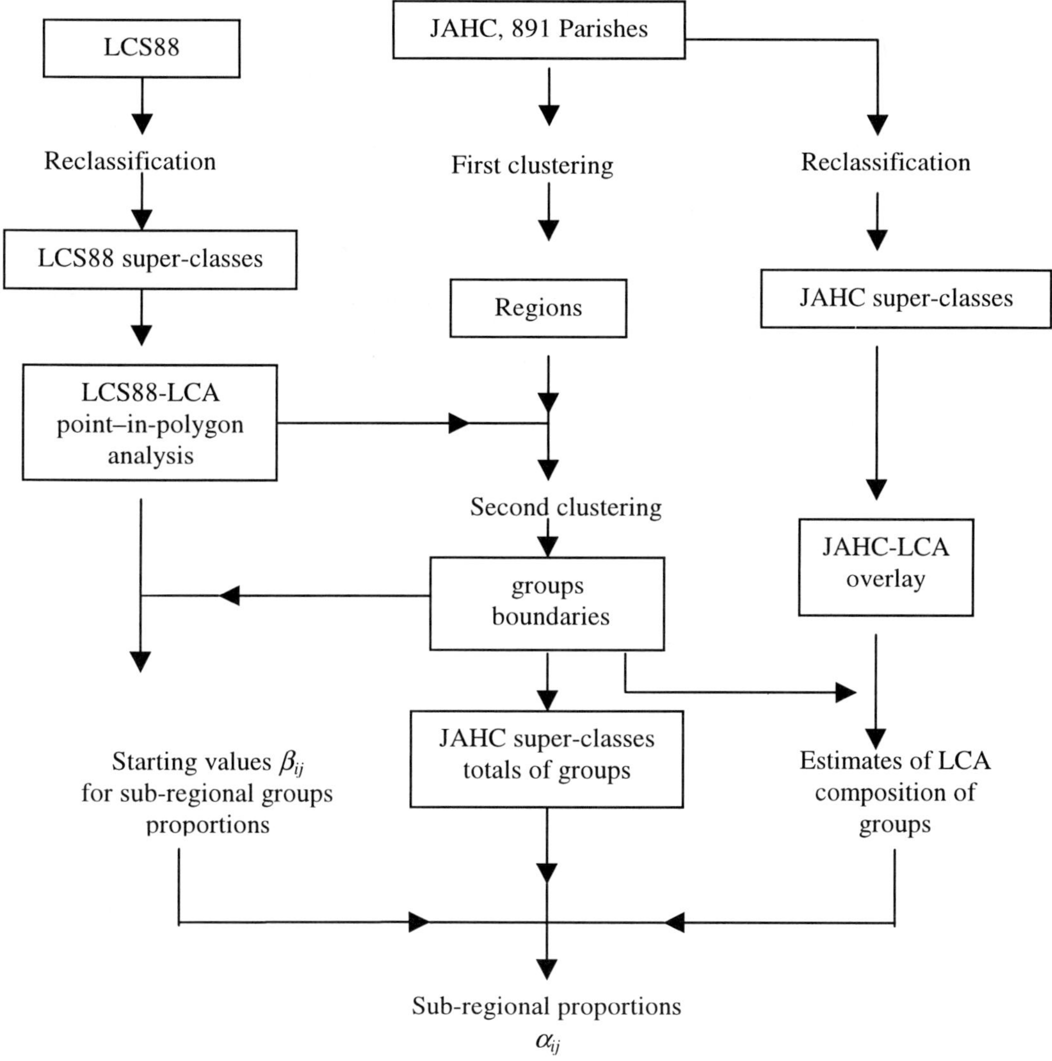

Figure 15.1 Schematic illustration of the disaggregation method

(p = 1,…,P) in the parish group. For each LCA_{ip}, a vector of proportions α_{ij} must be estimated, indicating how the area of land should be allocated across the summary types such that their total areas (expressed in hectares) summed for the parish group are preserved, and the proportions of land within each LCA class in the group sums to unity.

Convergence to a solution was based on minimising the differences between the predicted and observed summary areas for each group. Some parish groups were small, potentially giving rise to an ill determined problem. This is where the relationships that exist between the summary totals, derived using the LCS88 and the LCA class, was most useful. The individual cell values in the LCA-LCS88 cross-tabulation can be considered as offering prior knowledge about the, that is, to provide a first estimate of their value.

Therefore, they could be treated as starting values (denoted as β_{ij}) for the α_{ij}. To exploit this prior knowledge, a 'penalty function' ,W_{LCS88}, was added to the function to be minimised. This was equal to the sum of the squared differences between the prior estimate and the final estimated value for each parameter. This function aimed to minimise the differences between the estimated and starting parameters, α_{ij}, while satisfying the minimisation constraints. Formally, the expanded function is expressed in Equation (15.1) as:

Minimise

$$\Sigma_p \Sigma_j (AIRO_{jp} - \Sigma_i \alpha_{ij} * LCA_{ip})^2 + k * W_{LCS88} \quad (15.1)$$

subject to:

$$\alpha_{ij} >= 0$$

and

$$e = 0$$

where:

$$e = \Sigma_j (\Sigma_p AIRO_j - \Sigma_p \Sigma_i \alpha_{ij} * LCA_{ip})^2$$

In Equation (15.1) $W_{LCS88} = \Sigma_i \Sigma_j (\beta_{ij} - \alpha_{ij})^2$, denoting the differences between the estimated and starting proportions summed for all the parishes in the parish group; k is a constant allowing for different weights to be assigned to W_{LCS88} (equal weights were assigned to either side of the + sign in Equation (15.1), so $k = 1$).

Table 15.2 The disaggregation problem represented as a two-way frequency table of land area distribution by LCA class and summary group

$\alpha_{11p}\ LCA_{1p}$	$\alpha_{1jp}\ LCA_{1p}$	$\alpha_{1Np}\ LCA_{1p}$		LCA_{1p}
$\alpha_{i1p}\ LCA_{1p}$	$\alpha_{ijp}\ LCA_{ip}$	$\alpha_{iNp}\ LCA_{ip}$		LCA_{ip}
$\alpha_{M1p}\ LCA_{Mp}$	$\alpha_{Mjp}\ LCA_{Mp}$	$\alpha_{MNp}\ LCA_{Mp}$		LCA_{Mp}
$AIRO_{1p}$	$AIRO_{jp}$	$AIRO_{Np}$	......	A_p †

[† A_p is the total area of all parishes in the group.]

The problem was operationalised efficiently as a routine in a spreadsheet-based specialist optimisation package (Lindo Systems Inc., 1998). The final estimates α_{ij} quantify, for each LCA class, the proportion of the total class area to be assigned to each of the four summary land types.

15.4 RESULTS AND DISCUSSION

The total crop areas of the JAHC statistics for each sub-regional group were successfully reproduced by the disaggregation procedure, while spatial resolution was increased. Making the assumption that the class associated with each LCA sample point is representative of the one-kilometre (100 ha) square around the point, gridded raster surfaces can be generated for each of the four summary land types after multiplying the total area covered by a class in the parish group by its α_{ij} proportion (a less assuming alternative would be the production of a point lattice, which could be smoothed using kernel processes. This, however, may not preserve the parish totals).

Examples of mapped results for disaggregated JAHC classes are presented in Figures 15.2 and 15.3. These give an idea of the spatial pattern of the disagregated land cover. A more precise picture is obtained by examining the estimated parameters α_{ij}, that is, the proportions of LCA classes to be allocated to each of the four land cover "super-classes". An example of these proportions is shown in Table 15.3, which refers to the results for Region 1.

The raster surface representing the national mosaic of Arable production in Figure 15.2, clearly shows the concentration on the eastern lowland belt. Even in this zone, the transition to decreasing production is identifiable as progress is made up valley sides and onto hilltops. Production intensity appears highly localised inland of the north-east Banff and Buchan 'nose' and in the Central Lowlands, probably reflecting the mix of cropping and livestock farming in the former area and the competing uses of good quality land for urban and industrial uses in the latter.

The contrasts with the distribution pattern of *Improved Pasture* are shown in Figure 15.3. Improved Pasture occupies a greater proportion of each square kilometre in the warmer, wetter south-west, and notably also on the Orkney Isles. The large scale inset covers a section which details the decreasing production gradient from the coastline in the north-east transitionally along the coastal plain and the River Spey valley, running south-westwards, up to the northern limits of the Cairngorm and Grampian ranges.

Although local estimates could be obtained, it is problematic to obtain a global estimate of the spatial error of the disaggregation. This is because there is no data set that can truly and rigorously be regarded as 'correct' for the whole of Scotland. Sampling and misclassification errors, in fact, affect any data set collected by field survey or with remote sensing technology, especially when the extent is large. However, the results. appear consistent with topography derived from an Ordnance Survey digital elevation model. *Arable* and *Improved Pasture* areas have not been allocated to hill sides or mountains, and river valleys have contrasting values with the surrounding hills, (i.e. valleys have a higher concentration of pastures and agricultural land, see for example the inset in Figure 15.3). This is consistent with what it is reasonable to expect.

15.5 CONCLUSIONS

The method developed in this study can be used to integrate data sets based on administrative regions with higher resolution data, in particular raster data sets, including remotely sensed data.

Figure 15.2 Map of results for Arable. The brightness is proportional to hectares of Arable per 100 ha grid cell

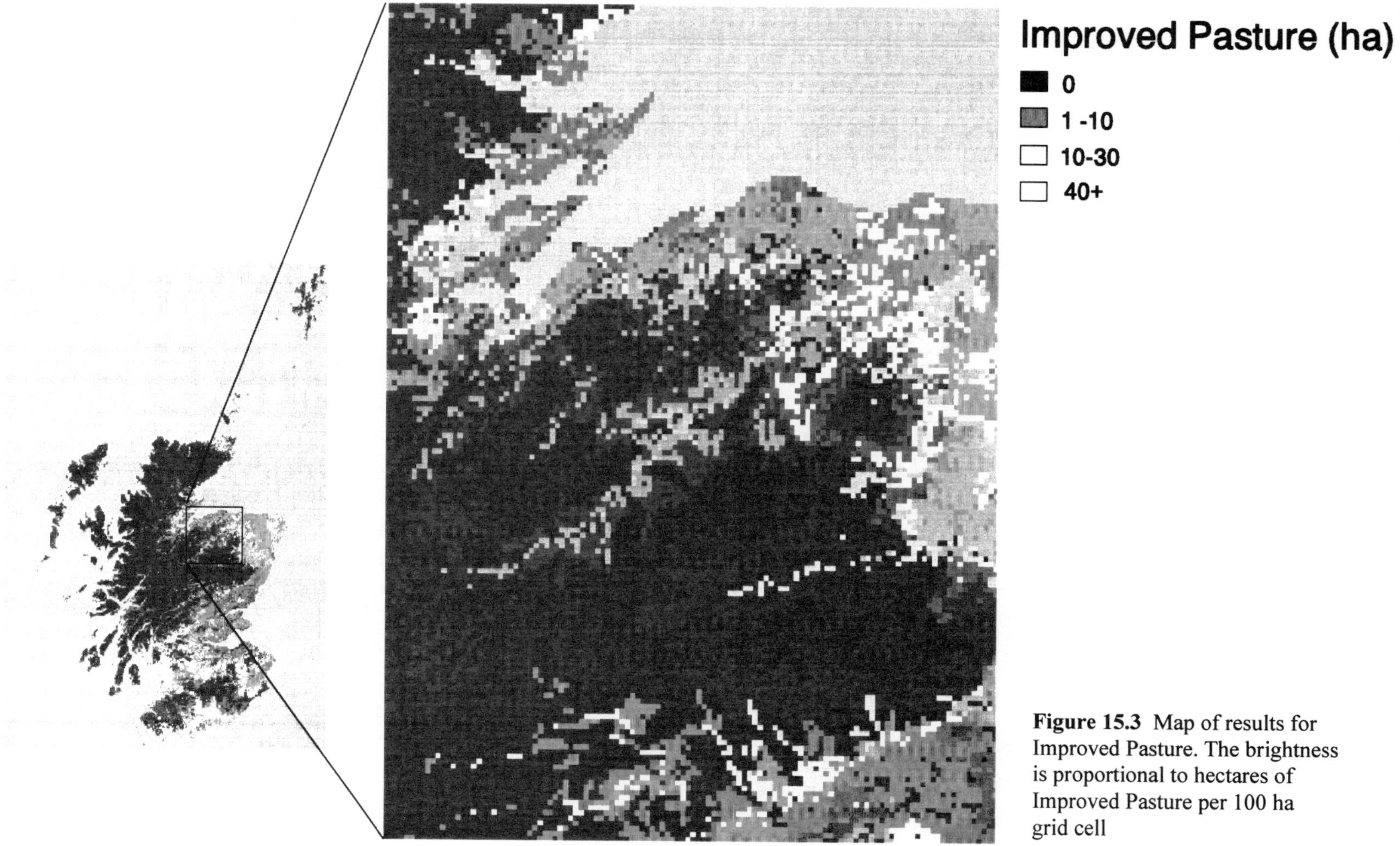

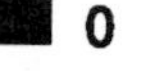

Figure 15.3 Map of results for Improved Pasture. The brightness is proportional to hectares of Improved Pasture per 100 ha grid cell

Table 15.3. Estimated proportions (α_{ij}) for Region 1. NA indicates that land in the corresponding LCA class is not represented in the region.

Sub regional group	*LCA class*	*Arable*	*Imp.past*	*Rough Grazing*	*Other*
1	1	NA	NA	NA	NA
1	2	NA	NA	NA	NA
1	3	NA	NA	NA	NA
1	4	0.014049	0.756659	0.070452	0.158841
1	5	0.071445	0.038249	0.574854	0.315451
1	6	0	0	0.128211	0.871789
1	7+	0.016786	0.103492	0.344469	0.535253
2	1	NA	NA	NA	NA
2	2	NA	NA	NA	NA
2	3	NA	NA	NA	NA
2	4	0.011598	0.243957	0.495462	0.248982
2	5	0.030255	0.200698	0.344316	0.424732
2	6	0.013406	0	0.294203	0.692391
2	7+	0.006743	0	0.630287	0.36297
3	1	NA	NA	NA	NA
3	2	NA	NA	NA	NA
3	3	NA	NA	NA	NA
3	4	0	0.310006	0.367138	0.322856
3	5	0.019221	0.045188	0.433797	0.501793
3	6	0.00892	0.009709	0.255022	0.726349
3	7+	0.003441	0	0.574845	0.421714
4	1	NA	NA	NA	NA
4	2	NA	NA	NA	NA
4	3	NA	NA	NA	NA
4	4	0.105982	0.409247	0.484772	0
4	5	0.083886	0.095976	0.318818	0.50132
4	6	0.032615	0	0.466095	0.50129
4	7+	0.022579	0.054816	0.390882	0.531723

This study tackled directly some of the long-standing drawbacks connected with limited spatial resolution of the agricultural census and the widening discrepancies between the parish system and the true geography of farming activity in Scotland. It exploited the official figures on actual production provided by the JAHC alongside knowledge of geography of limiting factors provided by the LCA and the comprehensiveness and relatively high spatial resolution of the LCS88.

The parishes provide a poor template for many researchers interested in using the JAHC, and many other rural survey data, and therefore the grid values which were output provide a more convenient 'zone neutral' framework for studying spatial variations in agricultural production which can be compared or combined with other data in a GIS.

15.6 ACKNOWLEDGEMENTS

The authors acknowledge the Scottish Office Agriculture, Environment and Fisheries Department (SOAEFD) for financial support and for the provision of data from the agricultural census. This paper is one of the output results of the SOAEFD funded project "Integration of land cover and agricultural information".

15.7 REFERENCES

Allanson, P., Savage, D. and White, B., 1992, Areal disaggregation of parish agricultural census data. In *Land Use Change: its Causes and Consequences*, edited by Whitby, M.C., (London: HMSO), pp. 82-100.

Aspinall, R.J. and Pearson, D.M., 1995, Describing and managing uncertainty of categorical maps in GIS. In *Innovations in GIS II*, edited by Fisher, P., (London: Taylor and Francis), pp. 71-83.

Bibby, J.S., Douglas, H.A., Douglas, Thomasson, A.J. and Robertson, J.S., 1982, *Soil Survey of Scotland. Land Capability Classification for Agriculture*, (Aberdeen: The Macaulay Institute for Soil Research).

Clark, G., 1979, Farm amalgamations in Scotland. *Scottish Geographical Magazine*, **95**, pp. 93-107.

Clark, G., 1982, *The Agricultural Census - United Kingdom and United States*, (Norwich: Geo Books).

Coppock, J.T., 1960, The parish as a geographical statistical unit. *Tijdschrift voor Economische en Sociale Geografie*, **51**, pp. 317-326.

Coppock, J.T., 1965, The cartographic representation of British agricultural statistics. *Geography*, **L(2)**, pp. 101-114.

Dragosits, U., Sutton, M.A., Place, C.J. and Bayley, A.A., 1998, Modelling the spatial distribution of agricultural ammonia emissions in the UK. *Environmental Pollution*, **102(S1)**, pp. 195-293.

Haining, R.P., Wise, S.M. and Ma, J., 1998, Exploratory spatial data analysis in a geographic information system environment. *Journal of the Royal Statistical Society, Series D*, **47**, pp. 457-469.

Hotson, J.M., 1988, *Land Use and Agricultural Activity: an Areal Approach for Harnessing the Agricultural Census of Scotland. Working Paper No. 11*, (Edinburgh: Economic and Social Research Council Regional Research Laboratory for Scotland).

Lindo Systems Inc., 1998, *What'sBest!*, (Chicago: Lindo Systems Inc).

Martin, D., 1997, Census geography 2001: designed by and for GIS?. In *Proceedings of the GIS Research-UK 5th National Conference* (Leeds: University of Leeds), pp. 18-21.

Martin, D., 1998, Optimising census geography: the separation of collection and output geographies. *International Journal of Geographical Information Science*, **12**, pp. 673-685.

MAFF (The Ministry of Agriculture, Fisheries and Food), 1997, *Digest of Agricultural Census Statistics United Kingdom 1997*, (London: HMSO)

MLURI (The Macaulay Land Use Research Institute), 1993, *The Land Cover of Scotland. Final Report on the Land Cover of Scotland Project*, (Aberdeen: The Macaulay Land Use Research Institute).

Moxey, A., McClean. C. and Allanson, P., 1995, Transforming the spatial base of agricultural census data. *Soil Use and Management*, **11**, pp. 21-25.

Openshaw, S., 1995, The future of the census, in Openshaw S., Census users' handbook. Cambridge: GeoInformation International, 389-411.

Robinson, G.M., 1988, Agricultural change: geographical studies of British Agriculture. Edinburgh: North British.

SOAEFD (The Scottish Office Agriculture, Environment and Fisheries Department), 1998, Economic report on Scottish agriculture 1998 edition. Edinburgh: The Stationery Office.

Towers, W., 1998, Personal communication.

Walker, P. and Mallawaarachchi T., 1998, Disaggregating agricultural census statistics using NOAA-AVHRR NDVI. *Remote Sensing of the Environment*, **63**, pp. 112-125.

Wise, S.M., Haining, R.P. and Ma, J., 1997, Regionalisation tools for the exploratory spatial analysis of health data. In Recent developments in Spatial Analysis: Spatial Statistics, Behavioural Modelling and Neurocomputing, edited by Fischer, M. and Getis, A. (Berlin: Springer), 83-100.

16

A fuzzy modelling approach to wild land mapping in Scotland

Steffen Fritz, Linda See and Steve Carver

16.1 INTRODUCTION

The use of GIS for wilderness mapping is a recent development, but several attempts have already been made that cover a range of different areas across the globe (; Kliskey and Kearsley, 1993, 1994; Lesslie, 1993, 1995; Carver 1996; Henry and Husby, *http://www.esri.com/recourses/userconf/proc95/to150/p113.html*). Methodologies range from the mechanistic and rigorous approach adopted by the Australian Heritage Commission (1988), which used deterministic, yet arbitrary parameters, to the approach of Kliskey and Kearsley (1993), which accounts for the subjective nature of wilderness using Stankey's wilderness purism scale (Stankey, 1977). However, there are drawbacks to the way in which Kliskey and Kearsley (1993) have translated the perceived levels of wilderness to the spatial domain. Moreover, none of these methodologies are directly applicable to Scotland. Here the term 'wild land' is proposed as a better representation of a landscape that has been dramatically altered due to its long history of settlement and rural land use (Aitken, 1977). In terms of biophysical naturalness, wild land has ceased to exist in nearly all areas of Scotland. Nonetheless, people still value the land according to factors such as solitude, remoteness and the absence of human artefacts and as such perceive it as wild. However not all factors can be measured easily in a quantitative sense. For example, solitude is highly dependent on the weather and personal ideals. Nevertheless, there are two main factors having a strong influence on wild land perception in Scotland that can be quantified. One factor is closely linked to the idea of the 'long walk in' and termed here as 'remoteness from mechanised access'. It can be measured as the minimum time it takes a walker to reach a particular destination from any origin (usually a road or car park). A second factor strongly influencing wild land perception is the impact of certain human-made features such as roads, hill roads, pylons, hydroelectric power plants etc. The presence of such features can detract from a 'wild land experience', particularly when the features are visible within the landscape.

This paper describes an approach to building a spatial mapping tool for wild land areas that captures qualitative perceptions of the factors affecting wild land quality. The methodology uses an Internet questionnaire designed specifically to collect softer, perceptual information such as naturalness, that is, forest and land cover, and artifactualism, which are important wild land indicators. This information is then translated to the spatial domain using a fuzzy logic modelling framework that combines these perceptions to produce a fuzzy wild land map. The method has been applied to the South West area of the Cairngorms.

16.2 METHODS OF WILDERNESS MAPPING

Several authors (Lesslie, 1985; Hendee, 1990; Countryside Commission, 1994; Carver, 1996) agree that there is no generally accepted definition of wilderness or wild land. A perceptual or sociological definition of wilderness can be found in Nash (1982). He defines wilderness from a personal perspective, that is, what people perceive it to be:

> *"There is no specific material thing that is wilderness. The term designates a quality that produces a certain mood or feeling in a given individual and, as a consequence, may be assigned by the person to a specific place. Wilderness, in short, is so heavily freighted with meaning of a personal, symbolic, and changing kind as to resist easy definition*". (Nash, 1982, p.5)

Due to the fact that the perception of wilderness quality by recreationalists differs widely amongst individuals and is influenced by a variety of personal factors, the establishment of a wilderness or wild land threshold is arbitrary. Therefore, it would be advantageous if wilderness areas could be identified in a relative way, either as a continuum or using fuzzy concepts.

Lesslie *et al.* (1988) define wilderness as undeveloped land which is relatively remote and relatively undisturbed by the process and influence of settled people. However, neither remoteness nor primitiveness can be assessed by a single wilderness quality indicator. Remoteness can be described as a proximity function to settled land and settled people whilst primitiveness is a more complex attribute since it has both a subjective and an objective component. Primitiveness might be defined in terms of biophysical or apparent disturbance. In order to consider the complexity of both remoteness and primitiveness, Lesslie *et al.* (1988) have identified four different indicators that define wilderness:

1. remoteness from settlement, i.e. points of permanent human occupation;
2. remoteness from access including constructed vehicular access routes (roads) and railway lines and stations;
3. apparent naturalness, which is the degree to which the landscape is free from the presence of the permanent structures of modern technological society; and,
4. biophysical naturalness, which can be defined as the degree to which the natural environment is free of anthropogenic biophysical disturbances.

A simple estimation of wilderness quality can be obtained by summing together the four wilderness indicator values at each grid point on a map, as shown in Figure 16.1. Although this method has been used to map wilderness in Australia at the national level, the approach has been criticised for being too mechanistic and for not taking the perceptual nature of wilderness into account (Kliskey and Kearsley, 1994; Bradbury, 1996). Moreover, the method assumes that each of the factors contribute equally to total wilderness quality since it consists of the simple addition of indicators, yet they are not necessarily comparable in a quantitative sense. However, despite these criticisms, the method is still seen as an effective and efficient way of deriving wilderness quality (Center for International Economics, 1998).

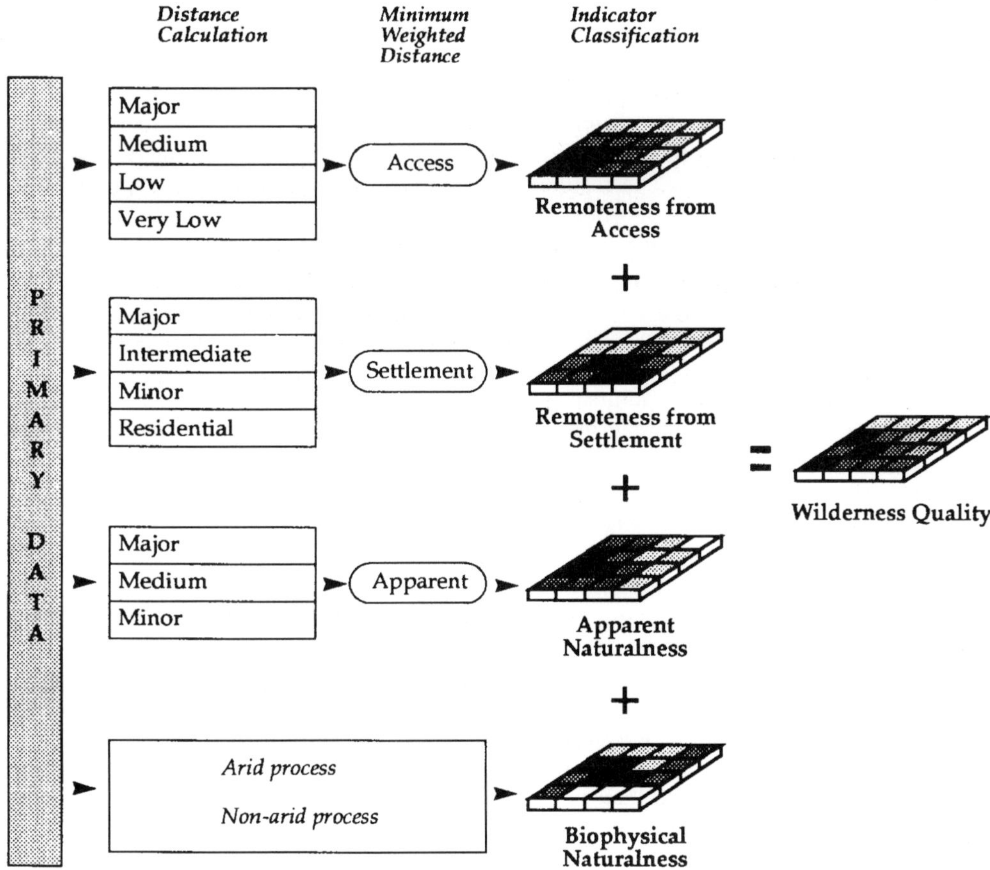

Figure 16.1 Method of deriving wilderness quality source (taken from Lesslie and Maslen, 1995, p. 8)

In a different approach, Kliskey and Kearsley (1993) mapped different peoples' perceptions of wilderness for the purpose of 'wilderness resource' zoning. Their study was based upon the premise that the concept of 'multiple perceptions of wilderness' is valuable as a theoretical notion and involves the following steps: 1. Acquire information on the variations in perceptions of wilderness; 2. Establish general properties defining each perception level; 3. Develop spatial criteria for each property; 4. Produce a wilderness perception map. Their method provides an approach to wilderness mapping that more closely approaches the definition of Nash (1982). It concentrates on the management of a national park and maps wilderness from the viewpoint of a backcountry user. In their study, wilderness was assessed for a national park area in the north of South Island, New Zealand. They formulated a wilderness perception survey, which attempted to measure four properties: artifactualism (absence of human impact); remoteness; naturalness (aspect of forest and vegetation); and solitude. Four backcountry user groups were then categorised with the use of a wilderness purism scale. This scale has been used to provide a mechanism that accommodates the variation of user definitions of wilderness (Stankey, 1973). Backcountry users were asked about their view towards desirability of various activities and experiential items in what they considered to be a wilderness setting. A value from 1 to 5 was assigned to each response (from strongly desirable to strongly undesirable) and each group in the wilderness purism scale had a range of scores (e.g. non purist 16-45). Contingency table analysis of purism groups and desirability of items in what is perceived as wilderness was used, supporting the use of these indicators for differentiating and determining variations in perception levels. The results were then translated into a spatial concept according to the four properties described above. The resulting maps revealed that differing user groups had entirely different perceptions of wilderness (Kliskey and Kearsley, 1993; Kliskey, 1994). Although their method clearly considered group perceptions of wilderness, the spatial criteria for determining the influences of human-made features were determined on an arbitrary basis for each group.

Using a similar type of approach to Kliskey and Kearsley (1993), a questionnaire was formulated and disseminated via the Internet. The questionnaire allows the respondents to establish their own spatial criteria, which determine the influence of a particular human-made feature. The questionnaire concentrates on assessing wild land impact in the South West area of the Cairngorms in Scotland but the methodology can be transferred easily to other areas.

16.3 THE INTERNET QUESTIONNAIRE

The Internet was chosen as the preferred method for dissemination to promote wider accessibility and to accommodate future plans, which will allow users to view and alter their resulting perceived wild land maps directly on the web. The questionnaire, which can be found at *http://www.ccg.leeds.ac.uk/steffen/questionnaire1.html*, asks participants to evaluate the spatial impact of human-made features, such as roads, isolated buildings, coniferous plantations, etc., on their perception of wild land in Scotland. Using a built-in visibility analysis, participants can also differentiate between the impact of features that are visible and those that are not. Instead of using simple buffers around the different features as used by Kearsley (1993), the factors influencing wild land are converted to fuzzy perceptual map layers and combined using a fuzzy logic approach. This enables people to establish their individual criteria and produce their own wild land map. The following study captures information on naturalness (forest and land cover) and artifactualism. Together these factors represent ‘apparent naturalness’ and are considered to be important wild land indicators. Another important indicator is ‘remoteness from mechanised access’, which will be considered in future versions of the questionnaire.

The current Internet questionnaire consists of three parts. Part one is used to gather personal information about the respondent whilst part two poses some general questions about hiking in Scotland and in the area covered by the questionnaire in particular. Eventually the information from these two parts will be used to classify the participants into different behavioural groups. Part three contains the main questions regarding the impact of certain features on the participant's perception of wild land. The respondents are first required to define what they feel constitutes being ‘near’, a ‘medium’ distance away and ‘far’ from visible features in either metres, kilometres or miles, where ‘near’, ‘medium’ and ‘far’ are fuzzy concepts. For example, a respondent might feel that they were 'near' if they were at a distance of up to 300 m or less while 500 m would represent a 'medium' distance away and 1 km or more would constitute being 'far' from the road. The respondents are then asked to define similar distances for the fuzzy terms 'close' and 'far away' for features that are not visible but which can still have an impact on the perception of wild land. The third part of the questionnaire is comprised of a series of questions regarding the impact of eleven different features on the perception of wild land when the respondent is 'near' to, a 'moderate' distance away and 'far' from a visible feature or 'close' and 'far away' from features that are out of sight. The questions are arranged in tabular format as shown in Figure 16.2 and the eleven different factors affecting wild land are listed in Table 16.1. Impact is also presented as a fuzzy concept ranging from a 'very strong impact' to 'no impact'. Question twelve invites the participant to enter any comments regarding any additional factors that might affect their perception of wild land or suggestions for improving the questionnaire.

1. What is the impact of a hill road on your perception of wild land when you are:

	near (visible)	medium (visible)	far (visible)	close (out of sight)	further away (out of sight)
very strong impact	○	○	○	○	○
strong impact	○	○	○	○	○
medium impact	○	○	○	○	○
low impact	○	○	○	○	○
very low impact	○	○	○	○	○
no impact	○	○	○	○	○

Figure 16.2 The first question of the Internet questionnaire on the impact of hill roads

16.4 A FUZZY LOGIC MODELLING APPROACH TO WILD LAND MAPPING

Fuzzy logic is one of several new alternative approaches to modelling that has emerged from the fields of Artificial Intelligence and process-based engineering. Originally formulated by Zadeh (1965), fuzzy logic replaces the crisp and arbitrary boundaries with a continuum, thereby allowing the uncertainty associated with human perception and individual concept definition to be captured. For this reason, fuzzy logic is particularly well suited to wild land mapping because it enables different factors influencing the perception of wild land to be integrated into a fuzzy wild land map, analogous to the way in which our brains might handle this information in a decision-making process. It also allows one to map different degrees of wild land quality, thereby eliminating the crisp boundary between wild and non-wild land. Moreover, this approach considers the spatial component explicitly by asking people to define their concept of distance and the subsequent impact of certain human-made features on their personal definition of wild land.

Table 16.1 Factors affecting the perception of wild land embedded in the Internet questionnaire. The * indicates the factors considered in producing the combined wild land quality impact maps

Factors/Impacts on Wild Land
*Surfaced road (paved)
*hill road (non-paved)
*Built up areas
*Isolated building
Pylons
Grazing sheep or cattle
Arable land
*Coniferous plantation
Hydroelectric power plant
Ski-lifts
Shielings

16.4.1 Visibility and distance analysis

A visibility map for the South West area in the Cairngorm Mountains in Scotland was produced using the ARC/INFO GRID module at a 50 m spatial resolution for five factors on the questionnaire including paved roads, hill roads, built-up areas, isolated buildings and coniferous plantations. A visibility analysis of the Digital Terrain Model (EDX data) was undertaken for each individual human-made feature. The distance to the closest visible feature for each factor was recorded. These factors were extracted from the LCS88 land cover map produced by the Macaulay Land Research Institute. The closest Euclidean distance was also calculated for each factor in order to acquire a data layer outlining those areas where a feature is not visible, but which may still have a potential influence on the perception of wild land quality. Figure 16.3 shows a map of the DTM with the features used for this study. In the future, the remaining factors listed in Table 16.1, which are currently part of the questionnaire, will be taken into account.

16.4.2 A fuzzy logic model for mapping wild land

The distances specified by the respondent were used to construct fuzzy sets for defining the concepts 'near', 'medium' and 'far' for visible features and 'close' and 'far away' for non-visible features. The fuzzy sets were built on the premise that the specified distances are considered by the respondent to completely represent the given concept of distance, which in fuzzy terms means a membership of 1.0. As one moves away from this distance, the membership value declines (or retains a value of 1.0 at the end points). To capture the uncertainty or fuzziness associated with these concepts, the fuzzy sets are overlapped, so it is possible to be both 'near' and a 'medium' distance from a feature, but to differing degrees. The output sets for wild land quality, which range from a 'very strong impact' to 'no impact' were evenly spread but overlapped across a continuum of 0 to 1. Example fuzzy sets are provided in Figure 16.4 using sample values of 200 m, 500 m, 1 km, 2 km and 6 km for the five distances to illustrate the fuzzy set construction procedure.

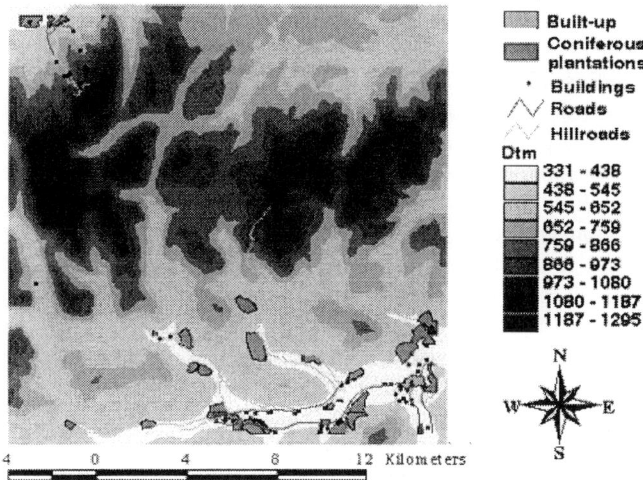

Figure 16.3 Human-made features overlaid on a Digital Elevation Model of the Cairngorm Mountains in Scotland

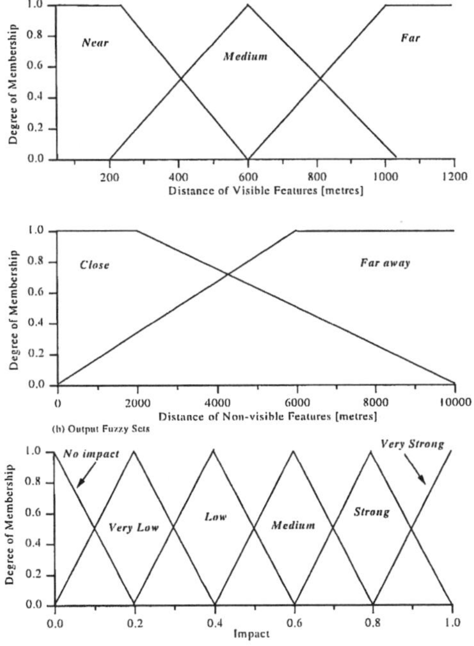

Figure 16.4 Fuzzy input and output sets defining distance and impact

Each question regarding the impact of a single factor produces a set of fuzzy rules. For example, if one clicks on the radio button at the intersection of the row 'very strong impact' and the column 'near' for the question on the impact of surface roads, the rule would take the form:

> IF you are 'near' to a surfaced road
> THEN this has a 'very strong impact' on wild land quality.

Each question yields 12 rules that link each of the six impact fuzzy sets to the five distances. 6 of the rules correspond to visible features while the other 6 cover non-visible features. Figure 16.5 provides a methodological outline of the procedure for processing the rules for each individual layer and then combining these layers to produce an integrated fuzzy wild land map of the effect of human-made features. For each grid cell, the visible and non-visible distances to a given feature are used to determine the membership values in each of the distance fuzzy sets. A classical fuzzy OR operator (Zadeh, 1965), which involves taking the maximum membership, is then used to combine the rules from the visible and non-visible rule bases. This means that the impact on wild land quality from a given factor will be determined by the highest value regardless of visibility. The fuzzy model is applied to each layer producing fuzzy wild land maps based on only one factor. To combine them, different Boolean and fuzzy decision-making operators can be used to integrate the layers into a single map. These operators represent various decision-making modes, for example, the fuzzy OR or maximum creates a map based on the factor that is the most important while the fuzzy AND or minimum represents a risk-adverse approach. In this paper, only the fuzzy OR has been applied to combine the layers.

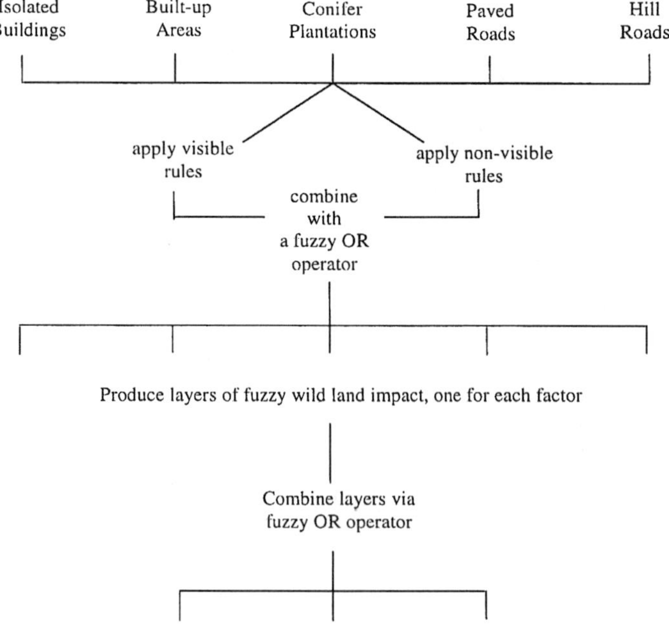

Figure 16.5 A methodological outline for the creation of fuzzy wild land maps

16.4.3 Sample results - Boolean vs Fuzzy logic

For each of the five factors considered in this paper, a Boolean and fuzzy logic approach were used to create single factor impact maps based on the answers of a single respondent. The Boolean approach involved creating buffer zones around the features based on the definitions of the 5 fuzzy distances. Figure 16.6(a) shows the impact of isolated buildings on wild land after applying the Boolean approach. Figure 16.6(b) contains the same map but created using a fuzzy logic model. A comparison of the two maps shows that the fuzzy logic-based wild land impact map changes more gradually as distance from isolated buildings increases compared to the map produced with the Boolean approach where the boundaries are far more distinct. The five single factor maps created with the Boolean approach were then combined using a Boolean OR operator to produce an impact map based on the five human-made features and the result is shown in Figure 16.7(a). Figure 16.7(b) is a plot of the same map but combined using the fuzzy OR operator. The crisp boundaries shown in the Boolean-derived map are replaced by a gradual decrease in wild land impact as the distance away from each feature increases in the map derived using the fuzzy logic approach. An abrupt change of values in the fuzzy impact map was noted only at the border between visible and non-visible areas which reflects clearly the rules specified by the respondent.

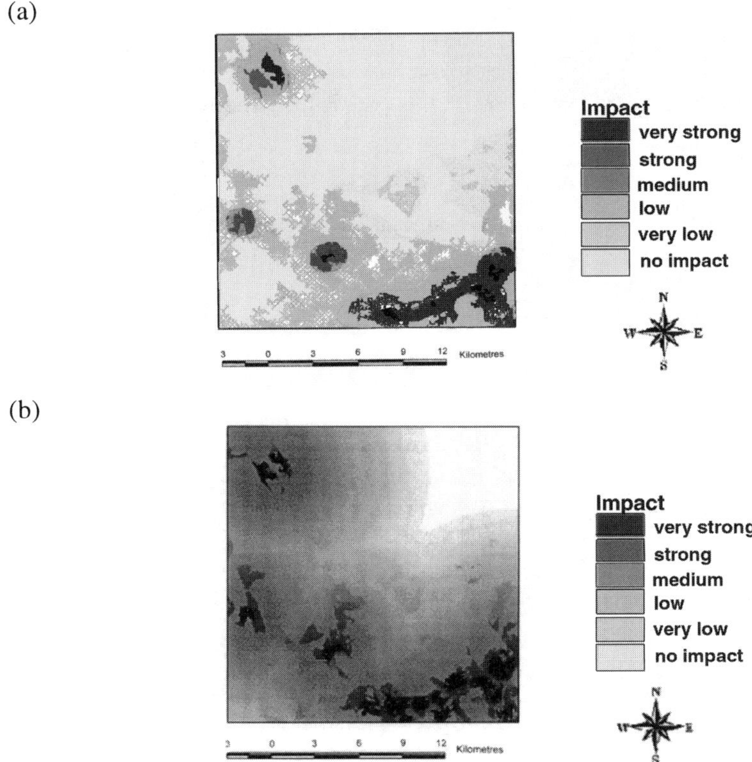

Figure 16.6 Perceptual impact map of isolated buildings on quality of wild land using (a) a Boolean and (b) a fuzzy approach

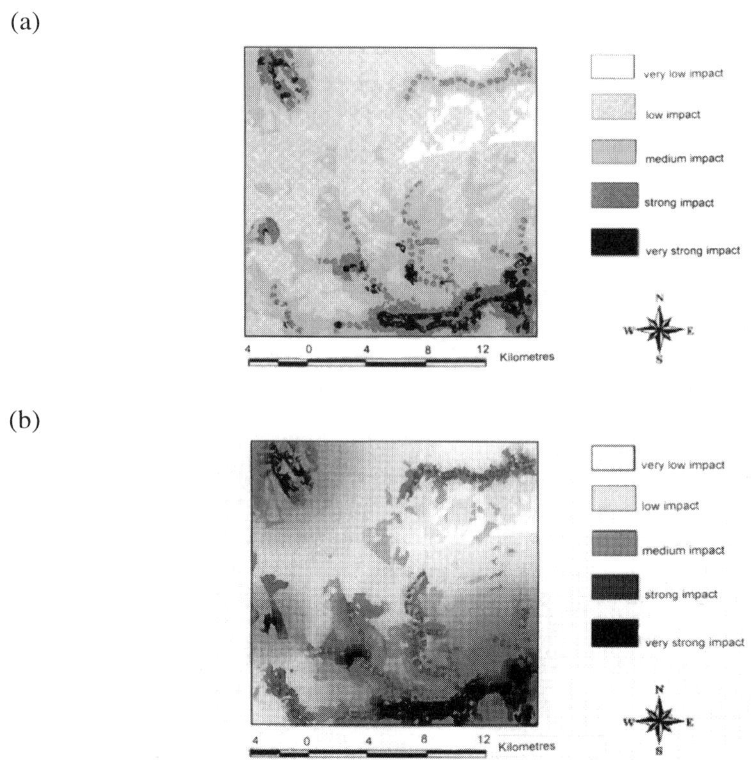

Figure 16.7 Perceptual impact map of human-made features on quality of wild land using (a) a Boolean and (b) a fuzzy approach

16.5 LIMITATIONS OF THE METHOD AND RECOMMENDATIONS FOR FURTHER RESEARCH

Although the fuzzy wild land maps appeared to produce a more natural graduation between impact categories as distance increased over the method employed by Kliskey and Kearsley (1993), there are still some unresolved problems with this approach. Firstly, visibility analysis is computationally a very intensive process. This method may require considerable computational time if it were to be applied to larger areas and if more factors were taken into account. However, the analysis needs to be performed only once. A second drawback is related to linear features. Although the visibility to the closest road has been recorded, the length of that feature is not taken into account, which may have an influence on the perception of wild land. A third limitation involves the use of the OR operator to combine the maps. This operator is problematic since it takes the strongest individual impact as the overall result rather than taking all the features into account. The application of more appropriate fuzzy operators that can handle a combination of factors such as the order weighted averaging operator (Yager, 1998) or the gamma operator (Bonham-Carter, 1994) will be investigated in the future. A final problem with the methodology is how to validate the results. One possible suggestion would be the addition of an interactive mapping component to the questionnaire, which would allow the respondent to view the resulting maps and then adjust their answers to better represent what they feel is their perception of wild land in that area.

Other improvements to the research will include incorporating the second important wild land indicator 'remoteness from mechanised access', also referred to as 'the long walk in', and adding a mapping tool onto the Internet questionnaire so that respondents can view their perception of wild land based on any combination of factors. Users will be able to change their definitions and answers to produce a map that better characterises their perceived idea of wild land quality. It will also be possible to generate an average wild land map based on all questionnaire respondents for a given behavioural group so that individuals can contrast their perception with that of the wider population of respondents. Photographs showing different distances to each of the features might also be incorporated to aid in defining the fuzzy distances in the future.

16.6 CONCLUSIONS

Mapping wild land areas is a difficult problem because it depends greatly on individual perception, and the mapping can only be achieved to a certain degree using measurable indicators. However, it is becoming increasingly important to try to quantify less tangible resources such as wild land, which can only be compared objectively if the perceptions of the public are quantified in some manner. For example, a stronger case might be made against the construction of a hydroelectric power scheme in a public enquiry if maps that incorporate people's perceptions could demonstrate that the wild land character of a vast area may be spoilt. To try and achieve this aim, a methodology was presented for the creation of wild land maps based on the answers from an Internet questionnaire, which was designed to collect perceptual information on how different human-made features affect an individual's overall perception of wild land. The questionnaire provided a useful method of collecting 'soft' information, which was subsequently utilised in a fuzzy logic model for the generation of single factor and combined fuzzy wild land maps. The fuzzy maps were compared to their Boolean equivalents and the results showed a more gradual distance decay effect rather than the sharp boundaries associated with the Boolean approach. The methodology was demonstrated on a small area in the Cairngorm Mountains of Scotland but has the potential to be extended to other regions in Scotland and, with minor adjustments, to other areas in Europe. The methodology can also incorporate a large range of influencing factors beyond the five considered in this paper, and it can be used to produce individual and average fuzzy wild land maps for certain behavioural groups or across the entire respondent population.

16.7 ACKNOWLEDGEMENTS

The authors gratefully acknowledge the help of the School of Geography, University of Leeds in funding this research.

16.8 REFERENCES

Aitken, R., 1977, *Wilderness Areas In Scotland*, unpublished Ph.D. Thesis, (Aberdeen: University of Aberdeen).

Bonham-Carter, G., 1994, *Geographic Information Systems for Geoscientists: Modelling with GIS*, (Kidlington: Elsevier Science Ltd.).

Carver, S., 1996, *Wilderness Britain? Using GIS and Multi-Criteria Evaluation Techniques to Map the Wilderness Continuum, Working Paper 96/16*, (Leeds: School of Geography).

Center for International Economics, 1998, *Evaluation of the National Wilderness Inventory Project,* (Canberra and Sydney: Center for International Economics).

Kliskey, A.D. and Kearsley, G.W., 1993, Mapping multiple perceptions of wilderness in southern New Zealand. *Applied Geography*, **13**, pp. 203-223.

Kliskey, A.D., 1994, A comparative analysis of approaches to wilderness perception mapping. *Journal of Environmental Management*, **41**, pp. 199-236.

Lesslie, R. G. and Taylor, S. G., 1985, the wilderness continuum concept and its implication for Australian wilderness preservation policy. *Biological Conservation*, **32**, pp. 309 - 333.

Lesslie, R. G., Mackey, B.G. and Preece, K. M., 1988, A computer based method of wilderness evaluation, *Environmental Conservation*, **15**, pp. 225 - 232.

Lesslie, R., and Maslen, M., 1995, *National Wilderness Inventory Handbook of Procedures, Content and Usage* 2nd edition, (Canberra: Commonwealth Government Printer).

Nash, R., 1982, *Wilderness and the American Mind*, 3rd edition, (London: Yale University Press).

Stankey, G.H., 1977, Some social aspects for outdoor recreation planning, *Outdoor Recreation: Advantages in Application of Economics*, US Department of Agriculture Forest Service, Gen. Tech. Rep, WOO, 2.

Yager, R.R., 1998, Including importances in OWA aggregations using fuzzy systems modeling. *IEEE Transactions on Fuzzy Systems*, **6**, pp. 286-294.

Zadeh, L.A., 1965, Fuzzy sets. *Information and Control*, **8**, pp. 338-353.

17

Evaluating the derivation of sub-urban land-cover data from Ordnance Survey Land-Line.Plus®

Michael Hughes and Peter Fisher

17.1 INTRODUCTION

As part of a project concerned with modelling sub-pixel components of land-cover in satellite imagery (FLIERS) it was necessary to provide detailed verification data for an area of sub-urban Leicester, UK. Aerial photograph interpretation (API) has been the traditional method for large-scale land-cover mapping since the 1940s (Lillesand and Kiefer, 1994) and is used widely for training, verification and validation of land-cover mapping from satellite remote sensing (e.g. Thunnissen *et al.*, 1992 and CORINE land-cover mapping project). Aerial photographs as a basic source of spatial data have become an important part of urban information systems (Cassettari, 1993). More recently, a new generation of high resolution satellite sensors (< 5 m pixel) promises to provide data suitable for large-scale land-cover mapping (Aplin *et al.*, 1997). Either way, image interpretation is an interactive process involving a trained interpreter. Objects in the image are delineated and identified with closed polygons, either by manual or digital means. The polygons are then labelled using the interpreter's knowledge of the area which may be enhanced by field survey or ancillary data (Avery & Berlin, 1985).

Although widely accepted, API has several disadvantages, namely: (i) Imagery, if available at all, may only be available for specific dates. If no imagery exists for a chosen area there may be high costs involved in planning and executing a flight; (ii) The interpretation process is time consuming and laborious, especially for complex areas such as the sub-urban landscape. The time taken depends on the level of classification which in this case is detailed; (iii) The interpretation is likely to contain uncertainty (both thematic and geometric) as a result of user subjectivity, digitising errors and properties of the photograph itself (e.g. scale, geometric registration). The use of orthophotos can help alleviate the latter problem. Interpretations of the same area by different observers may be quite different and (iv) Expert knowledge is required to correctly label the polygons (this may be provided by additional field survey or ancillary data).

Internationally, general topographic or cartographic databases at detailed scales are increasingly available. In Great Britain Ordnance Survey maintain the national topographic database which contains over 200 million geographically referenced features of the landscape. For sub-urban areas the most suitable products are Land-Line and Land-Line.Plus which have a vector data structure originally derived from maps at a scale of 1:1250 or 1:2500 (*http://www.ordsvy.gov.uk/*). Land-Line.Plus differs from Land-Line in that it contains additional landscape features. Features are names, points, lines or series of lines forming coherent units and are described using feature codes and text annotations (Ordnance Survey, 1997). Data coverage is extensive compared to aerial photography and under a forthcoming agreement between Ordnance Survey and the Joint Information Systems Committee 30% UK coverage of Land-Line.Plus data will be available to UK academic institutions by subscription (Ordnance Survey press release OS 18/99, *http://www.jisc.ac.uk/*).

If suitable land-cover classes can be extracted automatically from Land-Line.Plus then such a methodology would offer several benefits over traditional land-cover mapping by API. Firstly, the process could be automated using routines in a geographical information system (GIS). Large numbers of tiles could be processed quickly with minimal user intervention. The perceived geometric accuracy of Land-Line data is high compared to aerial photography and the data are revised periodically when new surveys have been carried out. Finally, the features are already classified thus providing the expert knowledge required for labelling. An automated routine, without user intervention, would eliminate the effects of user subjectivity and drastically reduce the time needed to produce an interpreted land-cover map.

The aim of this paper is therefore to answer the question: can land-cover classes comparable to those obtainable from API be extracted from Land-Line.Plus? For a sensible comparison between the two methods it was necessary to select classes which could be identified, and extracted as areal coverages, in both datasets. The sub-urban landscape is composed of a vast array of materials, many of which cover very small areas. For this study, however, three broad land-cover classes were chosen which dominate the sub-urban landscape and allow it to be characterised namely: *paved surface*, *woody vegetation* and *built surface*. Such classes are deemed to be useful in studies of urban systems, for example: energy budget calculations and surface runoff modelling (Fankhauser, 1999).

17.2 METHODS

The aim of both methods described below was to derive a land-cover interpretation for a 1 km^2 area of sub-urban Leicester. This interpretation was then used to derive land-cover proportions for imaginary satellite pixels (each of 25 m to simulate European Landsat TM pixel size) thus providing a suitable basis for comparison between the two methods.

17.2.1 Derivation of Land-cover Data from Aerial Photograph Interpretation

A digital colour aerial photograph mosaic, taken in November 1995 with a spatial resolution of 25 cm, was obtained for part of Leicester (from the Cities Revealed product range, GeoInformation Ltd). This product is derived from a map accurate digital ortho database and is supplied geo-referenced to the Ordnance Survey grid (*http://www.crworld.co.uk/*). Landscape features in the 1 km^2 study area were digitised using ERDAS Imagine vector utilities. Digitising at on-screen scales of between 1:500 and 1:2000 allowed the delineation of very small features (approximately 2 m across) and detailed vector coverages were obtained as a result. Originally, a more detailed land-cover classification scheme was used, but for the purposes of this work several classes were combined to form each of the three main classes for analysis: *paved surface*, *woody vegetation* and *built surface* (Table 17.1).

Table 17.1 Derivation of main classes from original API classes.

Land-cover Class	Original API Classes
paved surface	tarmac, concrete
woody vegetation	shrubs, deciduous trees coniferous trees, scrub
built surface	slate, tile, metal, asbestos, pitch, felt

17.2.2 Derivation of Land-cover Data from Land-Line.Plus

The derivation of areal coverages from Land-Line.Plus was complicated by the fact that it does not consist entirely of closed polygons with unique labels. Instead, the lines making up the vector coverage have a feature code attribute and label points are used only for certain features. Therefore, a certain amount of data preparation was necessary before land-cover polygons could be extracted. The following analysis was carried out in ArcInfo GIS after importing the original data (supplied in DXF format by Ordnance Survey).

The first step was to remove all non-physical entities (e.g. boundary information) from the coverage (Table 17.2) and then to join the four tiles covering the 1 km^2 area. The vectors were given polygon topology, so that each polygon had a unique identifier. At this stage it was assumed that only one class could be attributed to each polygon. The label points (seeds) were removed and kept as a separate coverage.

By combined analysis of the new polygon coverage and the original seeds it was possible to identify building polygons since each has a unique building seed. It was assumed that these polygons would equate to the *built surface* polygons identified by API. Polygons were identified as woody vegetation if they contained at least one of the seeds shown in Table 17.3. Land-Line also contains seed points representing individual trees and these could have been used together with an assumption of canopy size to provide an additional coverage of woody vegetation. The *paved surface* class, however, was less straightforward to assign since there are no specific seeds for paved features. The only component of the *paved surface* class that could be extracted reliably was road since road casings are feature coded lines. This left other paved areas such as pavements and car-parks which could not be identified automatically using feature codes. A coverage of road casings was produced by extracting line features with the relevant feature code. However, these lines do not form closed polygons (in our area at least) and there is no way of knowing (in an automated process) whether the road surface is to one side or the other of the lines. The only way of labelling the polygons was by user intervention. In two instances, polygons had to be closed manually by adding extra vectors — this is presumably a data error since the edges of metalled road surfaces are usually well surveyed. Fortunately, it is easy to identify road areas since they display a characteristic pattern.

Table 17.2 Land-Line.Plus features selected as physical entities

Feature code (type)	Feature code name
0001 (line)	building outline
0004 (line)	building pecks
0014 (line)	narrow gauge railway
0015 (line)	railway track
0021 (line)	road casings
0030 (line)	general line detail
0032 (line)	minor line detail
0035 (line)	vegetation/landform limit
0036 (line)	vegetation/landform limit
0059 (line)	water detail

Table 17.3 Derivation of woody vegetation class from Land-Line.Plus feature attributes.

Feature code (type)	Land-Line feature code name
0379 (point)	coniferous trees
0380 (point)	coniferous trees (scattered)
0381 (point)	coppice, osiers
0384 (point)	non-coniferous trees
0385 (point)	non-coniferous trees (scattered)
0386 (point)	orchard
0392 (point)	scrub

17.3 RESULTS

Land-cover interpretations derived using both methods can be seen in Figure 17.1. Figures 17.3 and 17.4 show class proportions mapped for 25 m pixels for API and Land-Line.Plus interpretations respectively. The pixel proportion approach provided a suitable basis for comparison between the two methods. Proportions per pixel per class for both methods were correlated to give an indication of their similarity. A summary of these results, including total estimated area per class, is shown in Table 17.4 and as scatter plots in Figure 17.2.

Table 17.4 Comparison of results for pixel proportions of land-cover classes obtained from API and Land-Line.Plus analysis.

	Total area comparison (% total)		Pixel based comparison	
Land-cover Class	API	Land-Line.Plus	Correlation coefficient	RMS
paved surface	26.2	16.6	0.761	0.201
woody vegetation	25.1	3.7	0.314	0.353
built surface	24.5	23.4	0.922	0.094

Figure 17.1 Land-cover classification of 1 km^2 area of sub-urban Leicester derived from a) aerial photograph interpretation and b) Ordnance Survey Land-Line.Plus.

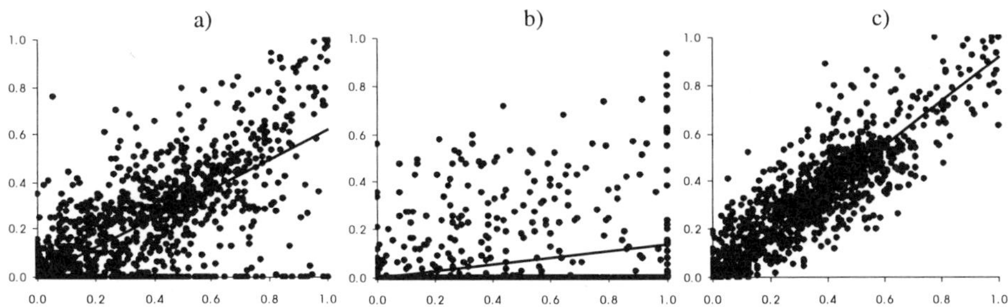

Figure 17.2 Scatter plots of land-cover proportions for 25 m pixels derived from Land-Line.Plus (y-axis) versus API (x-axis) for a) *paved surface*; b) *woody vegetation* and c) *built surface*.

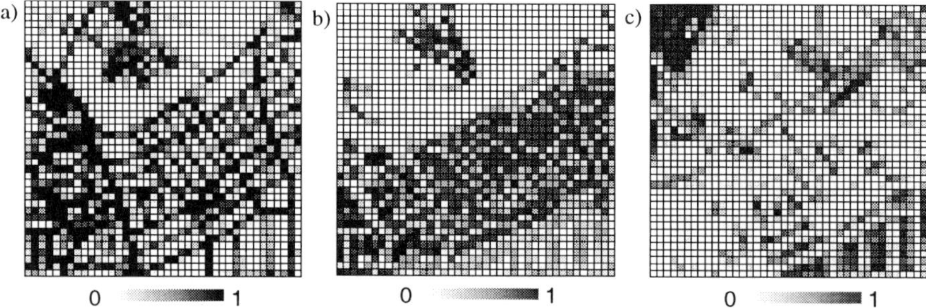

Figure 17.3 Class proportions in 25 m pixels derived from aerial photograph interpretation for: a) *paved surface*, b) *built surface* and c) *woody vegetation*.

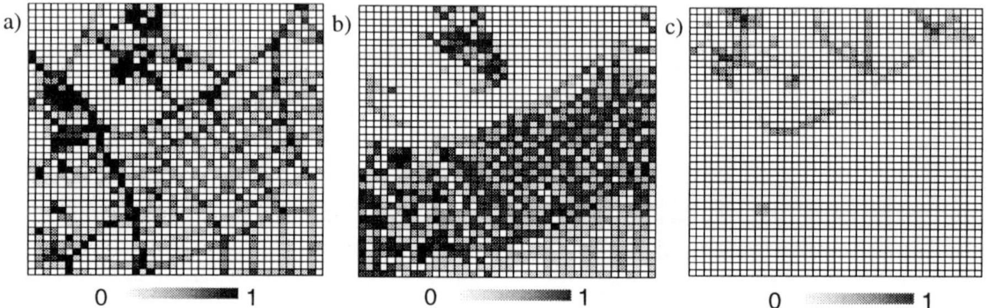

Figure 17.4 Class proportions in 25 m pixels derived from Ordnance Survey Land-Line.Plus for: a) *paved surface*, b) *built surface* and c) *woody vegetation*.

17.4 DISCUSSION

Before an analysis of the results it is important to first consider sources of uncertainty in the data. The two datasets can be regarded as representing two different properties of the sub-urban landscape. The aerial photograph represents land-cover as viewed from above (a true aerial view) as might be seen by a satellite sensor. Thus, if tree cover obscures a tiled roof, then the land-cover for that area will be tree and not tile (assuming the tree is in leaf and the interpreter doesn't try to guess the extent of the underlying building). Ordnance Survey topographic data, on the other hand, is a planimetric representation of the landscape, a plan of objects as they lie on the ground with some features symbolised for simplification (see Figure 17.5 for an example). In Land-Line, the roof area will be represented by its floor plan and the tree, if represented at all, may be a symbol placed next to the building outline. This example illustrates one potential problem of deriving land-cover from Land-Line, or indeed any cartographic product, for use in remote-sensing studies. That is, planimetric land-cover is not the same as aerial land-cover.

Another issue is the geometric fidelity of both datasets. Despite the accuracy statement accompanying Land-Line data and the fact that Cities Revealed photos are ortho-corrected using OS data there were observed locational shifts between obvious features in both datasets. The magnitude of these discrepancies (up to 8 m) suggests that they may be contributing to the observed variation between the two sets of pixel proportions. However, these errors would not be expected to greatly affect area-based statistics.

There are vital differences in the way the two data sources are processed to produce land-cover polygons. In API the placing of polygons is subject to digitising error and sometimes object boundaries are fuzzy (either in real-life, e.g. indistinct vegetation boundaries; or due to poor image quality in the photograph). Land-Line, on the other hand, is derived from extensive ground survey with a high degree of (stated) accuracy.

There is good agreement (both area-based and pixel-based) for the *built surface* class (Table 17.4). This is not surprising since buildings are easy to identify and digitise in aerial photography and are generally well surveyed by the Ordnance Survey. The API method gives a slightly larger total area of built surface class. This is probably due to features in the photo being identified which are not demarcated in Land-Line, for example, small garden sheds, out-buildings, garage extensions (mainly felt, pitch and asbestos cover types). The *paved surface* class too, gives relatively good visual agreement but the area-based estimate from Land-Line.Plus is much smaller. This is because, in the API method, pavements and other paved areas around buildings have been correctly identified as paved surface whereas for Land-Line.Plus, although road surfaces have been correctly labelled, the other areas are not sufficiently described (by feature codes) and have been left unclassified. Land-Line.Plus appears to be a very good data source for delineating metalled roads but not necessarily paved areas with other uses.

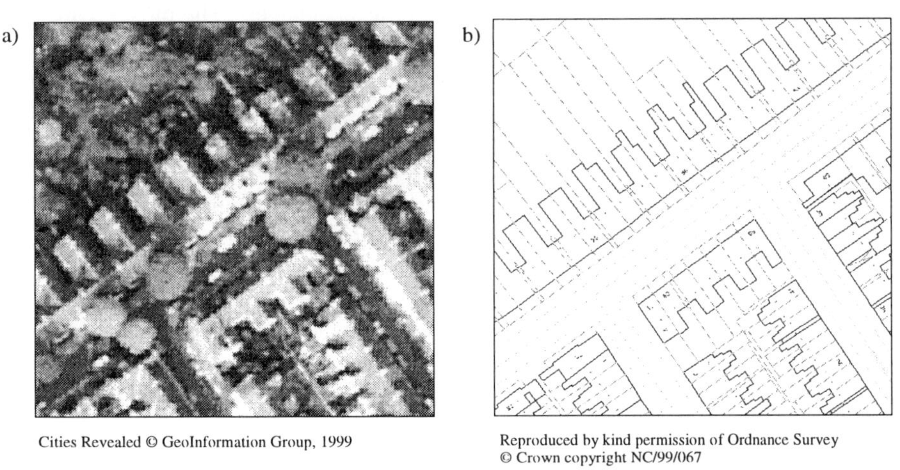

Cities Revealed © GeoInformation Group, 1999

Reproduced by kind permission of Ordnance Survey
© Crown copyright NC/99/067

Figure 17.5 Part of the study area (100 m x 100 m) as seen in a) digital aerial photography and b) Ordnance Survey Land-Line.Plus.

The woody vegetation class is a different story, with very poor agreement between the two methods. Areas identified as woody vegetation in Land-Line.Plus are also identified by API but there are many other areas which are only identified by API. In fairness to Ordnance Survey, it is not reasonable to expect large-scale mapping to adequately describe complex vegetation information, especially in urban and sub-urban areas where large changes may occur over short time-scales. The Ordnance Survey topographic database exists primarily to store information concerning the built environment and was not originally designed as a land-cover database. Interestingly, of the vegetation types currently distinguished on Ordnance Survey maps, there is no class that explicitly describes either grass or shrub — probably the two most common types of vegetated land-cover in these areas. This is certainly the cause of discrepancy between the two methods, since the woody vegetation cover derived from Ordnance Survey is really only showing areas with trees or scrub (although the Ordnance Survey definition of scrub extends to bushes and 'underwood' (Harley, 1975)). The remaining shrub areas, missed in the Land-Line.Plus classification, are mainly in gardens and around institutional buildings.

So far the discussion has concentrated on a comparison of the two methods. A more intuitive approach uses information from both cartographic and image sources (usually high resolution satellite imagery) to derive more accurate land-cover and land-use maps than using just one data source alone (e.g. Harris and Ventura, 1995; Aplin *et al.*, 1999). High resolution imagery of urban areas, including aerial photographs, can also be classified using other methods including maximum likelihood classification (Fankhauser, 1999), unsupervised texture analysis (Sali and Wolfson, 1992) and neural networks (e.g. Heikkonen and Varfis, 1998; Barnsley and Barr, 1996).

17.5 CONCLUSIONS

Land-cover information has a wide variety of uses and demand for high resolution land-cover data is increasing. Aerial photograph interpretation, although a tried and tested method for land-cover mapping, suffers from several disadvantages. The most significant of these is the amount of time needed to produce land-cover interpretations for large areas, such as might be needed in remote sensing studies. The current revolution in high resolution satellite imagery promises much but, so far, has delivered little. Even when data does become available it is expected to be relatively expensive although probably cheaper than digital aerial orthophotos.

The main drawback of Land-Line.Plus for land-cover mapping is the fact that it originates from a cartographic database (with feature-coded lines) which is biased toward the built environment. An ideal land-cover database would be area-based, with feature codes describing the surfaces within those areas. Furthermore, for indeterminate boundaries some way is needed of expressing fuzziness. With new products coming onto the GIS market to allow easier use of digital cartography it appears that Land-Line data in particular will become more widely used (Ordnance Survey Data News, Autumn 1998). For example, Lake *et al.* (1998) demonstrate a novel use of several Ordnance Survey digital products (Land-Line.Plus, ADDRESS-POINT and Land-Form PROFILE) in modelling negative impacts on property value.

In the UK, the Ordnance Survey's national topographic database is an important source of geographic data with the potential to facilitate development of land-cover datsets. Results from this work indicate that, for sub-urban areas at least, the Land-Line.Plus product can be used to extract reliable estimates of one land-cover type (*built surface*) with a second (*paved surface*) producing slightly divergent results compared to those obtained from API. However, Land-Line.Plus does not allow a more precise definition of land-cover types, such as different roofing materials, nor does it allow for delineation of different vegetation types (both of which are possible with API). We have also shown that it is possible to partially automate the process of feature extraction in a GIS, although some intervention is required to label certain land-cover classes and correct incomplete line-work. An approach combining analysis of digital imagery and cartographic data would seem to be the way ahead for urban and sub-urban land-cover mapping.

17.6 ACKNOWLEDGEMENTS

We would like to thank the co-operation of all those participating in the FLIERS project (*http://www.geog.le.ac.uk/fliers*) and especially Lucy Bastin. The work was funded by EU grant ENV4-CT96-0305.

17.7 REFERENCES

Aplin, P., Atkinson, P.M. and Curran, P.J., 1997, Fine spatial resolution satellite sensors for the next decade. *International Journal of Remote Sensing*, **18**, pp. 3873-3881.

Aplin, P., Atkinson, P.M. and Curran, P.J., 1999, Fine spatial resolution simulated satellite sensor imagery for land cover mapping in the United Kingdom. *Remote Sensing of Environment*, **68**, pp. 206-216.

Avery, T.E. and Berlin, G.L., 1985, Interpretation of Aerial Photographs, (Minnesota, USA: Burgess Publishing Company).

Barnsley M.J. and Barr S.L., 1996, Inferring urban land use from satellite sensor images using kernel-based spatial reclassification. *Photogrammetric Engineering and Remote Sensing*, **62**, pp. 949-958.

Cassettari, S., 1993, Geo-referenced image-based systems for urban information management. *Computers, Environment and Urban Systems*, **17**, pp. 287-295.

Fankhauser, R., 1999, Automatic determination of imperviousness in urban areas from digital orthophotos. *Water Science and Technology*, **39**, pp. 81-86.

Harley, J.B. 1975, *Ordnance Survey Maps: A Descriptive Manual*, (Southampton, UK: Ordnance Survey).

Harris P.M. and Ventura S.J., 1995, The integration of geographic data with remotely-sensed imagery to improve classification in an urban area. *Photogrammetric Engineering and Remote Sensing*, **61**, pp. 993-998.

Heikkonen J. and Varfis A., 1998, Land cover land use classification of urban areas: A remote sensing approach. *International Journal of Pattern Recognition and Artificial Intelligence*, **12**, pp. 121-136.

Lake, I.R., Lovett, A.A., Bateman, I.J. and Langford, I.H., 1998, Modelling environmental influences on property prices in an urban environment. *Computers, Environment and Urban Systems*, **22**, pp. 121-136.

Lillesand, T.M. and Kiefer, R.W. 1994, *Remote Sensing and Image Interpretation*, (New York, USA: John Wiley and Sons, Inc.).

Ordnance Survey. 1997, *Land-Line® User Guide - Format Section*, (Southampton, UK: Ordnance Survey).

Sali, E. and Wolfson, H., 1992, Texture classification in aerial photographs and satellite data. *International Journal of Remote Sensing*, **13**, pp. 3395-3408.

Thunnissen, H., Jaarsma, M.N. and Schoumans, O.F., 1992, Land cover inventory in the Netherlands using remote-sensing - Application in a soil and groundwater vulnerability assessment system. *International Journal of Remote Sensing*, **13**, pp. 1693-1708.

18

Interpolating elevation with locally-adaptive kriging

Christopher D. Lloyd and Peter M. Atkinson

18.1 INTRODUCTION

The tools of geostatistics are becoming increasingly important components of GIS (Burrough and McDonnell, 1998). Geostatistics can be used for (1) the characterisation of spatial variation, (2) interpolation, (3) simulation and (4) the design of optimal sampling strategies. In particular, interpolation of elevation values and other properties from samples is a major function of GIS. This chapter uses the geostatistical technique of ordinary kriging (OK) with the aim of making optimal estimates. A principal advantage of kriging-based algorithms over other interpolation techniques is that the spatial dependence (as represented by the variogram, for example) in the property of interest is used to inform the estimates. Also, kriging provides an estimation variance as part of the estimation procedure. In this chapter, a digital terrain data set is analysed using geostatistics.

Geostatistics have been used in several contexts relating to digital terrain data (Mulla, 1989; Bian and Walsh, 1993). Using OK, the weights assigned to the available observations are a function of the form of the model fitted to the variogram (or other other structure function) and the spatial configuration of the available data. To assign weights for kriging the variogram should be representative of the spatial variation modelled at all places within the region of interest. In geostatistical terminology, a stationary variogram model should be appropriate (Myers, 1989). Where spatial variation is markedly discontinuous across the region in concern (as it often is for digital terrain data) the variogram is not representative for all places equally. Then, weights assigned for kriging will be sub-optimal since changes in the coefficients of the variogram model may have a substantial effect on the kriged estimates (Isaaks and Srivastava, 1989; Herzfeld *et al.*, 1993; Goovaerts, 1997).

One possible solution is to segment the spatial variation on the basis of statistics or functions estimated for a moving window. An alternative is a locally-adaptive approach by which the variogram is estimated and modelled for a moving window (Haas, 1990). These approaches should, in principle, provide more accurate estimates than would be attained using a global variogram model. However, the automated fitting of models to variograms for use in kriging may be problematic, as it is necessary to ensure that the model fitted is valid and appropriate.

Segmentation of spatial variation and a locally-adaptive algorithm are both implemented in this chapter and some advantages and problems associated with these approaches are considered. A photogrammetrically-derived digital terrain model (DTM), representing an area of the Lake District in the UK, was sampled systematically and the complete DTM was used to quantify the errors obtained by estimating with a global model and with local models (segment by segment or within a moving window).

Applications that may benefit from segmentation or a locally-adaptive approach include (1) characterisation of spatial variation in terrain as an aim in itself, (2) estimation, (3) simulation and (4) the design of optimal sampling strategies that are based on geostatistical measures of uncertainty in estimates.

18.2 GEOSTATISTICS

18.2.1 Introduction

A variety of introductions to the most widely used tools of geostatistics are available (for example, Journal and Huijbregts, 1978; Isaaks and Srivastava, 1989; Webster and Oliver, 1990; Goovaerts, 1997; Kitanidis, 1997). Only a summary of the main tools used in this chapter will be given here. The principal tool of most geostatistical analyses is the variogram ($\gamma(\mathbf{h})$), a function which relates half the average squared difference between paired data values to the lag (distance and direction, where anisotropy is considered), $\mathbf{h}$, by which they are separated. The sample (or experimental) variogram is estimated for the $p(\mathbf{h})$ paired observations, $z(\mathbf{x}_i)$, $z(\mathbf{x}_i+\mathbf{h})$, $i=1,..., p(\mathbf{h})$:

$$\hat{\gamma}(\mathbf{h}) = \frac{1}{2p(\mathbf{h})} \sum_{i=1}^{p(\mathbf{h})} \{z(\mathbf{x}_i) - z(\mathbf{x}_i + \mathbf{h})\}^2 \qquad (18.1)$$

A mathematical model may be fitted to the variogram and the coefficients of the model may be used to assign optimal weights for interpolation using kriging. OK is the most widely used variant of kriging. A by-product of OK is the OK variance (or its square root, the OK standard error). It is a function of (1) the form of spatial variation of the data (modelled, for example, using the variogram) and (2) the spatial configuration of the observations. The OK variance serves as a measure of confidence in estimates. However, the OK variance is independent of the data values (Goovaerts, 1997) and its practical value is frequently limited.

18.2.2 Classification, segmentation and locally-adaptive kriging

Where the frequency and magnitude of spatial variation varies across the region of concern it may be profitable to sub-divide the data spatially. In practice, this is more likely to be the case where data are numerous. This may be achieved using classification or segmentation algorithms to identify sub-sets into which the data set should be divided

(St-Onge and Cavayas, 1997; Allard, 1998). One approach is to identify areas over which the coefficients of variogram models and summary statistics for a moving window, may be considered 'similar' (Lloyd and Atkinson, 1999a). This approach is useful when the aim is to determine sub-regions for which variogram model coefficients may be considered independent of location. If the variogram could be estimated locally and a model fitted automatically this problem could be reduced or even eliminated.

18.3 METHODS

18.3.1 Segmentation of spatial variation

The variogram (six 5 m lags) was computed for a moving window. The fractal dimension was derived from the slope of a linear model fitted to the double-log variogram. A region-growing segmentation algorithm (Haralick and Shapiro, 1985), written in Fortran 77, was used to segment the fractal dimension. This approach is detailed in Lloyd and Atkinson (1999b).

18.3.2 Locally-adaptive kriging

Kriging, as commonly implemented, is already adaptive in the sense that it takes into account the spatial configuration of data and the magnitude of data locally. The desire to adapt the variogram locally would arise if a global variogram model failed to adequately represent local spatial variation, leading to sub-optimal estimates of the property of concern and of the estimation variance, for at least part of the region of interest.

The fitting of variograms on an automatic basis may be extremely problematic and most geostatisticians caution against fully-automated fitting of models (Goovaerts, 1997). These reservations are based upon the central problems of fitting models to variograms: (1) the selection of the model coefficients, (2) the type of structured component(s) and (3) the choice of an isotropic or anisotropic model. Automation of these choices can be extremely difficult. The manual or semi-automated fitting of models can be very time consuming and it is not always possible to find any model that provides a 'satisfactory' fit. Furthermore, judgement of the success of the fit of a variogram model is problematic. There are no universally accepted measures representing how well a model fits a variogram.

The primary rationale behind this chapter is that the sub-optimal fitting of models locally may provide more accurate estimates than a well-fitted global variogram model that is not representative of large sub-sets of the region of concern (Haas, 1990 considered this issue but his results were unclear). Therefore, in this chapter, locally-adaptive kriging is compared to two other approaches: (1) to assume stationarity across the region in concern (to use a 'global' variogram model) and (2) to use variogram models for segments with internally similar variogram-derived fractal dimensions.

An approach to locally-adaptive kriging is presented in which a variogram is estimated for a moving window and three authorised models (together with nugget variances, c_0, the intercepts on the ordinate) are fitted automatically, one of which is selected for kriging. An algorithm was developed based on the first edition of the GSLIB variogram and kriging Fortran 77 subroutines (Deutsch and Journel, 1992). Least sum of squares was used to select one of the three models, which was then used to assign weights for kriging. The models were fitted using the generalised least squares functionality of ODRPACK (Boggs *et al.* 1989). The algorithm fitted a spherical, exponential and Gaussian structure (constrained such that all the coefficients should be greater than zero). These models are defined in turn:

spherical model:

$$\gamma(h) = c \cdot Sph\left(\frac{h}{a}\right) = \begin{cases} c \cdot [1.5\frac{h}{a} 0.5(\frac{h}{a})^3], & \text{if } h \leq a \\ c & \text{if } h \geq a \end{cases} \tag{18.2}$$

exponential model:

$$\gamma(h) = c \cdot Exp\left(\frac{h}{a}\right) = c \cdot \left[1 - exp\left(-\frac{h}{a}\right)\right] \tag{18.3}$$

Gaussian model:

$$\gamma(h) = c \cdot \left[1 - exp\left(-\frac{h^2}{a^2}\right)\right] \tag{18.4}$$

where a is the range (the limit of spatial dependence), c is the structured component and **h** is the lag (distance and direction, where anisotropy is considered, by which paired data are separated). One of the three models was then selected as the 'best fitting' in the sense that it minimised the weighted sum of squares (WSS). The model fitting was weighted by the number of pairs used at each lag and the inverse square of the semivariance (Pebesma and Wesseling, 1998):

$$w_i = \gamma(h_i)^{-2} N_i \tag{18.5}$$

Approximations of the variogram model coefficients were provided as initial values as follows: (1) nugget effect c_0 (semivariance at the first lag); (2) structured component c_1 (mean semivariance at all lags computed) (3) range a_1 (cut-off divided by two). These initial values were found to result in reasonable fits for most of the variograms examined. In all cases, a nugget variance or component (constrained to be greater than or equal to zero) and one structured component were fitted. These were found to provide reasonable estimates of the forms of the variograms in the majority of cases. In some instances, the addition of a second structure would have provided a better fit visually. However, because of the reasonable fits in most cases using only one structured component, as well as the increased difficulties in fitting automatically a nugget component and two structured components, only one component was fitted using the approach presented here.

Where the standardised WSS was greater than the predetermined threshold the coefficients of the global variogram model were used for estimation. The standardised WSS (as defined by Fernandes and Rivoirard, 1999) may be given as:

$$\gamma(h)_{WSS} = \frac{\sum_{h} w(h)[\gamma(h) - \hat{\gamma}(h)]^2}{\sum_{h} w(h)[\hat{\gamma}(h)]^2} \qquad (18.6)$$

where *w(h)* are the weights assigned to each lag as defined in equation 18.5. Where the standardised WSS was greater than a set figure the global variogram model was used for estimation. Five figures were used: 0.01, 0.02, 0.03, 0.04 and 0.05 and the accuracy of estimates made using the locally-adaptive approach was found to increase as the threshold increased. However, as the threshold increases the potential for poor fits of the models fitted automatically also increases, so a balance must be sought.

The coefficients of the selected models were input into an OK algorithm (based on the GSLIB OK code) and estimates made on a moving window basis. In some instances, a model could not be fitted locally because of the small number of data in some edge areas. In these cases, the coefficients of the model fitted to the global variogram were used for kriging, as noted above. An approach by which the variogram was re-computed each time the window moved was computationally expensive but served as a basis to demonstrate the applicability of the approach. A more efficient approach would entail re-computing or re-modelling the variogram in only some instances.

For the global and locally-adaptive approaches OK and OK with a trend model (KT) (Deutsch and Journel, 1992; Goovaerts, 1997) were used. Only KT was used for the segmentation approach, given the improvement in the global approach when KT was used. For KT the trend was modelled as quadratic in the *x* and *y* co-ordinates. OK and KT were both implemented using the algorithms provided in GSLIB (Deutsch and Journel, 1992).

18.4 DATA AND THE STUDY SITE

The analysis was based on a subset of a digital terrain model (DTM) derived photogrammetrically. The DTM represented an area in the Lake District, north-west England, falling primarily in the Ordnance Survey® (GB) (The Great Britain national mapping agency) tile NY11NE. The spatial resolution of the DTM was 5 m, and the cells were treated as discrete points (the sample spacing being, therefore, equal to the spatial resolution).

The DTM represented all terrain features (as opposed to just landform) and there are wooded areas in the north-east and central-east part of the region. All references to elevation in this chapter refer to heights of all surface features and not just to landform. The geographical limits of the subset DTM (Figure 18.1), defined in British National Grid References (for which the units are metres), were: north 517742.68; south 515902.68; west: 315953.79; east 317353.79. The maximum number of rows was 368 pixels (1840 m) and the maximum number of columns was 280 pixels (1400 m), although the edges of the DTM were not regular.

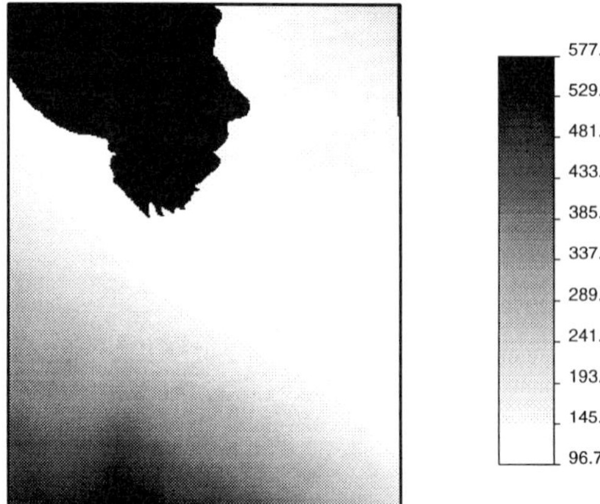

Figure 18.1 DTM (Ordnance Survey (GB) region NY11NE) with values estimated, using OK, to a 5 m grid. The black area in the north-west is water. Units are in metres

18.5 ANALYSIS

18.5.1 Introduction

Locally-adaptive kriging was compared to two other approaches: kriging using (1) a global variogram and (2) omnidirectional variograms computed on the basis of sub-sets of the DTM identified using region-growing segmentation of the fractal dimension (derived from the slope of a linear model fitted to the double-log variogram) for a moving window.

The DTM was sampled by extracting every other pixel, leaving a grid with a 10 m spacing. A 20 m and 30 m grid were also used to assess relative changes in estimation accuracy as the sample spacing increased. Then estimates were kriged to a 5 m grid, the aim being to reconstruct the original DTM from the sample. The estimates of the extracted data were used to assess the accuracy of the approaches presented by comparison with the complete DTM.

18.5.2 Variograms

Initially, a variogram was computed for the whole region. The variogram was fitted with a nugget component and a Gaussian component (Figure 18.2). In retrospect, a power model may have been more suitable, but the fit appeared reasonable. The presence of a low-frequency trend in the data was apparent from the form of the variogram (Starks and Fang, 1982). However, in this case the trend was ignored and variograms of the raw elevation values (rather than residuals from a polynomial trend) were used. The DTM represents an area of topography that is dominated by a gently-sloping valley so the presence of a low-frequency trend was expected. The coefficients for models fitted to the variograms estimated for the 17 segments identified with fractal-based region-growing segmentation are given in table 18.1.

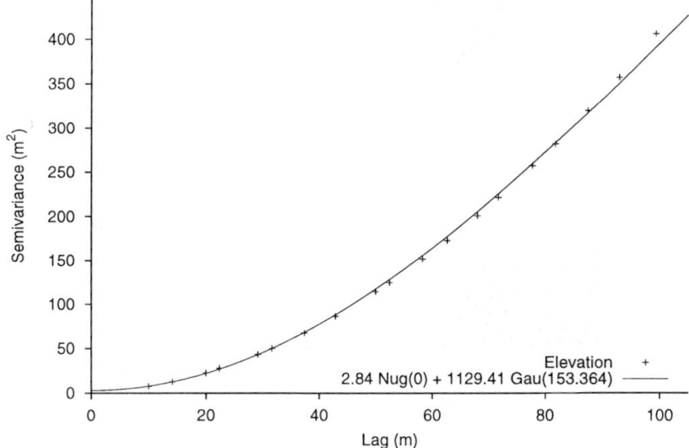

Figure 18.2 Global omnidirectional variogram

Table 18.1 Variogram model coefficients for segments identified with fractal-based segmentation. Seg. Is segment, str. is structure.

Seg.	c_0	str. 1	c_1	a_1	str. 2	c_2	a_2
1	8.80	pow.	0.10	1.55	–	–	–
2	0.88	Gau.	54.10	50.43	sph.	103.94	172.72
3	3.07	Gau.	51.08	120.09	sph.	54.95	182.08
4	–	pow.	0.12	1.863	–	–	–
5	0.54	Gau.	29.80	49.83	sph.	6.44	72.66
6	2.19	Gau.	57.12	92.02	–	–	–
7	4.89	pow.	0.25	1.11	–	–	–
8	–	pow.	0.00	1.17	–	–	–
9	1.07	Gau.	1311.92	167.29	sph.	115.87	201.82
10	0.17	pow.	0.00	1.74	–	–	–
11	–	pow.	0.21	1.67	–	–	–
12	1.00	Gau.	12.04	42.55	–	–	–
13	–	pow.	0.00	1.35	–	–	–
14	5.49	Gau.	964.62	102.28	–	–	–
15	–	Pow.	0.32	1.67	–	–	–
16	10.75	Gau.	897.41	119.01	–	–	–
17	0.35	sph.	0.61	18.608	sph.	0.71	81.25

18.5.3 Kriging

Summary statistics of the estimation errors for the global, segmentation and locally-adaptive approaches were compared (Table 18.2). The locally-adaptive approach with OK or KT provided an increase in estimation accuracy, as measured by the RMS error, over OK with the global model. However, KT with the global model provided greater accuracy than the locally-adaptive approach using either OK or KT. The segmentation-based approach provided clearly the highest accuracy (0.059 m less than the RMS for KT with the global model), as judged by the RMS error, though it involved by far the most effort to implement.

Although the locally-adaptive approach did not perform well for estimation from the 10 m grid, as sample spacing increased its estimation accuracy increased relative to the global model approach employing KT. For a 20 m grid the RMSE for the global approach was 2.30 m, for the locally-adaptive approach it was 2.31 m. Thus, the global approach RMSE was 0.01 m less than that for locally-adaptive kriging. When the grid spacing was increased to 30 m the relevant figures were 2.696 m and 2.688 m, an RMSE 0.008 m smaller for locally-adaptive kriging than for the global approach.

Table 18.2 Errors for global variogram, segmentation and locally-adaptive approaches. Seg. is segmentation, LA is locally-adaptive

Approach	Max. neg. error (m)	Max. pos. error (m)	Mean error (m)	RMSE (m)	$\gamma(h)_{WSS}$ threshold
Global OK	–17.115	15.352	–0.061	2.024	–
Global KT	–17.433	16.041	–0.049	1.789	–
Seg. KT	–18.342	16.600	–0.093	1.730	–
LA OK	–17.115	15.352	–0.068	2.003	0.01
LA OK	–17.115	15.088	–0.067	1.997	0.02
LA OK	–17.115	15.088	–0.068	1.995	0.03
LA OK	–17.115	15.088	–0.068	1.995	0.04
LA OK	–17.115	15.088	–0.068	1.994	0.05
LA KT	–17.428	15.088	–0.074	1.830	0.05

18.5.4 Distribution of errors

The estimation errors for both the global and locally-adaptive approaches centred around zero. The histogram of estimation errors for the locally-adaptive approach (figure 18.3), as well as that for a global variogram model, was approximately normally distributed.

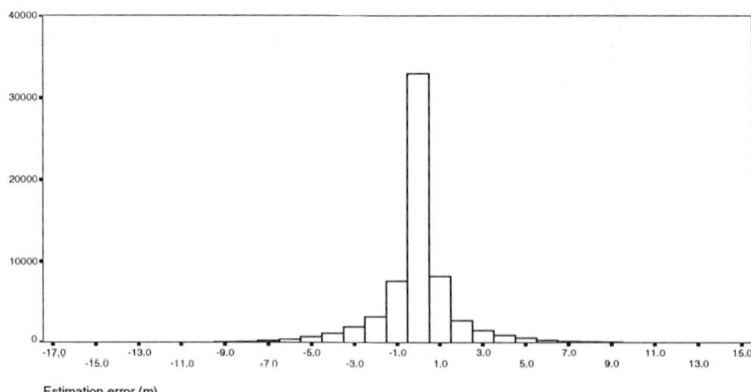

Figure 18.3 Histogram of estimation errors for locally-adaptive approach. Units are in metres

18.5.5 Mapped errors

Summary statistics, such as the RMS error, are of limited value. It is useful to look at the spatial distribution of errors to assess the performance of the approaches used. Figure 18.4 shows the spatial distribution of estimation errors for the locally-adaptive approach (employing OK). The largest estimation errors, visible as patches in Figure 18.4, occurred in areas with extensive vegetation cover and, in particular, forested areas. In relation to this issue, the approaches outlined in this chapter may be useful in two specific ways. Firstly, where the researcher is only interested in surface heights characterisation of spatial variation may help to identify, and remove in some way, features such as vegetation. Secondly, if all terrain features are of interest, adapting to the spatial variation within areas covered with vegetation will be particularly important since the spatial variation within and outside these areas may be markedly different.

Figure 18.5 is a map of the absolute difference between estimation errors made using the global variogram model and the locally-adaptive approach. Figure 18.6 shows the same image but with the lower and upper limits set to -0.5 m and 0.5 m respectively to illustrate smaller differences. In Figures 18.5 and 18.6 positive values indicate that the locally-adaptive approach has provided the most accurate estimates and negative values indicate that the global approach gave more accurate estimates. The largest differences are apparent in the north-east of the region, where the frequency of spatial variation in land form is highest and there is extensive vegetation cover. Differences, though less obvious, are apparent across the rest of the region. In particular, a series of ridges may be observed in the south and south-west. The majority of differences are close to zero but it is difficult to ascertain visually the relative proportions of negative and positive differences. The summary statistics indicate, of course, that where OK was used the locally-adaptive estimates were more accurate on average than the global variogram based estimates. However, the reverse was true for KT.

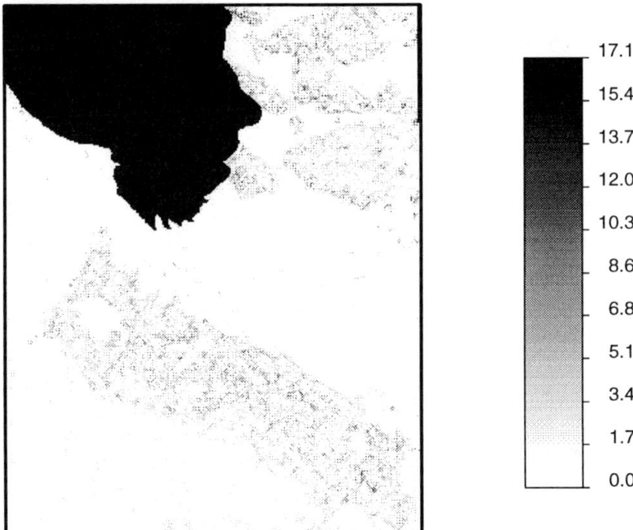

Figure 18.4 Map of estimation errors for the locally-adaptive approach. Units are in metres. The black area in the north-west is water

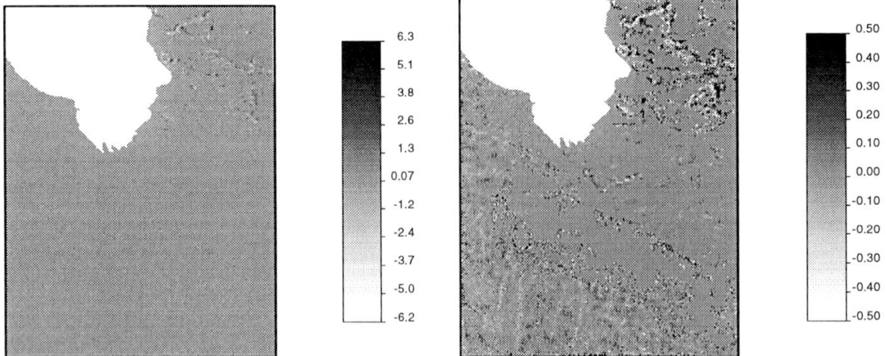

Figure 18.5 Map of differences between estimation errors for the global variogram model and the locally-adaptive approach. Positive values indicate that locally-adaptive estimates were more accurate, negative values indicate that the global model estimates were more accurate. Units are in metres. The white area in the north-west is water

18.6 As Figure 18.5, but scale restricted to –0.5 m to 0.5 m to illustrate small differences more clearly

18.5.6 Sampling design

One of the primary concerns of the research of which this chapter is part was to use geostatistics to assess uncertainty in estimates and to design optimal sampling strategies. For the global variogram model the r^2 for the absolute error against the kriging standard error was 0.0002 for OK and 0.0009 for KT. The largest r^2 for any approach was 0.1824, for the segmentation approach. For the locally-adaptive approach the r^2 was 0.0025 for a $\gamma(h)_{WSS}$ threshold of 0.01, rising to 0.0318 for a threshold of 0.05. Where KT was used with a threshold of 0.05 the r^2 was 0.0496. Since the kriging standard error represents an estimate of the spread of errors, rather than actual errors, it is not surprising that the relationships are not linear. The magnitude of the differences between standard errors and mean kriging standard errors do not provide a satisfactory picture for the locally-adaptive approach (Table 18.3). The OK/KT standard errors for the locally-adaptive approach clearly better represent the maximum errors than do those for the global variogram model approach.

18.5.6 Summary

From the results presented in this chapter it is clear that the form of the variogram model may have a substantial impact on the accuracy of estimates. Previous work has illustrated that the effect on the estimation (kriging) variances was even greater (Lloyd and Atkinson, 1998) and this chapter substantiates that work. In this analysis, the strength of the correlation between estimation errors and the kriging standard errors (square root of the kriging variances) was observed to increase when the locally-adaptive approach was used. The need to adapt the variogram to change in spatial variation across the region in concern is clear, though a segmentation-based approach was more successful than the locally-adaptive approach presented here.

Table 18.3 Errors and the kriging standard error for global variogram, segment variograms and locally-adaptive approaches. Seg. is segmentation, LA is locally-adaptive, AE is absolute error, SE is standard error

Approach	SE (m)	Max. AE (m)	Mean OK SE (m)	Max. OK SE (m)	AE vs. OKSE(r^2)	$\gamma(h)_{WSS}$ threshold
Global OK	2.01	17.12	1.79	2.49	0.0002	-
Global KT	1.78	17.43	1.90	2.94	0.0009	-
Seg. KT	1.72	18.34	1.85	4.39	0.1824	-
LA OK	2.00	17.12	2.71	23.86	0.0025	0.01
LA OK	1.99	17.12	2.91	23.86	0.0150	0.02
LA OK	1.99	17.12	2.98	23.86	0.0242	0.03
LA OK	1.99	17.12	3.00	23.86	0.0300	0.04
LA OK	1.99	17.12	3.00	23.86	0.0318	0.05
LA KT	1.82	17.43	4.26	25.96	0.0496	0.05

18.6 CONCLUSIONS

The accuracy of estimates of elevation made using kriging were assessed through three specific approaches. Weights were assigned using: (1) coefficients of a global variogram model; (2) coefficients of segment-based variogram models; (3) a locally adaptive variogram. Large differences between estimates made using the three approaches were noted and support the use of segmentation. Where estimates were made from 20 m and 30 m grids the locally adaptive kriging results improved with relation to the global model estimates. The principal conclusions of the chapter may be summarised:

- A semi-automated segmentation approach may enable more effective characterisation of spatial variation and significantly greater accuracy of estimation than an entirely arbitrary approach.
- Estimates derived using the fractal-based segmentation approach were more accurate than global approach (using KT) with a reduction in the RMS error of 0.059 m.
- A locally-adaptive approach may provide more accurate estimates than a global approach, where the sample spacing is sufficiently large (RMSE 0.008 m less than the global model with KT for a threshold of 0.05 for estimation from a 30 m grid). There is much room to increase the success of automated model fitting.
- Where sampling design is the concern, a locally-adaptive approach may be valuable as the kriging variance is affected by the magnitude of data values where the variogram is adapted locally. This was illustrated clearly in this chapter. The gain is achieved for no extra effort on the part of the user, but there in an increase in computational intensity.

18.7 ACKNOWLEDGEMENTS

The authors would like to thank the Department of Geography, University of Southampton and the Ordnance Survey for funding. We are also grateful to Dr. David Holland of the Ordnance Survey for arranging the supply of the data on which this analysis was based.

18.8 REFERENCES

Allard, D., 1998, Geostatistical classification and class kriging. *Journal of Geographic Information and Decision Analysis*, **2**, pp. 87-101. *ftp://ftp.geog.uwo.ca/SIC97/Allard/Allard.html*

Bian, L. and Walsh, S. J., 1993, Scale dependencies of vegetation and topography in a mountainous environment of Montana. *Professional Geographer*, **45**, pp. 1-11.

Boggs, P. T., Donaldson, J. R., Byrd, R. H. and Schnabel, R. B., 1989, Algorithm 676. ODRPACK: Software for Weighted Orthogonal Distance Regression. *ACM Transactions on Mathematical Software*, **15**, pp. 348-364.

Burrough, P A. and McDonnell, R. A., 1998, *Principles of Geographical Information Systems*, (Oxford: Oxford University Press).

Deutsch, C. V. and Journel, A. G., 1992, *GSLIB: Geostatistical Software Library and User's Guide*, (New York: Oxford University Press).

Fernandes, P. G. and Rivoirard, J., 1999, A geostatistical analysis of the spatial distribution and abundance of cod, haddock and whiting in North Scotland. In *GeoENV II: Geostatistics for Environmental Applications*, edited by Gómez-Hernández, J., Soares, A. and Froidevaux, R. (Dordrecht: Kluwer Academic Publishers), in press.

Goovaerts, P., 1997, *Geostatistics for Natural Resources Evaluation*, (New York: Oxford University Press).

Haralick, R. M. and Shapiro, L. G., 1985, Image segmentation techniques. *Computer Vision, Graphics, and Image Processing*, **29**, pp. 100-132.

Haas, T. C., 1990, Lognormal and moving window methods of estimating acid deposition. *Journal of the American Statistical Association*, **85**, pp. 950-963.

Herzfeld, U. C., Eriksson, M. G. and Holmlund, P., 1993, On the influence of kriging parameters on the cartographic output — A study in mapping subglacial topography. *Mathematical Geology*, **25**, pp. 881-900.

Isaaks, E. H. and Srivastava, R. M., 1989, *An Introduction to Applied Geostatistics*, (New York: Oxford University Press).

Kitanidis, P. K., 1997, *Introduction to Geostatistics: Applications to hydrogeology*, (Cambridge: Cambridge University Press).

Lloyd, C. D. and Atkinson, P. M., 1998, Scale and the spatial structure of landform: optimising sampling strategies with geostatistics. In *Proceedings of the Third International Conference on GeoComputation, University of Bristol, 17-19 September 1998*, (Manchester: GeoComputation CD-ROM). *http://www.geog.port.ac.uk/geocomp/geo98/15/gc_15.htm*

Lloyd, C. D. and Atkinson, P. M., 1999a, The effect of scale-related issues on the geostatistical analysis of Ordnance Survey® digital elevation data at the national scale. In *GeoENV II: Geostatistics for Environmental Applications*, edited by Gómez-Hernández, J., Soares, A. and Froidevaux, R. (Dordrecht: Kluwer Academic Publishers), 537-548.

Lloyd, C. D. and Atkinson, P. M., 1999b, Increasing the accuracy of kriging and the kriging variance through fractal-based segmentation: Application to a photogrammetrically-derived DTM. In *RSS'99: Earth Observation: From Data to Information. Proceedings of the 25th Annual Conference and Exhibition of the Remote Sensing Society, University of Wales, 8–10 September 1999.* (Nottingham: Remote Sensing Society), pp. 291-298.

Mulla, D. J., 1988, Using geostatistics and spectral analysis to study spatial patterns in the topography of southeastern Washington State, U.S.A. *Earth Surface Processes and Landforms*, **13**, pp. 389-405.

Myers, D., 1989, To be or not to be ... stationary? That is the question. *Mathematical Geology*, **21**, 347–362.

Pebesma, E. J. and Wesseling, C. G., 1998, Gstat, a program for geostatistical modelling, prediction and simulation. *Computers and Geosciences*, **24**, pp. 17–31.

St-Onge, B. A. and Cavayas, F., 1997, Automated forest structure mapping from high resolution imagery based on directional semivariogram estimates. *Remote Sensing of Environment*, **61**, pp. 82–95.

Starks, T. H. and Fang, J. H., 1982, The effect of drift on the experimental semivariogram. *Mathematical Geology*, **14**, pp. 309–319.

Webster, R. and Oliver, M. A., 1990, *Statistical Methods in Soil and Land Resource Survey*, (Oxford: Oxford University Press).

19

Assessing spatial similarity in geographic databases

Alia Abdelmoty and Baher El-Geresy

19.1 INTRODUCTION

One of the main functions of spatial information systems such as GIS is the unification and integration of different data sets and making them available for coherent manipulation and analysis by different applications. Integrating data in spatial information systems involves the integration of diverse types of information drawn from a variety of sources requiring effective matching of similar entities in these data sets and information consistency across data sets. Typically, spatial information can be provided in different forms by a number of sources. Data sources in GIS can include maps, field surveys, photogrammetry and remote sensing. Data sets may be collected at different scales or resolutions at different times. They may be collected in incompatible ways and may vary in reliability. Some details may be missing or undefined. Incompatibilities between different data sets can include incompatibilities between the spatial entities for which data are recorded, including differences in dimension, shape and positional accuracy.

For example, it may be required that a schematic representation of a certain region in space be stored in a GIS besides a more faithful representation (a schematic representation can be useful as an interactive tourist map). The two data sets are different. Many objects may be omitted from the schematic representation. The positional accuracy of the objects may not be maintained. However, both data sets hold the same relative position and orientation for the common subset of objects they hold.

A pre-requisite for the effective use and manipulation of several diverse spatial data sets is the understanding of the contents of the data sets and how they compare to each other. In this paper, a systematic approach is proposed for studying spatial similarity of geographic data sets. The approach involves the following steps:

- Analysing the different aspects of equivalence between the data sets. A range of spatial equivalence classes are identified which can be checked in isolation.
- Studying measures of spatial equivalence which can be applied to every class. Different levels of equivalence are proposed, namely, total, partial, conditional and inconsistent. Data sets can then be ranked as being consistent in which class to which level. This provides the flexibility for two data sets to be integrated without necessarily being totally consistent in every aspect.

- Developing methods for checking and representing explicitly the different equivalence classes and levels in the spatial database.
- Explicit representation of ambiguity or uncertainty resulting from the inconsistency of the data sets studied.

A qualitative representation scheme is proposed where the spatial content of the data sets is encoded. A simple scheme is first presented for handling topological information which is then extended for handling both topological and orientation information.

The use of a common representation scheme for different sets allows the direct (qualitative) comparison of those sets and the detection of any (qualitative) inconsistencies among them. This approach can, in some cases, alleviate the need for the expensive, error-prone, operation of transformation of data sets from one form to another (e.g. from raster to vector) which is the process commonly used for comparing spatial data sets. Automatic spatial reasoning techniques can be incorporated for the derivation of spatial relationships which are not represented explicitly.

The rest of this paper is structured as follows. Section 2 gives an overview of related work. In section 19.3 the different aspects of spatial equivalence are identified and classified between object-based and relation-based types. Section 19.4 presents a simple approach to the explicit representation of topological equivalence which is also extended to handle orientation equivalence. Conclusions are given in section 19.5.

19.2 RELATED WORK

Methods for checking consistency in spatial databases have been limited to checking topological consistency of pairs of spatial objects and not to whole map scenes (Egenhofer and Sharma 1992; Egenhofer *et al.*, 1994; Kuijpers, 1995, 1997).

In (Egenhofer and Sharma, 1992), consistency networks were used to check the consistency of a spatial scene containing regions with holes. In Tryfona (1997) consistency of topological relations between multiple representations of objects, specifically between parts and aggregate representations, is given. Approaches to the qualitative representation of images or maps can be classified into two categories. In the first category, spatial relations are studied and defined between pairs of objects, for example, defining relationships between two simple regions or two linear objects, etc. In the second category, approaches attempt to describe continuous spaces by describing sets of objects and relationships in these spaces.

In the first category, several methods were proposed, namely, the work of Cohn *et al.* (1993a, 1993b, 1996) and the work of Egenhofer *et al.* (1990, 1993a, 1993b) and Jen and Boursier (1994). The set of topological relations between two spatial objects, e.g. convex regions, are first defined. Then those are used to define relationships between more complex objects such as regions with holes. In Egenhofer and Sharma (1992), eight topological relations between simple regions were used to represent composite and non-composite fields using a method similar to consistency networks.

In the second category, the main approaches proposed define spatial scenes using symbolic arrays and minimum bounding rectangles (Chang *et al.*, 1987; Glasgow, 1990; Papadias, 1994). However, it is recognised that approximating objects by their minimum bounding rectangles may produce misleading results. In a different approach, Lundell (1996) used graphical illustrations to represent the adjacency between composite and non-composite physical fields. Composite fields are represented by drawing connected lines between the different representations of data layers or themes. The representation of change is depicted through a sequence of diagrams. A computational model for this method is however not envisaged directly. Glasgow and Papadias (1995) showed how a symbolic array can represent whole map scenes schematically.

19.3 ASPECTS OF SPATIAL EQUIVALENCE

In checking the similarity of two geographic data sets which relate to the same area in space, two consecutive steps are needed,

1. Object matching: where corresponding objects in both sets are identified using some equivalence tests. The result of this procedure is the identification of which objects in both sets can be considered to be the same, for example, matching two sets of land parcels in an old and up to date map or matching two road networks in maps with different scales, etc. Note that those objects could differ with regard both to positional information and geometric structure.
2. Spatial Equivalence representation: where the explicit representation of the relationship between the data sets is needed to allow for the intelligent manipulation of both sets by the system and to project to the user a clear view of the nature of the data used.

The equivalence of two representations of a spatial object can be studied from three points of view: relative to a fixed frame of reference, relative to the principal object studied, or, with reference to relationships with other objects in the data sets. Thus spatial equivalence can be studied using an absolute frame of reference, an object-based frame of reference and a relation-based frame of reference. Three classes of spatial equivalence can therefore be identified as follows.

19.3.1 Positional Equivalence

Objects are represented by the specific coordinates describing their spatial extents. Under this reference, two objects from two different data sets match only if their representative sets of coordinates match exactly and two data sets can be considered as locationally consistent if any position (x,y,z) corresponds to the same object in both sets.

19.3.2 Object-Based Equivalence Classes

A spatial data set consists of the spatial properties of a set of objects in a defined space. These properties include a description of spatial extent, from which the dimension and the shape of the object can be derived. An object in the data set can be composite, that is, consisting of, or containing other objects. Object-based consistency can be classified using the above properties. Two spatial data sets can be said to be object-based consistent of a certain class if for each object in both sets this consistency is achieved.

1. Object Existence Equivalence: two data sets are existentially equivalent if all the object classes and instances in one data set exist in the other data set.

2. Object Dimension Equivalence: two data sets are equivalent with reference to object dimension, if every object in one set has the same spatial dimension as that of the corresponding object in the other set.
3. Object Shape Equivalence: equivalence based on object shape can be as flexible as needed. On a strict level object shapes can be defined using equations of the curve or set of curves defining its boundary. On a less precise level object shapes can approximate well-known geometric shapes, for example, a circle, a square, a T shape, zig-zag, etc. Two data sets are said to be equivalent with reference to object shape if every object in the set can be described as shape equivalent to the corresponding object in the other set.
4. Object Size Equivalence: several measures of size exist including, length of boundaries, areas and volumes of shapes. Two data sets may be considered as equivalent with reference to object size if every object in one set has a similar size to the corresponding object in the other set.
5. Spatial Detail Equivalence: objects in the data sets may be composite, that is, containing other objects or made up of several connected or non-connected objects. Two data sets can be considered to be equivalent with reference to object detail if corresponding composite objects in both sets can be considered to be equivalent. An example of object-based equivalence is shown in Figure 19.1.

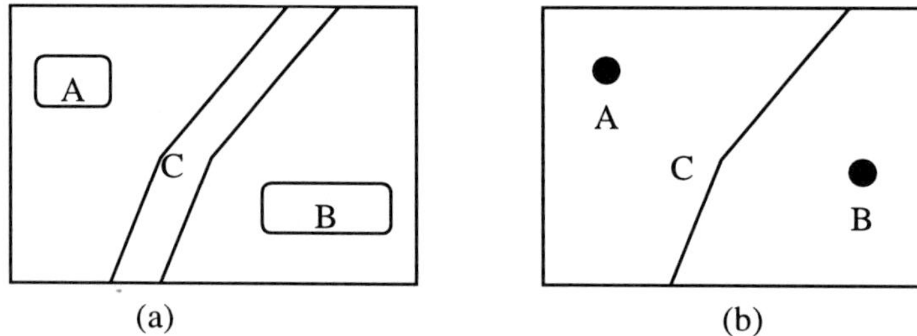

Figure 19.1 Non-consistent data sets with reference to object dimension.

19.3.3 Interdependency between Equivalence Classes

Other classes of object-based equivalence may exist. The above set of classes are possibly the most important from a general point of view. Note that the above classes may not be mutually exclusive. In particular, the positional consistency implies every other type of consistency and is by default the strictest measure of spatial equivalence. Shape and size imply dimension and all equivalence classes imply existence equivalence. Shape equivalence may imply spatial detail consistency if the object is composed of non-connected sets, etc. Also, it is assumed that a certain degree of inaccuracy can be acceptable in the measurement of some of the properties, for example, size and shape. However, this depends on the applications intended for these data sets. Note, that non-spatial equivalence is assumed here. Measuring non-spatial equivalence is part of the overall problem and is not discussed in this paper.

19.3.4 Relation-Based Equivalence Classes

The third type of consistency measures is based on the spatial relationships between objects in the data sets considered. Three classes of equivalence can be classified according to the types of spatial relationships (Abdelmoty and El-Geresy, 1994; Abdelmoty and Williams, 1994).

1. Topological Equivalence: two data sets can be regarded as topologically consistent if the set of topological relationships derived from one set are the same as those derived from the other. For example, the two sets in Figure 19.2 are not topologically consistent.

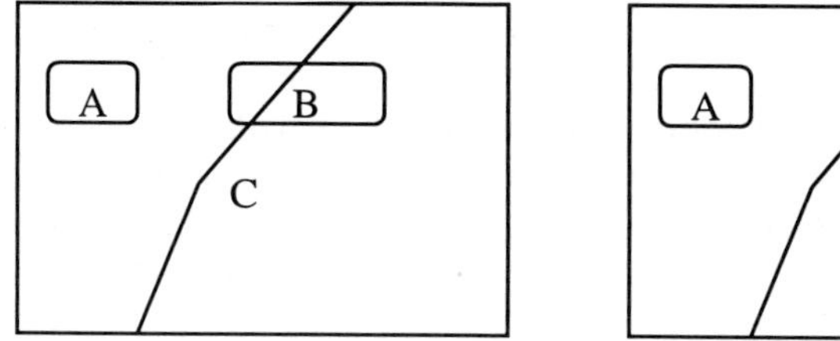

Figure 19.2 Topological inconsistency. (a) Object B crosses object C. (b) Object B is disjoint from C.

2. Direction or Orientation Equivalence: two data sets can be regarded as directionally consistent if the relative direction relationship in one set is the same as the other set.
3. Relative Size Equivalence: two data sets can be regarded as consistent with reference to relative size relationships if the qualitative size relations of larger and smaller are maintained between corresponding sets of objects in the two sets.

19.3.5 Different Levels of Spatial Consistency

Two spatial data sets can be consistent in more than one class of those defined above. For example, the data sets can be topologically and dimensionally equivalent, or consistent with reference to dimension, detail and category, etc. As noted earlier some consistencies do assume others. For example, topological equivalence may assume spatial detail. Up till now, the discussion is based on one level of consistency, namely, when all objects in the data sets conform to the consistency class studied. In reality, this is not always the case. Ranking the level of consistency for the different classes identified is important as it would provide the user of the GDB with an initial measure of the nature of the data sets in use. Further processing of this ranking would be used to identify how the data sets compare and which parts of the data sets are consistent, that is, the nature of such consistency. Let S1 and S2 represent the set of knowledge present in two data sets. This knowledge consists of all the different types of information that can be derived from every data set. It can be classified according to the object-based and relation-based classes. Let $S1_i$ and $S2_i$ represent the subsets of the set of knowledge $S1$ and $S2$ respectively, which belong to a certain class i, for example, shape properties or directional or topological relationships, etc. Four different levels of consistency can be identified.

1. Total Consistency: two data sets S_1 and S_2 can be said to be totally consistent with reference to a certain consistency class i, if $S_{1_i} \cap S_{2_i} = S_{1_i} \vee S_{2_i}$, that is, $S_{1_i} = S_{2_i}$. In this case a query to the GIS involving only properties of class i shall return identical results if posed to either S_1 or S_2.
2. Partial Consistency: two data sets S_1 and S_2 can be said to be partially consistent with reference to a certain consistency class i, if $S_{1_i} \cap S_{2_i} = C_i$ and $C_i \subseteq S_{1_i} \vee C_i \subseteq S_{2_i}$. In this case, only part of class i knowledge is consistent in the two sets. If the two data sets are to be used together, then it is important to know which subsets of the different classes of knowledge can be manipulated interchangeably between sets.
3. Conditional Consistency: two data sets S_1 and S_2 are said to be conditionally consistent with reference to a certain consistency class i, if there exists a set of functions F which when applied to S_{1_i} makes it totally consistent with S_{2_i}, that is, $S_{2_i} = F(S_{1_i})$. This can also represent the case where S_{1_i} is consistent with S_{2_i} but S_{2_i} is not consistent with S_{1_i}, i.e. $S_{1_i} \cap S_{2_i} = S_{1_i} \wedge (S_{1_i} \subset S_{2_i})$, (an asymmetric consistency). The set of functions F must be non-ad-hoc, that is, predefined. For example, the set of cartographic generalisation rules used to produce maps at different scales or a set of predefined rules used to produce a schematic from a faithful representation of a map.
4. Inconsistency Level: two data sets S_1 and S_2 can be said to be inconsistent with reference to a certain consistency class i, if $S_{1_i} \cap S_{2_i} = \phi$, that is, they do not share any piece of knowledge from that class. In this case, a query to the GIS involving properties of class i shall return non-identical results if posed to S_1 and S_2.

In most cases, the data sets studied relate to a combination of classes and levels. For example, two data sets can be partially consistent in terms of shape and dimension but are totally consistent topologically, or are conditionally consistent with respect to object detail as well as partially consistent topologically. Figure 19.3 shows the integration of different sets of knowledge which are consistent in different classes and levels.

19.4 REPRESENATION OF DIFFERENT LEVELS OF CONSISTENCY FOR DIFFERENT CLASSES

Determining the class and level of consistency between two data sets involves the extraction and comparison of the set of properties or relationships for that class. Although it is useful for the user and the system to be informed of the class and level of consistency in general, it may not be enough for certain application domains. In those cases explicit representation of the consistent set of knowledge is needed. A closer look at the different classes of consistency reveals that they are mostly qualitative measures (apart from location, size and shape). Hence, the common set of spatial knowledge between data sets can be represented qualitatively. A structuring mechanism can be envisaged which can be applied on a geographic data set to allow the explicit representation of some of the qualitative properties and relationships and the derivation of others. Multiple spatial representations can exist for the same geographic objects. However, properties and relationships are always related to objects and not to their underlying representations. Hence the structuring mechanism envisaged should be based on the geographic objects level and not on the geometrical representations. This structure can then be built for any data set irrespective of its underlying form of spatial representation.

Manipulation of such qualitative structure could make use of spatial reasoning techniques (Cui, *et al.*, 1993; Egenhofer, 1994; Hernandez, 1994; El-Geresy, 1997). For example it would be possible to store only some of the topological relationships and derive others using composition tables for similar and mixed types of spatial relations. Explicit representation of this knowledge would allow comparisons between data sets, seamless manipulation of existing sets, integration of new sets and consistent update of existing ones. In developing the proposed structuring mechanism, several questions need to be answered, including:

- What are the types of knowledge that can be represented explicitly and which can be derived?
- How can the different classes of knowledge be structured?

In this section, the representation of the class and level of topological consistency is first given and then extended to include orientation relationships.

19.4.1 Representation of Topological Equivalence with Adjacency Relationships

Checking topological equivalence between two geographic data sets is the process of checking that the same set of topological relationships between objects in one set exist for the corresponding objects in the other set. This process involves the explicit extraction and representation of topological relationships. Several approaches to checking the topological consistency of two spatial scenes have been proposed (Egenhofer and Sharma, 1992; Egenhofer *et al*, 1994; Kuijpers *et al.*, 1995; 1997). However, they do not consider the issue of integrating both scenes and hence do not provide a means of representing the common set of consistent knowledge. In this section, a simple structure for storing the adjacency relationships between objects in the data sets is proposed from which topological relationships can be derived. The structure can then be used to represent the common set of consistent knowledge between data sets as well as the ambiguity or uncertainty in the knowledge derived from both sets. The structure is based on the following assumptions.

19.4.2 Assumptions

Given a space S and a set of spatial entities $O_1 \ldots O_n$ embedded in it.

- Space S is dense and infinite.
- The spatial entities are connected. If an entity is not connected, each of its components will be considered separately.
- The entities jointly cover the whole space, that is, $S = O_1 \cap \ldots O_n \cap S_0$ where S_0 is the complement of the entities in space S. The inclusion of S_0 is necessary for two reasons: a) to avoid mis-interpretation of space topology and b) to provide an explicit representation of the edges of the scene (or map).
- The spatial entities don't overlap, that is, $O_i \cap O_n = \phi$ for all $1 \leq i \leq n$.

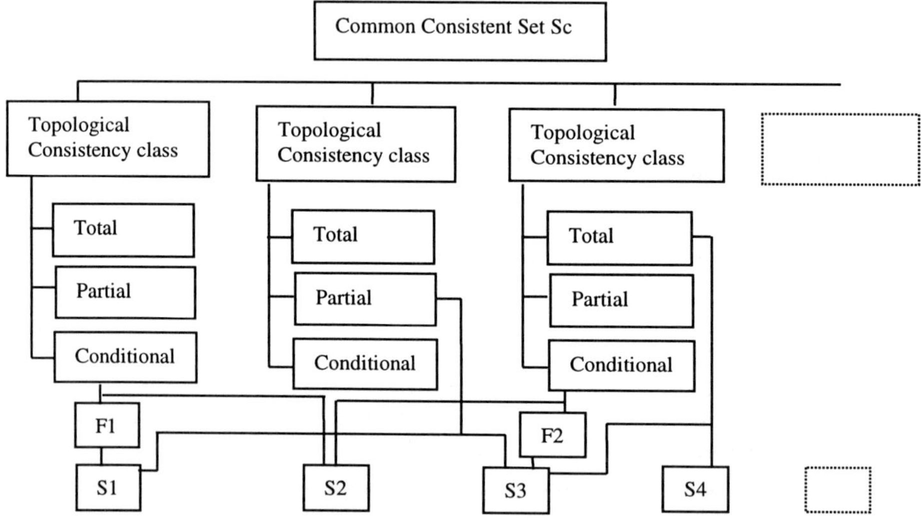

Figure 19.3 Integrating different data sets with different classes and levels of consistency to produce a common set of consistent knowledge. F1 and F2 represent sets of predefined functions for conditional consistency.

19.4.3 Capturing Topology- The Adjacency Matrix

The adjacency matrix is a qualitative spatial structure which captures the adjacency relations between different spatial objects. The adjacency relation (a binary symmetric relation) can be used for capturing the topological distribution of objects. In Figure 19.4(a) a map is shown with five entities A, B, C, D and E. In 19.4(b) the adjacency between the entities are encoded in a matrix. The fact that two entities are adjacent is represented by a (1) in the matrix and by a (0) otherwise. For example, A is adjacent to B, C and D but not to E, and D is adjacent to all others. Since adjacency is a symmetric relation, the resulting matrix will be symmetric around the diagonal. Hence, only half the matrix is sufficient for the representation of the space topology and the matrix can be collapsed to the structure in Figure 19.4(c). The complement of the objects in question shall be considered to be infinite. The suffix 0 (S_0) is used to represent this component.

As seen in the figure, the map edges are represented explicitly by the adjacency relations of S_0 (complement of objects in S). Objects B and E do not touch any of the map edges.

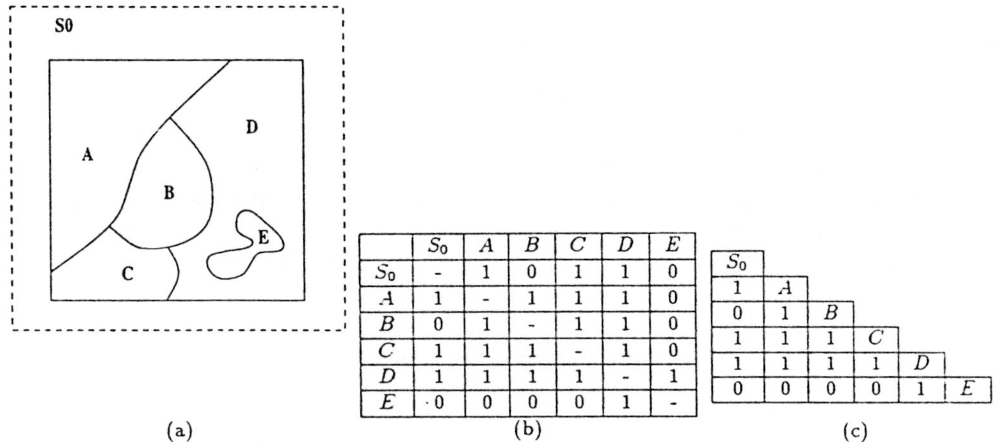

	S_0	A	B	C	D	E
S_0	-	1	0	1	1	0
A	1	-	1	1	1	0
B	0	1	-	1	1	0
C	1	1	1	-	1	0
D	1	1	1	1	-	1
E	·0	0	0	0	1	-

S_0					
1	A				
0	1	B			
1	1	1	C		
1	1	1	1	D	
0	0	0	0	1	E

Figure 19.4 (a) Space containing five objects. (b) Adjacency matrix for the scene in (a). (c) Half the symmetric adjacency matrix is sufficient to capture the scene representation.

19.4.4 Checking Topological Equivalence

The adjacency matrix can be used to check the topological equivalence of two scenes. Figure 19.5 shows a different data set of the same geographical area as Figure 19.4 and its corresponding adjacency matrix. There are two differences between the two scenes as can be seen from the structures. These are: in 19.4 object A is connected to C while it is not in 19.5, and object E in 19.4 does not exist in 19.5.

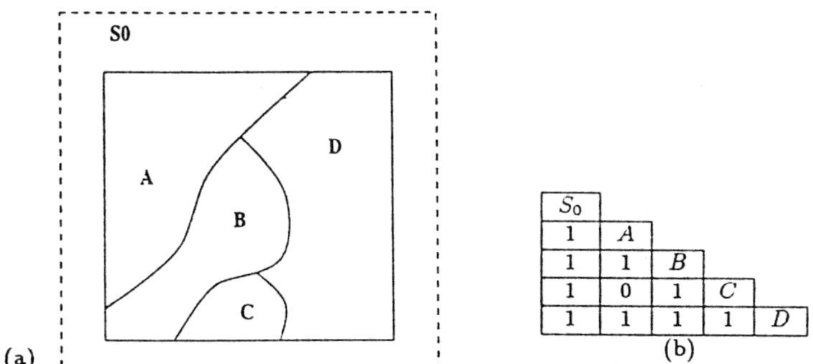

S_0				
1	A			
1	1	B		
1	0	1	C	
1	1	1	1	D

Figure 19.5 (a) Different data set for the same area in Figure 19.4. (b) Its corresponding adjacency matrix.

The only relationship stored explicitly in the above structures is *adjacency* and other topological relationships can be derived simply. For example, in Figure 19.4 object E is adjacent only to D and hence it is topologically inside D. Also, the relationship for complex objects can be realised from the grouping of relationships between its constituting parts, and so on. Hence, using this structure alone we can redraw the topological equivalences of the two scenes (obviously the exact shape of each object is not meant to be represented here). The adjacency structures can be organised in a tree structure representing different levels of detail in the data sets. Also, an explicit reference to object dimension will enable a (schematic) reproduction of the topological equivalence of the data sets. However, object dimension in both data sets need not be consistent.

19.4.5 Representing the Common Consistent Set of Knowledge

The scenes in 19.4 and 19.5 are partially topologically consistent. The set of common knowledge in both data sets can be grouped in an adjacency structure as shown in Figure 19.6. The structure in Figure 19.6 is informative of the consistent topological common knowledge between the two data sets. In this case, the adjacency between objects A and C is unknown, represented by a (-), and object E does not exist in both data sets and hence it is deleted from this set. Using this structure one can recreate the common knowledge in both scenes with the ambiguity of the relation between A and C.

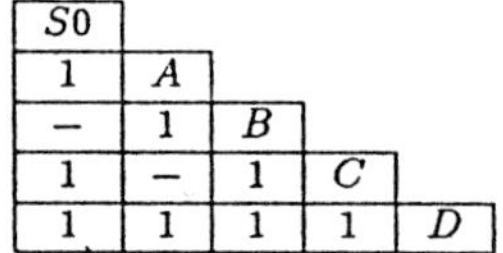

S0				
1	*A*			
–	1	*B*		
1	–	1	*C*	
1	1	1	1	*D*

Figure 19.6 The adjacency structure representing the common set of consistent knowledge in the structures of Figures 19.4 and 19.5.

19.4.6 Capturing Orientation: The Matrix Map

The adjacency matrix captures the topology of space under consideration. Orientation relations can be added to the cells of the matrix. Orientation relations have converses and therefore half the matrix is still enough to capture these relations. The matrix can be kept compact by exploiting the transitive property of the relations by qualitative reasoning. Thus those relations shall be explicitly defined between adjacent spatial entities only. Other relations between non-adjacent entities can then be deduced using qualitative reasoning. The convention of orientation relations is R(column,row). For example, in Figure 19.7, West(A,B) and South(A,C). Different granularities of the orientation relations can be defined, for example, south-west(A,D). Consider the example in Figure 19.7. The following orientation relations are defined between adjacent objects.

$W(A,B), N(A,C), S(A,D) \wedge SW(A,D) \wedge S(B,D), W(B,E), W(C,D), W(D,E)$.[19.1] The matrix in Figure 19.7(b) contains the orientation relations between adjacent objects only. The rest of the orientation relations can then be derived using the rules:

$$W(A,B) \wedge W(B,E) \rightarrow W(A,E)$$
$$S(A,D) \wedge W(D,E) \rightarrow S(A,E) \vee SW(A,E) \vee W(A,E)$$
$$SW(A,D) \wedge W(D,E) \rightarrow W(a,E) \vee SW(A,E)$$

Note that more than one reasoning path exists. For example, from the above rules we conclude that $S(A,E) \vee W(A,E) \vee SW(A,E)$. If an object is surrounded partly or fully by another object, such as in the case of part-whole relations, a notation is used to represent both relations, for example, $IE(A,B)$ denotes the relations $Inside(A,B) \vee East(A,B)$ as shown in Figure 19.8. The matrix structure is given in 19.8(b) and the converse relations are used if the order of objects is reversed in the matrix as in 19.8 (c). Whether A is totally inside B or shares its boundary can be inferred by examining the rest of the matrix cells for A and B. (A is totally inside B if it is only adjacent to B, that is, its corresponding row and column contain the value 1 only with object B).

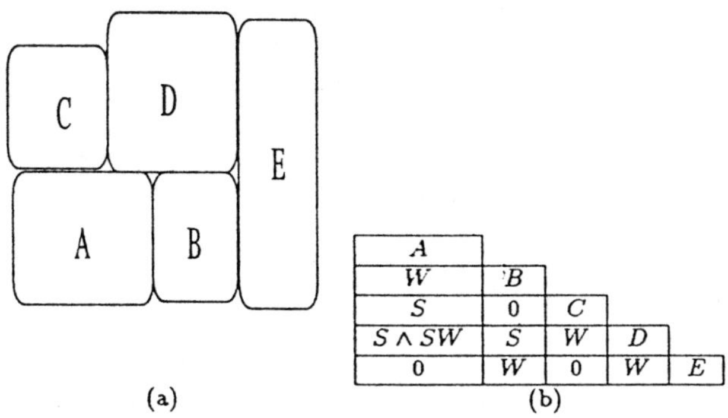

Figure 19.7 (a) Set of adjacent regions. (b) Corresponding adjacency matrix including orientation relations between adjacent regions only.

The combined adjacency and orientation relations and the explicit edge representation can be denoted the **Matrix Map**. A sketch map can be recreated from the matrix.
The matrix map can be further enriched with the size relations by specifying an ordered set of size relations between objects. The set D > A > B > C will capture the complete size relations between objects in the map.

[19.1] S denotes South, W denotes West, etc

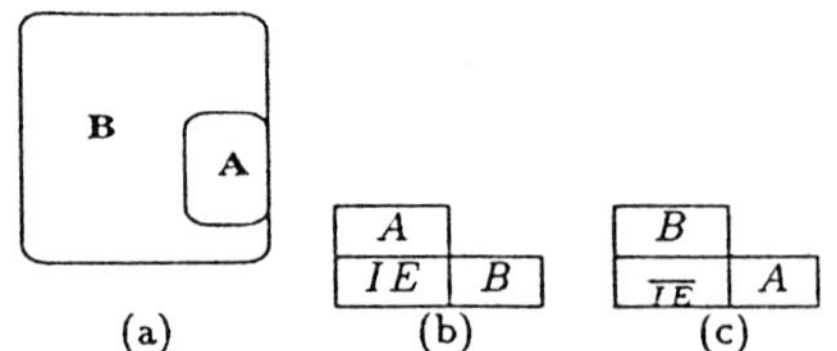

Figure 19.8 (a) Representing part-whole relationships - $Inside(A, B) \wedge East(A, B)$. (b) its matrix representation, (c) the matrix with the order of the objects reversed.

19.4.7 Example

Consider the data sets in Figure 19.9. The difference between the two sets is only apparent when their matrix maps are considered. The two data sets are totally equivalent topologically but partially equivalent directionally. Note that in defining the orientation relations, a specific consistent frame of reference has to be adopted. Different approaches exist for the representation of orientation relations (Abdelmoty and Williams, 1994; Hernandez, 1994). In this example, a simple conic division of the orientation space is adopted.

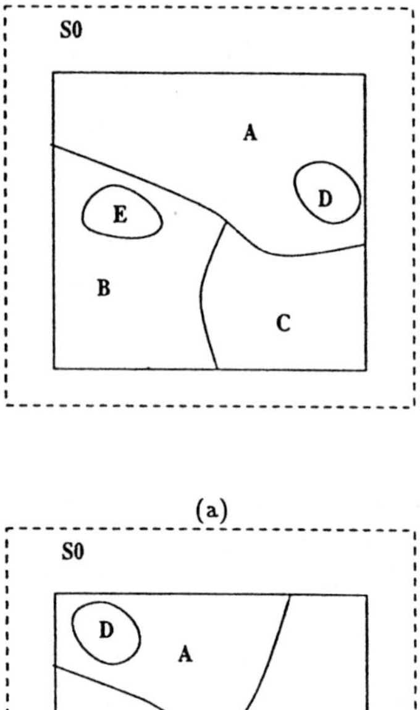

SO A D E B C

(a)

S_0					
1	A				
1	1	B			
1	1	1	C		
0	1	0	0	D	
0	0	1	0	0	E

(b)

S_0					
1	A				
1	$N \wedge E$	B			
1	N	W	C		
0	$\overline{IE}$	0	0	D	
0	0	$\overline{IN}$	0	0	E

(c)

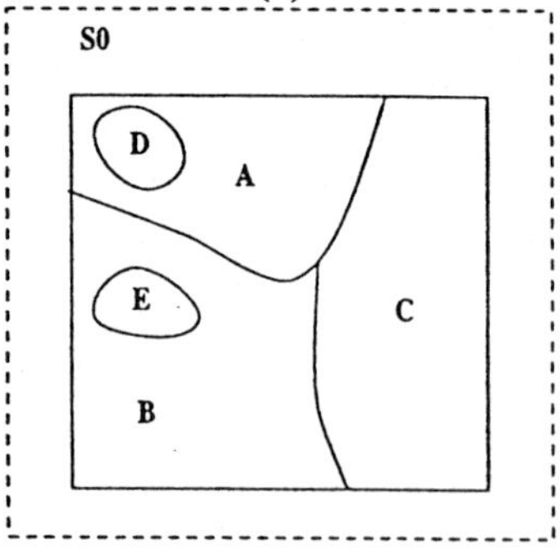

SO D A E C B

(d)

S_0					
1	A				
1	1	B			
1	1	1	C		
0	1	0	0	D	
0	0	1	0	0	E

(e)

S_0					
1	A				
1	N	B			
1	$N \wedge W$	W	C		
0	$\overline{IE}$	0	0	D	
0	0	$\overline{IN}$	0	0	E

(f)

Figure 19.9 (a) and (d) Two data sets of the same geographic area. (b) and (e) Their corresponding, equivalent, adjacency matrices. (c) and (f) Their corresponding (different) matrix maps.

19.5 CONCLUSIONS

In this paper, a study of the nature of equivalence between spatial data sets is presented. The proposed approach can be summarised as follows:

- Equivalence of data sets is broken down into two main categories: comparison of basic properties of objects and relationships between those objects. Different equivalence classes were identified which can be checked in isolation.
- For every class identified, data sets can be equivalent to a certain level or degree. Four levels of equivalence are proposed, namely, total, partial, conditional and inconsistent. Data sets can be ranked according to those levels, for example, totally consistent topologically but partially consistent with reference to object dimension and so on.
- Explicit representation of the different equivalence classes and levels of consistency is needed in the spatial database when different data sets are to be used together.
- The common set of consistent knowledge in the data sets needs to be expressed explicitly. A qualitative structure is proposed to hold different types of knowledge on the geographic feature or object level (as opposed to the geometric level).

As an example, the representation of topological equivalence is presented using a simple structure which stores adjacency relationships. Topological relationships can be derived from the structure and ambiguity in the relationships can be derived. It was also shown how the structure can be extended to incorporate orientation relationships. Further work needs to be done to devise representation methods for the different consistency classes and for their coherent integration. The work in this paper was done in the context of an ongoing research project which aims to develop methods for the modelling and manipulation of hybrid data sets in a GIS (Jones *et al.*, 1996).

19.6 REFERENCES

Abdelmoty, A.I. and El-Geresy, B.A., 1994, An intersection-based formalism for representing orientation relations in a geographic database. In *Proceedings of the 2nd ACM Workshop on Advances In Geographic Information Systems*, (New York: ACM press), pp. 44-51.

Abdelmoty, A.I. and Williams, M.H., 1994, Approaches to the Representation of Qualitative Spatial Relationships for Geographic Databases. In *Advanced Geographic Data Modelling (AGDM'94), Publications on Geodesy No. 40*, (Delft: Netherlands Geodetic Commission), pp. 204-216.

Cui, Z., Cohn, A.G. and Randell, D.A., 1993, Qualitative and topological relationships in spatial databases, In *Design and Implementation of Large Spatial Databases, SSD '93, LNCS 692*, (Berlin: Springer Verlag), pp. 396-315.

Cohn, A.G. and Gotts, N.M., The "egg-yolk" representation of regions with indeterminate boundaries, In *Geographic Objects with Indeterminate Boundaries, GISDATA*, edited by P.A. Burrough and A.U. Frank, (London: Taylor & Francis), pp. 171-187.

Cohn, A.G., Randell, D.A., Cui, Z. and Bennet, B., 1993, Qualitative spatial reasoning and representation, In *Qualitative Reasoning and Decision Technologies,* edited by Carrete, P. and Singh, M.G..

Chang, S.K, Shi, Q.Y. and Yan, W., 1987, Iconic indexing by 2-D strings. *IEEE Transactions on Pattern Analysis and Machine Intelligence*, **PAMI9**, pp. 413-428.

Egenhofer, M.J., Clementini, E. and Di Felice, P., 1994, Evaluating inconsistencies among multiple representations. In *Proceedings of the 5th International Symposium on Spatial Data Handling,* Vol. 2, (Charleston : IGU Commission of GIS), pp. 901-918.

El-Geresy B.A and Abdelmoty A.I., 1997, Order in space: a general formalism for spatial reasoning. *International Journal of Artificial Intelligence Tools*, **6**, pp. 423-450.

Egenhofer, M.J., 1993, Definitions of line-line relations for geographic databases. *Data Engineering*, **16**, pp. 40-45.

Egenhofer, M.J., Deriving the composition of binary topological relations. *Journal of Visual Languages and Computing*, **5**, pp. 133-149.

Egenhofer, M.J. and Herring, J.R., 1990, A mathematical framework for the definition of topological relationships. In *Proceedings of the 4th International Symposium on Spatial Data Handling*, Vol. 2, (Ohio: IGU Commission of GIS), pp. 803-13.

Egenhofer, M.J. and Sharma, J., 1992, Topological consistency. In *Proceedings of the 5th International Symposium on Spatial Data Handling,* Vol. 2, edited by Bresnahan, P., Corwin, E. and Cowen, D., (Charleston: IGU Commission of GIS), pp. 335-343.

Egenhofer, M.J. and Sharma, J., 1993, Topological relations between regions in R^2 and Z^2, In *Design and Implementation of Large Spatial Databases, SSD '93, LNCS 692*, (Berlin: Springer Verlag), pp. 316-336.

Glasgow, J.I., 1990, Artificial intelligence and imagery, In *Proceedings of the 2nd IEEE Conference on Tools with Artificial Intelligence, TAI'90*, (Washington: IEEE Computer Society), pp. 554-563.

Glasgow, J. and Papadias, D., 1992, Computation imagery, *Cognitive Science*, **16**, pp. 355-394.

Hernandez, D., 1994, *Qualitative Representation of Spatial Knowledge*, (Berlin: Springer Verlag).

Jen, T.Y. and Boursier, P., 1994, A model for handling topological relationships in a 2D environment, In *Proceedings of the 6th International Symposium on Spatial Data Handling,* Vol. 1, (London: Taylor & Francis), pp. 73-89.

Jones, C.B., Kidner, D.B., Luo, L.Q., Bundy, G.L, and Ware, J.M., 1996, Database design for a multiscale spatial information system. *International Journal of Geographical Information Systems*, **10**, pp. 901-920.

Kuijpers, B., Paredaens, J. and den Bussche, J.V., 1995, Lossless representation of topological spatial data, In *Proceedings of the 4th International Symposium on Large Spatial Databases, SSD'95*, edited by Egenhofer, M.J. and Herring J.R., (Berlin: Springer Verlag), pp. 1-13..

Kuijpers, B., Paredaens, J. and den Bussche, J.V., On topological elementary equivalence of spatial databases, In *Proceedings of 6th International Conference on Database Theory, ICDT '97*, edited by Afrati, F. and Kolaitis, P., (Berlin: Springer Verlag), pp. 432-446.

Lundell, M., A qualitative model of physical fields, In *Proceeding of the 14th National Conference on Artificial Intelligence, AAAI-92*, (Cambridge: AAAI Press, The MIT Press), pp. 1016-1021.

Papadias, D., 1994, *Relation-Based Representation for Spatial Knowledge*, PhD Thesis, (Athens: National Technical University of Athens).

Tryfona, N.and Egenhofer, M.J., 1997, Consistency among parts and aggregates: a computational model. *Transactions in GIS*, **1**, pp. 189-206.

Index

accuracy 176, 195
ADDRESS-POINT 109
adjacency 263–4
matrix 262–4
aerial photograph interpretation (API) 231, 234, 237–8
AGENT project 185
agents 56–61, 63–6, 68, 97
aggregation effects 103, 115, 118, 123
agricultural census 205–6, 216
agricultural land use 210
agriculture 163, 189
air photography survey 207
algorithm 59–61, 63, 66–7, 106, 111, 119, 122, 140, 143, 147–8, 151, 156, 175
AML *see* Arc Macro Language
anisotropy 243–4
ANN *see* artificial neural network
API *see* aerial photograph interpretation
Arc/Info 80, 233
GRID 224
Arc Macro Language (AML) 80, 82
architecture
centralized 18–19
cooperative 19
federated 20
functional 21–2
artificial life 87
artificial neural network (ANN) 76, 88, 189, 193, 195–6
architecture 200
backpropagation 195–6
astronomy 1
Australian Map Grid 197
automated zoning procedure 105–6, 121
AZP *see* automated zoning procedure

BDCARTO 136, 143
BDTOPO 136, 143, 145
Bellman-Dijkstra 61
Birmingham 132
Boolean approach 227
boundary coverage 199
boundary placement 105, 109, 111
Bristol 132
Britain 120
buffer 148–52, 157
buildings 137, 175, 177–8, 180, 183–5
Byzantine situation 20

CA *see* cellular automata
Cairngorms 220–4
canonical variate analysis 192
CASE tools 171
categorical maps 161, 163, 166, 170–1
cellular automata (CA) 55–7, 73–7, 82, 84
cellular space 56, 58
census 103, 106–8, 111, 115, 117, 120, 132
surveys 205
Chi-square 82
choropleth 103, 163
cindynics 16
classification 192
maximum likelihood 192
misclassification 209, 213
parametric 196
unsupervised 197
closed polygon 233–4
cluster 90, 92–3
cluster analysis 209–10
clustering 91, 94, 104
composition relationships 138
conceptual modelling 32
conflation 137
consistency 256–7
constraints 121, 127, 132, 142, 147–8, 161, 164–5, 168–71
contiguity 104, 109, 122
contrast 175–6
control centre 11, 18–9

correlation relationships 138–41
correspondence relationships
  138, 143–4
crime 51, 87

data
  aggregation 103
  cube 36
  infrastructure 111
  flow 115–9, 121–2, 132
  generation 91–3
  mining 1, 87–9, 95–6, 98
  reduction 88
  reference 135, 138–9
  simulated 87
  space 27–8, 32–4
  spatio-temporal 41
  structure 115–6, 166
  synthetic 89, 90–1
  warehouse 87
  visualisation 1
database
  consistency 135–6, 139,
   141–2, 144
  derived 135–6, 139–40
  multi-scale 136–7
  multi-scale reference 127, 135,
   138, 140
  schema 139
  topographic 176
  topographic query 29, 30, 33–4
decision support system (DSS) 15, 20
density contrasts 175, 177–80, 182–4
dependency relationships 138
derivation 136, 139–41, 143, 161
differencing 108
digital terrain model (DTM) 64, 224,
  242, 245–6
digitisation 176
digitising 197, 233, 237
disaggregation procedure 206, 208–10,
  213
distance 55, 58, 60–3, 67, 148–9,
  152, 154, 157, 179
Dordogne 154
Douglas-Peucker 148, 153
DSS *see* Decision Support Systems
DTM *see* Digital Terrain Model
DUEM 58

dynamic spatial modelling 2

ED *see* Enumeration District
EDA *see* exploratory data analysis
embarked computers 11
England 108, 120, 125–7, 132
enumeration district (ED) 90, 92–3,
  96, 106–7, 108, 110
environment 56–7, 65, 161
environmental data space 194
epidemiology 50–1, 89–90, 107
ERDAS Imagine 233
error 208, 248–50
exploratory data analysis (EDA) 88
exploratory geographical analysis 90
exploratory spatial analysis 125

feature history 36
fleet management 11, 13–4
FLIERS 231
flock 97–9
flood-fill 67
flow 115–6, 118, 120, 122, 125, 128
Fortran 77 244
fractal 65, 243
fuzzy 161
  concept 220
  logic modelling 219, 223, 227
  set 224
  wild land map 226

GA *see* genetic algorithm
GAM *see* Geographical Analysis
  Machine
Game of Life 74
Gaussian 244
GDM *see* geographical data miner
generalization 136, 161–4, 170–1,
  175–6, 178, 180, 183–5
genetic algorithm (GA) 97–8
geoBoids 97
Geocomputation 1–3, 5–7, 69,
  103, 111
  conference 1
geographic objects 261
geographic space 194
geographical information systems (GIS)
  revolution 115

geographical analysis machine (GAM) 87, 94, 97, 99
geographical data matching 137
geographical data miner (GDM) 95–6, 97, 99
geographical information producers 135
geographically weighted regression 2
geology 162
geometric fidelity 236
geometric relationships 138–9
georeferencing 108, 115
geostatistics 241–2
gerrymandering 107
GIS *see* geographical information systems
global positioning system (GPS) 11–2, 14–6, 18
goodness of fit 121
GPS *see* global positioning system
gradient 64–5
graph theory 104, 119
gravity models 115, 125
GSLIB 243, 245
Guadalquivir 155

Hausdorff distance 149, 154–6
hazmat (hazardous materials) 14–5
heuristics 74, 77
high resolution satellite sensors 231, 238
homogeneity 107, 110, 177, 180, 183–4
Humberside 90
hydrological modelling 191
hyperspaces 89

IGN 135, 137, 181
interaction 43, 56, 89, 93–4, 98, 107, 115–6, 121, 123, 143
Internet 161
  questionnaire 219, 222
interpolation 241
intersection operation 35–6
intramax 121, 123–4, 129, 132
intramin 124
isovist 66, 67

JAHC *see* June Agricultural and Horticultural Census
Java 165
join operation 36
journey to work 115, 116, 120, 125, 132
June Agricultural and Horticultural Census (JAHC) 206

kernel 77, 92
  smoothing 213
kriging 241–242, 247
  ordinary 242, 245, 247, 250
  locally adaptive 242–3, 245–7, 249–51
k-means classification 210

label 147–9, 153–4, 156
LAD *see* local authority district
LAMPS2 165
land capability 161, 163, 166–8
land cover 195, 206–7
  map database 206
  mapping 231–2
  proportion 232, 234
Land Cover of Scotland (LCS) 206–7
land resource analysis 205
land use pattern 77, 79
Land-Line.Plus 232–3, 237–8
landscape 66, 69, 161–2
LCS *see* Land Cover of Scotland
Leeds 127, 132
linkage rules 162
Liverpool 127
Liverpool Plains 190–1, 194
local authority district (LAD) 125–7
logistic
  distribution 76
  regression 78–9
Logo 58
London 132

Manchester 127, 132
map algebra 56
MAPEX 94, 96–7, 99
MAS *see* simulation, multi-agent
MATPAC 91, 120
Matrix Map 265–6
MAUP *see* modifiable areal unit problem

Merseyside 107
metadata 139–40, 142
micro-simulation 55
migration 115–6
minimum bounding rectangle 256
minimum spanning tree (MST) 180–1
mixel 199
model/modelling
    database 41, 163, 167
    environmental 191–2
    fitting 242–3, 246
    gravity 115, 125
    simulation 55
    temporal space 41–3, 45–6, 50, 52
    variogram 242–5, 248, 250–1
modifiable areal unit problem (MAUP) 103, 115, 118, 206
Monte Carlo simulation 76, 79
Moore neighbourhood 58, 62, 64
Moran's I 82
MST *see* minimum spanning tree
multi-agent systems 55, 60
multi-criteria evaluation 76
multinomial logit model 76
Murray-Darling Basin 189

national mapping agencies 135
navigation 61, 63
neighbourhood 56–8
network 62, 64, 143
neural network *see* artificial neural network
NeuralWorks v. 2.5 196
Newcastle 127, 132
nugget 244, 246

OA *see* Output Area
object constraint language 161, 164–5, 170–1
object matching 257
object-orientation 161, 165, 167
obstacle 59–61
OCL *see* object constraint language
ODRPACK 244
optimisation 125
OR operator 227–8
Ordnance Survey 109, 178, 232, 245
output area (OA) 106, 110, 115

parallel 106
parcel 178–9, 182–3, 185
path 59, 64
pattern 87–92, 94–6, 116, 125, 175, 184
Pearson's correlation coefficient 208
phenomenological control 175
pixel 232
Poisson distribution 76
Poisson process 91
political districting 107
pollution 15–6
polygon 233–4
polyline 147–9, 155, 157
population 90, 92, 103–5, 125
positional accuracy 255
positional equivalence 257
postcode 108–10, 120
precision 195
probability/potentiality surface 79–80
problem space 27
process-oriented models 36, 38
propagation 135–7, 139–145

quadratic programming 206

raster 195, 199
    data 74, 76
Rational Rose 171
ray tracing 66
real time 11
regionalisation 119, 123–4, 126
remote sensing *see also* satellite remote sensing 87, 115, 191, 194–5
research 161, 171
    design 194
    possibilities 23
resolution *see also* spatial resolution 176
risk 16, 90
rivers 15, 55, 64–5
RMS error 248–9
road network 77, 79
route
    finding 55
    network 55
    shortest 61–3, 69

SAGE 107, 210
salinisation 189, 191–2, 194–5, 201
sampling 241–2, 250, 251
satellite 62
  imagery *see also* SPOT imagery 189
  remote sensing *see also* remote sensing 231
scale 103, 117, 123, 135, 176
schematic map 255
Scotland 120, 167
search space 29–31
self-organising growth 74, 76–7, 80
semantic classification 32
semantics 45, 49, 138, 144, 162, 165, 167, 169, 171, 176
sensor 12, 15, 18–9
shape statistics 110, 132
Sheffield 107, 127, 132
shortest path 60
signal detection theory 189
simulated annealing 106, 149
simulation 242
  multi-agent (MAS) 56–7
sliver polygons 110
smart geographical analysis tools 87
SMS *see* special migration statistics
SMSTAB 120
soils 161–2, 166, 201
space-time 55, 87, 89–91, 93–6, 98
sparsity 115, 125
spatial
  autocorrelation 82
  behaviour 55, 116
  calibration 121
  change 28, 35
  consistency 259–60
  decision support system 20
  diffusion 51
  equivalence 255–9
  error 213
  information 192
  interaction 56
  relationships 44–5
  resolution 216, 233
spatio-temporal GIS 27
spatio-temporal query 29–30, 38
special migration statistics 120
spectral space 194
spontaneous growth 74, 76
SPOT imagery *see also* imagery 197, 200–1
Starlogo 58
statutory boundaries 109
street 59
  frontage 178–9, 182–3, 185
STREETS 57
sub-pixel components 231
sub-urban landscape 236, 238
Sugarscape 58
surface 64–5
SWARM 57
symbolic array 257
Synoptics 12, 20, 23

tabu 106
Tate Gallery 55, 67–9
telegeomonitoring 11, 23
temporal
  GIS (TGIS) 27, 42–43
  relationships (TSR) 43–6, 48–52
  space 41, 45–7, 50–1
terrain 241, 249
TGIS *see* temporal GIS
Thematic Mapper (TM) 192
Thiessen polygons 109–10
time 41–43
time clustering 91
time-event interaction 89
TM *see* Thematic Mapper
TOP25 145
topographic maps 136–7, 170, 176
topological
  consistency 260, 264
  equivalence 259, 261–3
  relations 256, 261
traffic management 12
TRANSIMS 57
transition rule 74, 77
travel to work areas (TTWA) 120
TRIAD model 35
triangle inequality 154
TSR *see* temporal relationships
TTWA *see* travel to work areas

UML *see* unified modelling language
uncertainty 231, 236, 256, 261

unified modelling language (UML) 161, 163, 165, 170–1
update 135, 136–7, 139–45
urban
  areas 177
  environment 66
  growth 73–4
  land use change 73–4
  morphology 76
  simulation 73–4, 77, 80
  systems 56

vehicles 12–14, 18
viewpoint 66
viewshed 66
visibility map 224
visual fields 66–7
visualisation 89, 118, 132, 147
Voroni cell 179–80
Voroni diagram 152, 154, 178, 180

Wales 108, 120, 125–6, 132
walking 58–60, 62, 64, 67
ward 120
watershed 64–5
wild land 219–20, 222
  perception 219, 221–2, 224
  quality 224
wilderness mapping 219–22

Yorkshire 90

ZDES *see* zone design system
ZDeSi *see* zone design system for interaction data
zone design 101–2, 106–7, 109, 115–21, 130
zone design system (ZDES) 104–5, 130
zone design system for interaction data (ZDeSi) 115, 119, 121–3, 130

CARDIFF UNIVERSITY
PRIFYSGOL CAERDYDD